Archimedes

New Studies in the History and Philosophy of Science and Technology

Volume 74

Archimedes has three fundamental goals: to further the integration of the histories of science and technology with one another; to investigate the technical, social and practical histories of specific developments in science and technology; and finally, where possible and desirable, to bring the histories of science and technology into closer contact with the philosophy of science.

The series is interested in receiving book proposals that treat the history of any of the sciences, ranging from biology through physics, all aspects of the history of technology, broadly construed, as well as historically-engaged philosophy of science or technology. Taken as a whole, Archimedes will be of interest to historians, philosophers, and scientists, as well as to those in business and industry who seek to understand how science and industry have come to be so strongly linked.

Submission / Instructions for Authors and Editors: The series editors aim to make a first decision within one month of submission. In case of a positive first decision the work will be provisionally contracted: the final decision about publication will depend upon the result of the anonymous peer-review of the complete manuscript. The series editors aim to have the work peer-reviewed within 3 months after submission of the complete manuscript.

The series editors discourage the submission of manuscripts that contain reprints of previously published material and of manuscripts that are below 150 printed pages (75,000 words). For inquiries and submission of proposals prospective authors can contact one of the editors:

Please find on the top right side of our webpage a link toour *Book Proposal Form.*

Francesca Rochberg
Editor

Studies in Premodern Sciences in Memory of Noel M. Swerdlow

Antiquity to the Renaissance

Springer

Editor
Francesca Rochberg
Department of Middle Eastern Languages and Cultures
University of California
Berkeley, CA, USA

ISSN 1385-0180 ISSN 2215-0064 (electronic)
Archimedes
ISBN 978-3-032-04039-8 ISBN 978-3-032-04040-4 (eBook)
https://doi.org/10.1007/978-3-032-04040-4

This Springer imprint is published by the registered company Springer Nature Switzerland AG
The registered company address is: Gewerbestrasse 11, 6330 Cham, Switzerland

zà-mí-zu dug-ga-àm
your praise is sweet

Preface

This volume presents the proceedings of the 2023 Bacon Conference "Thematic Continuities in Astronomy and Astrology from Antiquity to the Renaissance: A Conference in Memory of Noel M. Swerdlow," which took place on May 18–21, 2023. The conference was planned as a tribute to Noel Swerdlow, our colleague and friend.

Noel Swerdlow's work represents a singular contribution to our historical vision and understanding of the premodern astronomical sciences, and all of us were the beneficiaries, one way or another, of his work, his unparalleled command of the history of astronomy, and his kindness. The conference focused a new lens on the ancient, medieval, and early modern sciences of the heavens, with particular attention to some of the themes of interest to Noel throughout his scholarly career. Such themes included the transmission of astronomical knowledge, the development of the Babylonian theory of the Moon and planets, the Almagest, and medieval and Renaissance astronomy, including sources in Greek, Latin, Arabic, and Sanskrit. The papers published in this collection represent nearly all those presented at the conference.

Berkeley, CA, USA Francesca Rochberg

Acknowledgments

It is a singular honor to be able to call Noel M. Swerdlow my teacher, mentor, and friend, and to have had the privilege of the occasion of the 2023 Francis Bacon Award in History and Philosophy of Technology and Science at Caltech to remember him and celebrate him with this memorial volume. I thank all the contributors for their work.

I am indebted to Jed Buchwald and the Francis Bacon Foundation for the exceptional honor of the Bacon Award at Caltech. I thank Henry Gibbons and Karl Swaidan of the Bacon Foundation for their generous support of the 2023 Caltech conference to commemorate the consummate scholarship of Noel M. Swerdlow. Thanks are also due to Francine Tise of the Division of Humanities and Social Sciences at Caltech for her organizational support of the conference. For copyediting support, I gratefully acknowledge Kelly Burch.

Contents

Contributors

Alan C. Bowen Institute for Research in Classical Philosophy and Science, Baysville, ON, Canada

Lis Brack-Bernsen Wissenschaftsgeschichte, Universität Regensburg, Regensburg, Germany

Margaret Gaida Department of History, Occidental College, Los Angeles, CA, USA

Anthony Grafton Department of History, Princeton University, Princeton, NJ, USA

Florence C. Hsia Department of History, University of Wisconsin, Madison, WI, USA

Clemency Montelle School of Mathematics and Statistics, University of Canterbury, Christchurch, New Zealand

Robert G. Morrison Department of Religion and Program in Middle Eastern and North African Studies, Bowdoin College, Brunswick, ME, USA

Betsey Price Department of History, Glendon College, York University, Toronto, ON, Canada

Francesca Rochberg Department of Middle Eastern Languages and Cultures, University of California, Berkeley, CA, USA

John M. Steele Department of Egyptology and Assyriology, Brown University, Providence, RI, USA

Liba Taub Department of History and Philosophy of Science, Newnham College, University of Cambridge, Cambridge, UK

Abbreviations

AA	*Astronomical Almanac*
ACT	Neugebauer 1955 *Astronomical Cuneiform Texts*, 3 vols.
Alm.	Ptolemy, *Almagest*
BYY	Burnett et al. 2004, *The Introduction to Astrology: Editions of the Arabic and Latin Texts and an English Translation*
CAD	*The Assyrian Dictionary of the University of Chicago*
Cael.	Cleomedes, *Caelestia*
De nat. deor.	Aristotle, *De natura deorum*
EAE	*Enūma Anu Enlil*
HAMA	Neugebauer 1975, *A History of Ancient Mathematical Astronomy*, 3 vols.
Harm.	Ptolemy, *Harmonica*
HM	Caspar (1619) 1940, *Harmonices Mundi*
Intro. ast.	Geminus, *Introductio astronomiae*
KGW	Dyck et al. 1937–2012, *Kepler Gesammelte Werke*
Legg.	Plato, *Laws*
LSJ	Liddell et al. 2001, *The Online Liddell-Scott-Jones Greek-English Lexicon*
MC	Kepler 1596, *Mysterium Cosmographicum*
Meta.	Aristotle, *Metaphysica*
MUL.APIN	An Astronomical Compendium in Cuneiform
Phaedr.	Plato, *Phaedrus*
Phys.	Aristotle, *Physica*
SAA	State Archives of Assyria, Helsinki 1987 ff.
TU	Thureau-Dangin 1922, "Tablettes d'Uruk"

List of Figures

List of Tables

Part I
Babylonia

Chapter 1
Early Babylonian Astronomy: Observations and Numerical Fits

Lis Brack-Bernsen

> *"The one duty we owe to history is to rewrite it."*
>
> —*Oscar Wilde*

One goal of this paper is to illustrate what we can learn about the development of Babylonian lunar astronomy by controlling the precision of the astronomical numerical schemes in EAE 14[1] and MUL.APIN[2] against nature. Predictions, based on those schemes, shall also be controlled and matched with the empirical know-how of the Babylonians. Their empirical knowledge at the time around 500 BCE can be found in the Diaries, which for half a year list observation undertaken by the Babylonian astronomers (Sachs and Hunger 1988, 1996) and in predicting rules written on TU 11.[3]

Another goal is to show that the schemes for day lengths found in the advanced Mathematical (or ACT) Texts most likely origin in the observation of ŠÚ+NA and ME+GE$_6$, which measure the daily delay of the setting and rising full Moon,

Noel Swerdlow in Memoriam.

[1] EAE 14 is the 14th tablet of the astronomical/astrological series Enūma Anu Enlil (Al-Rawi and George 1991–1992).

[2] MUL.APIN is a Babylonian astronomical compendium consisting in two cuneiform tablets (Hunger and Pingree 1989; Hunger and Steele 2019).

[3] TU 11 is a procedure text written in Uruk toward the end of the third century BCE, and published as No. 11 in "Tablettes d'Uruk" (Thureau-Dangin 1922). The tablet is a copy of an older tablet, but we do not know when that was composed. Parallel tablets dating back to the fifth century BCE have been found in the British Museum. TU 11 contains a variety of early prediction rules and methods for finding lunar events. Most methods were introduced by means of calculated examples. Some were based on EAE numbers, others give advice of how to make predictions based on observations (Brack-Bernsen and Hunger 2002).

L. Brack-Bernsen (✉)
Wissenschaftsgeschichte, Universität Regensburg, Regensburg, Germany
e-mail: lis.brack-bernsen@psk.uni-regensburg.de

F. Rochberg (ed.), *Studies in Premodern Sciences in Memory of Noel M. Swerdlow*, Archimedes 74, https://doi.org/10.1007/978-3-032-04040-4_1

respectively. These time differences have been regularly observed and recorded since the seventh century BCE. The daily retardation of the Moon was already treated in EAE and MUL.APIN.

1.1 Introduction

The mathematical astronomical table texts fit nature quite well. These texts constitute the end product of the Babylonian astronomers. They stem from the last four centuries BCE: The Mathematical Table- and Procedure-Texts were published by Neugebauer (1955) in his "ACT," Astronomical Cuneiform Texts volumes I–III. Therefore, such astronomical texts are often called ACT texts.

Ossendrijver (2012) published an enlarged and revised version of the Procedure Texts, and here he introduced the new denotation "Babylonian Mathematical Astronomy." The ACT or mathematical functions constitute quite good numerical fits to nature. This is true for the lunar models as well as for the planetary models: periods, phase, maximum, and minimum of the numerical functions are well chosen. Therefore, it is not evident which observational data were the basis for the construction of the numerical functions. There are several proposals; but it is difficult to detect how the Babylonians developed their different advanced numerical models. We learn more about the developing astronomy, when the Babylonian parameters do not describe the astronomical phenomena that well. Therefore, the focus in this paper shall be on the earlier development of Babylonian Astronomy. In EAE 14 and MUL.APIN, periodic astronomical phenomena are reproduced in number schemes. These schemes utilized the schematic year of 12 months with 30 days each. They recorded, e.g., the length of day and night during the year (as a function of the date), or starting at new Moon, they listed day by day when, with respect to sunset or sunrise, the Moon would set or rise. According to their schemes, from day to day the Moon rose and set later by the same amount of time, which was depending on the month. This delay, the daily delay of the Moon, could also be observed in the days around full Moon, and it was the basis for precise predictions. We shall look at the astronomical schemes in EAE 14 and MUL.APIN and investigate how precise they are and how they were used for calculating lunar phenomena. We call this type of calculations "schematic astronomy," and we shall look at their connection to the precise empirical predicting rules which also occur on early cuneiform tablets. EAE14 and MUL.APIN stem from the seventh century BCE or earlier, but they were copied and used until the last century BCE. Other tablets show that the Babylonian astronomers during the same time had found elegant methods for the prediction of astronomical phenomena, based on observations. The late procedure text TU 11 collects many different methods for the prediction of astronomical events. The methods are given either through extremely short advices or through calculated examples, which sometimes are based on observed values, sometimes on the schematic values from EAE 14 or MUL.APIN. One very important prediction method, the Goal-Year method, is recorded on TU 11 (Brack-Bernsen 1999). It enabled the Babylonians to give precise predictions of the times between rising or setting of Sun and Moon in the days around full Moon. We know that this method was known already in the seventh century BCE, since there exist complete lists of such lunar data from that time—evidently not all such values can have been observed: some must have been predicted.[4] This chapter investigates the schematic astronomy and shall look at its connection to the precise empirical predicting rules, which also occur on early cuneiform tablets. And we shall see that the daily retardation of the Moon, which through centuries was observed, listed and used for predictions, also can be used to determine the lengths of day and night as found in the ACT tables. We shall see that the day lengths determined by such observable data show those variations and internal symmetries which we find in the day-length schemes, as reconstructed by Neugebauer (1955, 47, 72; 1975, 368, 712–717).

What we call "observed" Babylonian data are computer-simulated data: From observed data collected in the Diaries and other cuneiform tablets, we know which phenomena were observed by the Babylonians, and we also know the accuracy of such observations, since they can be tested by modern programs able to compute old astronomical events. For example, the observed "Lunar Six" time differences are astonishingly accurate, and the same is true for planetary observations. The Lunar Six are time differences between risings and settings of Sun and Moon, measured in the days around full and new Moon (see Sect. 1.4). The accuracy of recorded observations implies that we can produce computer-simulated "observed data" and compare these data with the numerical models developed by the Babylonians. In this paper, we concentrate on lunar phenomena and compare the Babylonian numerical descriptions with "nature," i.e., with computer-simulated data. These data are partly calculated by S. L. Moshier's code "AA," and partly found by means of Peter Huber's "Creslong.dat."[5]

1.2 The Lunar Schemes from EAE 14 and MUL.APIN

EAE 14 and MUL.APIN describe astronomical phenomena in very similar ways. In both compendia the schematic year consisting of 12 months of 30 days serves as an independent variable. The duration of day and night as a function of the date is given in number schemes. The daily delay of the Moon, being derived from the duration of the night, is hence also a function of the date. The values for day and night lengths are the same in the two compendia, however, their calendars are slightly

[4] Huber and Steele 2007 have been able to date collections of Lunar Six data back to 643 BCE.

[5] Moshier (1996). Computer code "AA" (v. 5.4, public domain) was used for ephemerides of the solar system, using algorithms published in the *Astronomical Almanac* (AA) of the US Government Printing Office. For the Moon, special use is made of Chapront-Touzé and Chapront 1988. Peter Huber has in his creslong.dat (29 November 1997) computed the important lunar data for all full and new Moons from July 758 BCE to October 68 BCE. He has generously made this huge list of data available to those who can use it.

e	1	[DIŠ *sin*	ud 1.ka]m		12	*šiqil*(gín)
e	2	[DIŠ	ud 2.ka]m	1/3 *ma*	4	*šiqil*
de	3	[DIŠ	ud 3.ka]m	½ *ma*	6	*šiqil*
de	4	[DIŠ	ud 4.ka]m	2/3 *ma*	8	*šiqil*
de	5	DIŠ	[ud] ⸢5⸣.[ka]m	1 *ma-na*		
de	6	DIŠ	ud 6.kam	1 *ma*	12	*šiqil*
de	7	DIŠ	ud 7.kam	1 1/3 *ma**	4	*šiqil*
de	8	DIŠ	⸢ud⸣ 8.kam	1½ *ma*	6	*šiqil*
de	9	DIŠ	ud 9.kam	1 2/3 *ma*	8	*šiqil*
de	10	DIŠ	ud 10.kam	2 *ma-na*		
de	11	DIŠ	ud 11.kam	2 *ma*	12	*šiqil*
de	12	DIŠ	ud 12.kam	2 1/3 *ma*	4	*šiqil*
de	13	DIŠ	ud 13.kam	2½ *ma*	6	*šiqil*
de	14	DIŠ	ud 14.kam	2 2/3 *ma*	8	*šiqil*
de	15	DIŠ	ud 15.kam,.	3 *ma-na*		

Fig. 1.1 The first half of Table B in EAE 14 gives the duration of lunar visibility for the first 15 days of the equinoctial Month XII. Note. The durations are here given in units of Mina and Shekel. We see that the lunar visibility increases by 12 Shekel from day to day, and that 60 (= 5 × 12) Shekel equals 1 Mina. On day 15, the Moon is visible over the duration of 3 Mina, i.e., the whole night

different: In EAE 14 spring equinox is put at day 15 in month XII (with length of day and night equal to 3 Mina), while MUL.APIN has the equality of day and night on day 15 of month I. This means that the values of the different astronomical quantities are the same but shifted by one month. In Fig. 1.1, the first half of EAE 14, Table B (Al-Rawi and George 1991–1992, 56) is reproduced.

We see that on day 1 of month XII, the new crescent is described as being visible for 12 shekel. According to the scheme, from day to day the lunar visibility will increase by 12 shekel until day 15, where the Moon is visible during the whole night. Knowing that 1 Mina equals 60 shekel, we can convert the visibility durations into units of shekel:[6] On day 5 the visibility time is set to 1,0 = 60 shekel, on day 6 to 1,12 = 72 shekel, on day 7 to 1,24 = 84 shekel, and so on until the Moon should be visible for 3, 0 = 180 shekel, i.e., the whole night on day 15. Another scheme in EAE 14, Table A (Al-Rawi and George 1991–92, 55), also list the visibility duration of the Moon in the days of the equinoctial month. In this Table A, the time is, however, measured in units of UŠ. An UŠ is a time degree: During 24 hours, the celestial sphere performs a whole revolution of 360°. Therefore, 360 UŠ equal 24 hours, so

[6] The Babylonians used the sexagesimal system in their astronomical tables. It is a position number system like ours, just does it have 60 as basis, instead of 10. In this chapter, I also use the sexagesimal system as it is generally used and illustrated by the following examples: the sexagesimal number 1,24 equals 84 and the number 7, 13, 2; 28,30 equals $7 \times 60^2 + 13 \times 60 + 2 + 28/60 + 30/60^2$.

b	1	DIŠ *sin*	ud 1.kam	3,45 (3¾)			*izzaz*(gub
b	2	DIŠ	ud 2.kam	7,30 (7½)			*izzaz*
b	3	DIŠ	ud 3.kam	15			*izzaz*
b	4	DIŠ	ud 4.kam	30			*izzaz*
b	5	DIŠ	ud 5.kam	1,0 (60)			*izzaz*
bc	6	DIŠ	ud 6.kam	1,12 (72)			*izzaz*
bc	7	DIŠ	ud 7.kam	1,24 (84)			*izzaz*
bc	8	DIŠ	ud 8.kam	1,36 (96)			*izzaz*
bc	9	DIŠ	ud 9.kam	1,48 (108)			*izzaz*
bc	10	DIŠ	ud 10.kam	2,0 (120)			*izzaz*
bc	11	DIŠ	ud 11.kam	2,12 (132)			*izzaz*
bc	12	DIŠ	ud 12.kam	2,24 (144)*			[*izzaz*]
bc	13	DIŠ	ud 13.kam	2,36 (156)			[*izzaz*]
bc	14	DIŠ	ud 14.kam	2,48 (168)			[*izzaz*]
bc	15	DIŠ	ud 15.kam	3,0 (180)			[*izzaz*]
bc	16	DIŠ	ud 16.kam	2,48 (168)	*izzaz*	12	[gi_6.zal]

Fig. 1.2 The first part of Table A in EAE 14 lists the durations over which the Moon is present on the sky for the first 16 days of the equinoctial Month XII. Note. The durations are here given in units of time degrees, UŠ. We see that from day 5 on, the lunar visibility durations are the same as in Table B (Fig. 1.1) increasing by 12 UŠ from day to day. On day 15, the Moon is visible over 180°, i.e., the whole night. On day 16, the scheme tells that the Moon is only visible for 168 UŠ, indicating that the night has passed by over 12 UŠ before the Moon rises. The duration from moonrise to sunset is set to 180° – 12° = 168°

that 1 UŠ will correspond to 4 min, and the daily delay of the Moon 12 UŠ (given in Table A) equals 48 min, which is a reasonable value. The first part of Table A (Al-Rawi and George 1991–1992, 55), is reproduced in Fig. 1.2. We see that for all the days 5–15 of the equinoctial month, the visibility durations given in shekel in Table B are equal to the durations given in time degrees in Table A. This means that the Babylonian unit shekel, measuring time, is equal to the unit UŠ and that 60 UŠ equal 1 mina. The unit mina, used in early times for measuring time is discussed in the appendix. In Old Babylonian mathematical texts, the unit mina was the general unit—used for all kinds of objects—and this unit was at that time divided into 60 shekels or 60 UŠ.

The visibility times of the Moon given in Table B Fig. 1.1 are plotted in Fig. 1.3—together with the visibility durations of the Moon for the months of winter and summer solstices. Based on the insight that the time unit shekel equals time degees UŠ and that 1 mina = 60 UŠ, the durations in Fig. 1.3 are given in units of UŠ.

Other tables from EAE 14 and MUL.APIN list approximations to additional astronomical quantities: (1) the lengths of day and night on the 15th and 30th day of each of the schematic months are listed in EAE 14 Table C, and (2) the daily retardations of the Moon on the first and 15th of each schematic month are given in EAE 14 Table D as well as in MUL.APIN II ii 43–II iii 15, were also the length of the night is given. We recapitulate that the lengths of day and night are found as functions of the date, and that these lengths are used as generating functions for other

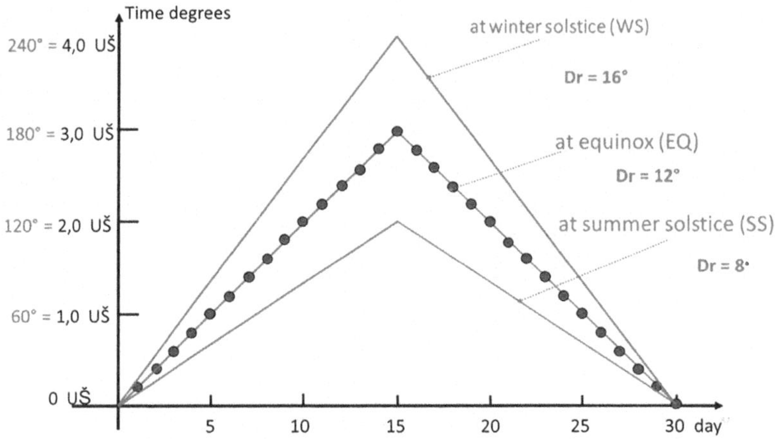

Fig. 1.3 The visibility durations of the Moon. Note. Visibility durations of the Moon in the equinoctial month (EAE 14 Table B) are visualized through the heavy dots. They are connected by straight lines. We see that the visibility durations of the Moon are approximated by a "linear zigzag" function. The same is true for the other months—here we have also drawn the zigzag functions for the months of summer and winter solstice. For these four cardinal points of the year, the daily retardation, Dr, of the Moon, which equals 1/15 night, is also given

astronomical functions. For instance, the daily delay of the Moon is found as 1/15 of the whole night of full Moon (Table 1.1).

It is a reasonable approach to determine the daily retardation of the Moon as the length of the full Moon night divided by 15: at full Moon on day 15, the schematic Moon is visible the whole night. The Moon is supposed to rise at sunset and set at the following sunrise, while on day 0 it is not visible at all (visible for 0 UŠ). Over 15 days, the visibility duration of the Moon goes from 0 UŠ to the whole night N UŠ, hence the increase pr. day is N/15.

As we shall see, the lengths of the night as well as the daily retardations of the Moon are used for other predictions. Therefore, let's examine how well the schematic duration of the night approximates nature: For the 13 months starting at 20 February 747 BCE and ending with 12 March 746 BCE, i.e., year 1 of Nabonassar (747–733 BCE), the **schematic duration of the night** on day 0 and 15 of each Babylonian month are compared to the **real night length** in Fig. 1.4. The actual night length is given in hours, and the schematic night length, given in UŠ, is converted to the unit of hours utilizing that 1 hour = 15 UŠ.

Table 1.1 The length of the night as given in EAE 14 Table C and in MUL.APIN in units of mina and shekel

EAE 14 date table C	MUL.APIN date	Length of night	Length of night given in UŠ	
I 15	II 15	2 2/3 minas	2, 40 UŠ	
I 30	III 1	2 1/2 minas	2, 30 UŠ	
II 15	III 15	2 1/3 minas	2, 20 UŠ	
II 30	IV 1	2 minas 10 shekel	2, 10 UŠ	
III 15	IV 15	2 minas	2, 0 UŠ	Summer solstice
III 30	V 1	2 minas 10 shekel	2, 10 UŠ	
IV 15	V 15	2 1/3 minas	2, 20 UŠ	
IV 30	VI 1	2 1/2 minas	2, 30 UŠ	
V 15	VI 15	2 2/3 minas	2, 40 UŠ	
V 30	VII 1	2 5/6 minas	2, 50 UŠ	
VI 15	VII 15	3 minas	3, 0 UŠ	Fall equinox
VI 30	VIII 1	3 minas 10 shekel	3, 10 UŠ	
VII 15	VIII 15	3 1/3 minas	3, 20 UŠ	
VII 30	IX 1	3 1/2 minas	3, 30 UŠ	
VIII 15	IX 15	3 2/3 minas	3, 40 UŠ	
VIII 30	X 1	3 5/6 minas	3, 50 UŠ	
IX 15	X 15	4 minas	4, 0 UŠ	Winter solstice
IX 30	XI 1	3 5/6 minas	3, 50 UŠ	
X 15	XI 15	3 2/3 minas	3, 40 UŠ	
X 30	XII 1	3 1/2 minas	3, 30 UŠ	
IX 15	XII 15	3 1/3 minas	3, 20 UŠ	
IX 30	I 1	3 minas 10 shekel	3, 10 UŠ	
XII 15	I 15	3 minas	3, 0 UŠ	Spring equinox
XII 30	II 1	2 5/6 minas	2, 50 UŠ	

Note. In column 4, the length of the night is also given in units of UŠ. Note that the cardinal points of the year in EAE 14 are given as the 15th, i.e., full Moon day of month XII, III, VI, and IX, respectively, while in MUL.APIN the cardinal points are given on day 15 of month I, IV, VII, and X, respectively. The length of the night ascribed to day 30 of a month in EAE 14 can also be seen as ascribed to day 0 of the next month. In MUL.APIN, the same length is ascribed to the first day of the next month

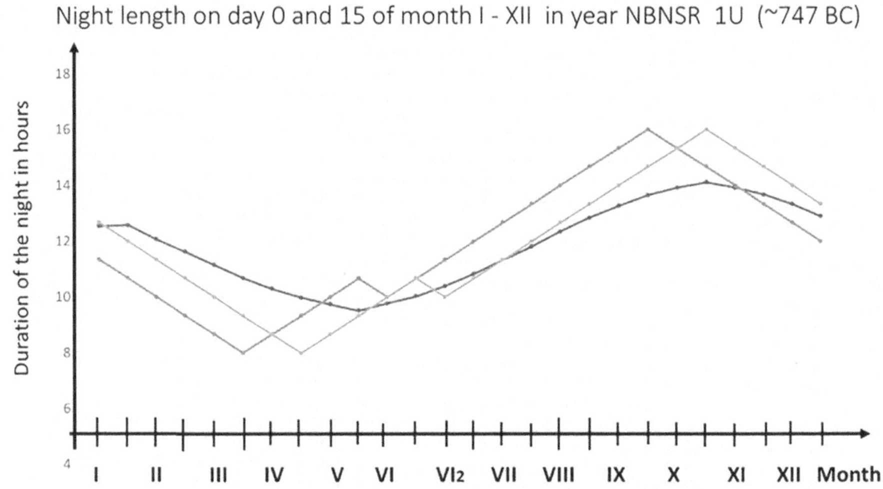

Fig. 1.4 For the first year of Nabonassar, NBNST 1 U, the duration of the night has been found (the blue curve) and compared to the schematic night length given in EAE 14 (the orange zigzag function) and in MUL.APIN (the gray zigzag function). Note. The U of year 1 U indicates that a second month VI, namely month VI_2, is inserted, resulting in a year of 13 months. This extra month is the reason for the little extra zigzags. The schematic duration of the night is in tune with the real night length. The nights are long in the winter and short in the summer. There is, however, a clear discrepancy between the schematic and real night times: the fraction between the longest and shortest night at the latitude of Babylon is 3/2, but the schemes in EAE and MUL.APIN use the fraction 4/2: here the longest night is listed as 4, 0 UŠ = 16 hours and the shortest night as 2, 0 UŠ = 8 hours. This discrepancy has been registered long time ago, and the Babylonians themselves must have noticed it, since they used the correct fraction 3/2 in their later mathematical astronomical texts

1.3 Astronomical Predictions Based on the Lunar Schemes from EAE 14 and MUL.APIN

In the important procedure text TU 11,[7] we find predicting rules based on **observations** as well as predicting rules based on the astronomical schemes from EAE and MUL.APIN, and a strange mixture of both plus some omen astrology. The scribe, Anu-uballiṭ wrote TU 11 around 215/214 BCE. At several places, he writes "he-pi" = broken, which indicates that the tablet is a copy of an older tablet. In addition, some parallel passages have been found in a tablet from the fifth century BCE, and we know that the Goal-Year method was known since the seventh century BCE. This means that at least several paragraphs of TU 11 are from that time. A very important section of TU 11 is Section 19, written in the lines rev. 8–15. This was the only

[7] Brack-Bernsen and Hunger 2002.

section of TU 11 which was translated by Neugebauer[8] who evidently also realized its importance. The first Part of TU 11, Section 19, reads in the translation of Hermann Hunger:

In order for you to find the time of invisibility (of the Moon). (A.) If in the **month I** of your new year (on the) 27th, the **last visibility is 25°,** 3;20(mana) the daylight, 2;40 (mana) the night, multiply 3;20 by 4; (the result is) 13;20. Subtract 13;20 from 25, and 11;40 remains; on the 28th day (the Moon) remains 11;40° behind the Sun. Subtract 13;20 from 11;40; on the 29th day the Moon has passed the Sun 1;40° Add 13;20 to 1;40, and 15 (is the result); on the 30th day the Moon has passed the Sun 15.

Comments to the text. We find here a model calculation: The method how to find the day at which the new Moon will become visible is exemplified by model calculations for month I. Aim of the calculations is to find the day (30, or 1) when the new crescent is visible for the first time after conjunction. The lengths of day and night during month I are listed here to be 3;20 mines and 2;40 mines, respectively. According to the scheme in EAE we have here the duration, of day and night on day 15 of month I. The daily retardation of the rising old Moon (on morning 27) is found as 1/15 of the day length:

$$3;20\,\text{mina times}\,4 = 3,20\,\text{U}\check{\text{S}} \times 1/15 = 13;20\,\text{U}\check{\text{S}}$$

This value for the daily delay of the rising old Moon is then used to find how far the Moon is away from the Sun in the following mornings. The text assumes the value of KUR, the time between last visible Moon rise and sunrise, to be 25 UŠ = 25°. The time between the invisible moonrise and sunrise on the following mornings is found by subtracting 13;20° from this value, and so on. We see, that the **time difference** between rising of Moon and Sun is identified with the **difference in position** along the ecliptic: "11;40 remains; on the 28th day (the Moon) remains 11;40° behind the sun." This means that the time measured along the equator is identified with the lunar movement along the ecliptic. This is quite a daring approximation. It means that in these calculations no differentiation is made between equator and ecliptic. In Fig. 1.5, the calculations of section 19 are reproduced together with a model of the celestial sphere at the latitude of Babylon, where the positions of the Moon at sunrise on the days 27, 28, and 29 are indicated. The horizontal thin great circle is the horizon, the heavy great circle is the equator, and the dotted great circle is the ecliptic. Both Sun and Moon are moving along the ecliptic in the direction indicated by the thin arrow. The Sun ⨀ at the horizon is just about to rise while the Moon at these three consecutive mornings is situated at the positions ☽KUR, ☽28, and ☽29 (the lunar latitude is neglected). On morning 27, the Moon rises at the same time as point A of the equator, while the Sun rises simultaneously with point O. Therefore, the time KUR (from last visible Moon rise to sunrise) is given by the length of arc (A, O). On morning 28, the Moon is at position ☽28 at the moment of sunrise. It is too close to the Sun to be visible. The time interval KUR(28) between the risings of

[8] Neugebauer 1947, 37 n. 4.

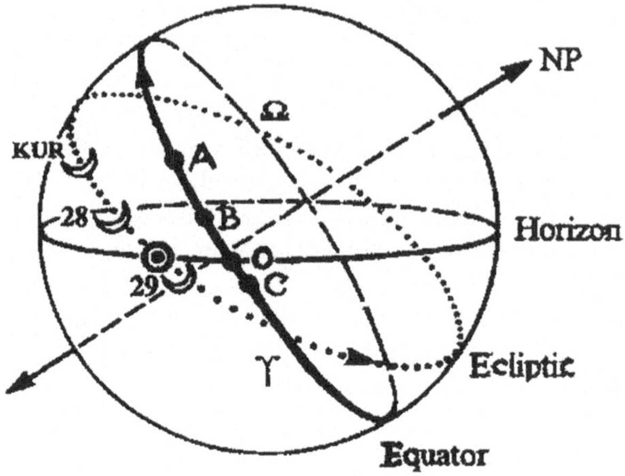

Fig. 1.5 The situation on the celestial sphere for Babylon at sunrise on the mornings of day 27, 28, and 29. Note. The Sun, symbolized by ⨀, situated at the eastern horizon, is about to rise. In the positions 28 and 29, the Moon is too close to the Sun to be observable

Moon and Sun is given by the length of arc (B,O). Sometimes between mornings 28 and 29, the Moon will pass the Sun; the conjunction has taken place, so that the Moon on morning 29 will rise after sunrise. The calculations in Section 19 ignore the difference between the ecliptic and the equator and just identify the ecliptic arc between ☽KUR and ☽28 with arc AB, and arc ☽28, ☽29 with arc BC. According to these calculations, the Moon has passed the Sun on morning 29 by 1;40°, and on morning 30 by 15°.

Schematic reproduction of the calculations for "KUR(m)," the time interval from moonrise to sunrise on morning m:

KUR(27) = KUR = 25°
KUR(28)) = KUR – 13;20 = 11;40°
KUR(29)) = KUR(28) – 13;20 = –1;40°
KUR(30)) = KUR(29) – 13;20 = –15°

In spite of these shortcomings, this method, and thus the idea behind these calculations, is very important. It shows us how the Babylonians used the daily retardation of the Moon for predictions, and it led me to discover and understand the elegant and precise Goal-Year method for the prediction of the observable Lunar Six.

But how accurate is the method? In order to estimate this method (TU 11§ 19), where ΔKUR is approximated by 1/15 day length, the value of the real daily retardation ΔKUR of the rising old Moon has been calculated from Peter Huber's "Creslongdat." It turns out that during the 18 years of the Saros (−459 to −441), ΔKUR (month I) varies between 7.1° and 11.7°, i.e., with a mean value of 9.5° as compared to 13;20° used in § 19. The method seems to be quite inaccurate. In order to check the method directly, **ΔKUR** has been calculated for all months of year Artaxerxes 5 (−459/−458) (Fig. 1.6).

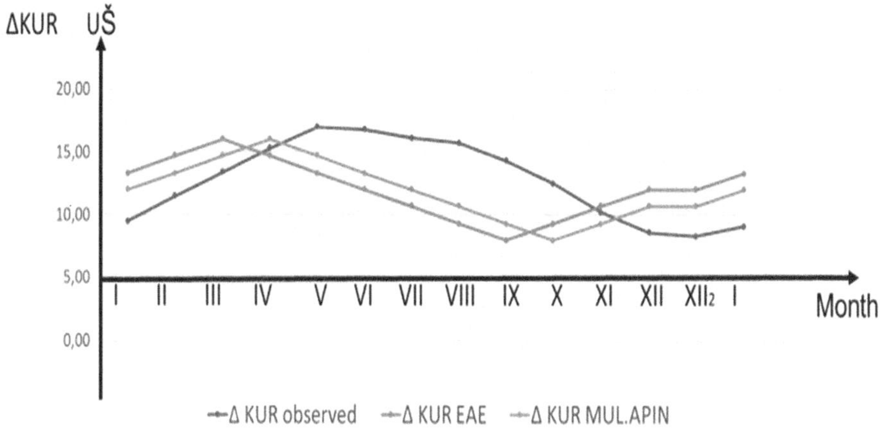

Fig. 1.6 A comparison between the schematic values of ΔKUR for all months I to XII and the exact values of ΔKUR found by Huber's computer simulations, here called "ΔKUR observed." Note. In TU 11, general rules or model calculations are formulated for month I, but they are meant to apply for all months. Therefore, the method must be checked for all months. We see that there is a huge difference (up to 7°) between the real retardation of the rising old Moon and the values from the schemes of EAE and MUL.APIN. The use of schematic values for extrapolation over several days will give completely wrong results. This means that the method found in the calculations in TU 11, section 19, really does not work! It gives, however, important insight into astronomical practice and considerations

1.4 Predicting Astronomy Based on Observations

In contrast to the schematic astronomy, which ignored the difference between the ecliptic and the equator, and in some sense also the difference between East and West, the Babylonian observers surveyed the full Moon when it was setting in the West and rising in the East. They measured the times between risings and settings of Sun and Moon in the days around full Moon. Such observations evidently differentiate between East and West and disclose the changing obliquity of the ecliptic at the horizon. Figure 1.7 shows the situation at the **western horizon** at the **moment of sunrise** on the mornings just before and after full Moon; being near spring equinox, the ecliptic is flat to the horizon. The Babylonians observed the time interval ŠÚ from moonset to sunrise at one morning and the time interval NA from sunrise to setting of the full Moon on the next morning. Their sum, **ŠÚ+NA, is the daily retardation of the setting full Moon.** At sunset, near fall equinox about half a year later, the ecliptic will be much steeper. The time between Moon set and sunrise, e.g., the time NA it takes the ecliptic arc between anti-Sun and ☽NA to set, is obviously heavily dependent of the inclination of the ecliptic. Therefore, the daily retardation ŠÚ+NA of the setting Moon will generally be smaller when observed in spring times than when observed in the fall. This must have been noticed by the astronomers who observed and analyzed these events and use them for predictions.

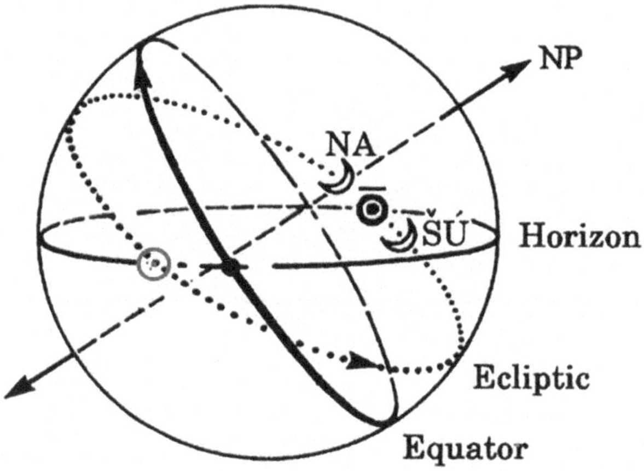

Fig. 1.7 The celestial sphere at sunrise in Babylon shortly before spring equinox. Note. The positions of the full Moon at sunrise on two consecutive mornings is shown (the Sun ☉ is rising at the eastern horizon, the "anti-Sun" is setting at the western horizon). The sphere is turning 360° around the NP axis each day, in the direction indicated by the heavy arrow on the equator. At one morning, the full Moon, in position ☽ŠÚ, has set shortly before sunrise, while the Moon on the next morning in position ☽NA will soon set. The black-dotted ecliptic is situated flat to the horizon at this time

In case of the daily retardation of the rising full Moon, ME+GE_6, it is the other way around: the ecliptic is steep at sunset in the spring (illustrated in Fig. 1.8) and flat at fall time. Therefore, the daily retardation ME+GE_6 of the rising full Moon will generally be larger when observed in the spring than when observed in the fall. Evidently, these observables reveal a different behavior of the Moon when observed by setting or rising in the spring from that observed in the fall. In contrast, the schemes in EAE and MUL.APIN derived the daily retardation of the Moon from the night length, which was 3,00 US at spring as well as fall equinox. The daily retardation of the Moon was assumed to be the same, namely 12°, in the month of spring equinox (XII or I, respectively) as that in the month of fall equinox (VI and VII, respectively).

In Fig. 1.8, the anti-Sun indicates the position in the ecliptic, where the opposition of Moon and Sun took place. The time interval ME between the rising of the Moon in position ☽ME and anti-sunrise (= sunset) was measured. On the next morning, the time GE_6 from sunset to moonrise, with the Moon in position ☽GE_6, was also measured. Obviously, ME+GE_6 is the daily retardation of the rising full Moon.

Around full Moon, these four time intervals could be measured, while around new Moon, only two such time intervals were observable: KUR and NA_1. We call the six observable time intervals between risings of Sun and Moon around full and new Moon the "Lunar Six." They have been observed regularly and recorded by the Babylonians since 700 BCE.

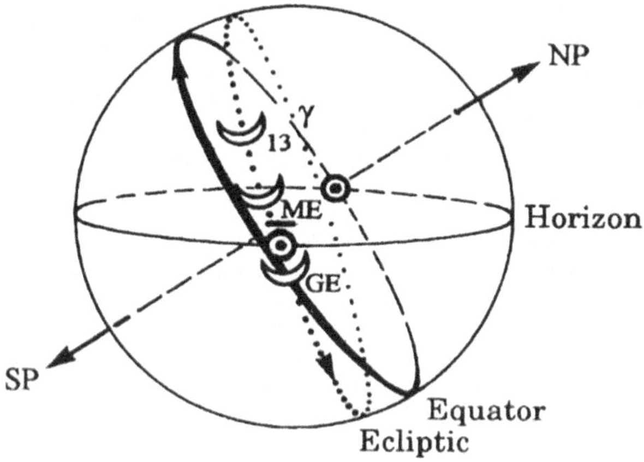

Fig. 1.8 The celestial sphere at sunset in Babylon shortly before spring equinox. Note. The situation at the eastern horizon on evenings shortly before and after opposition is shown. The "anti-Sun" is rising at the eastern horizon. The full Moon moves along the ecliptic as indicated by the thin arrow. On evening 13 the Moon is at the position ☽13, on the next evening it is at position ☽ME, and on evening 15 it is at position ☽GE_6

The Lunar Six:

Shortly before and after conjunction, KUR and NA_1 can be observed:
KUR is the time from last visible moonrise to sunrise ☾↑⊙↑
NA_1 is the time from sunset to first visible moonset after sunset ⊙↓☾↓
Around opposition, the "Lunar Four" can be observed:
ŠÚ is the time from last moonset before sunrise to sunrise ☾↓⊙↑
NA is the time from sunrise to first moonset after sunrise ⊙↑☾↓
ME is the time from last moonrise before sunset to sunset ☾↑⊙↓
GE_6 is the time from sunset to first moonrise after sunset ⊙↓☾↑

The so-called Goal-Year Tables collect all Lunar Six values for a whole year in addition to the sums ŠÚ+NA and ME+GE_6 for half a year. Test calculations have shown that such data were astonishingly accurate, and they were used for predictions of ŠÚ and NA or ME and GE_6, respectively. This shows that the Babylonians differentiated between the daily retardation of the setting full Moon, ŠÚ+NA, and the daily retardation of the rising full Moon, ME+GE_6.[9] How these sums were used for prediction can be read in TU 11 §16.

The Goal-Year method for the prediction of the Lunar Six in TU 11 §16:

Rev.11′) In order for you to calculate ŠÚ and NA, [you return] 18 years behind you[… you add ŠÚ and NA together,]

12′) you take one-third of them, and [you subtract] it from NA of the middle of month I, [and what remains of NA you predict as the NA of your new year;]

[9]Their sum $\Sigma = \check{S}\acute{U} + NA + ME + GE_6$ measures the movement of the Moon relative to the Sun over two days.

13′) and you add one-third of ŠÚ and NA to the ŠÚ, and [predict it as the ŠÚ of your new year]
14′) If you subtract one-third of ŠÚ and NA from the NA and it goes beyond the NA. …

This predicting method is composed so scarcely that it is difficult to be understood by non-insiders. But our knowledge from § 19 helped to understand it. Here again, the daily retardation of the Moon ŠÚ+NA was used for interpolation: The Lunar Four are predicted on the basis of the "old" Lunar Four values observed one Saros = 18 years earlier.

$$\check{S}\acute{U}(\text{new}) = \check{S}\acute{U}(\text{old}) + 1/3\left(\check{S}\acute{U} + \text{NA}\right)(\text{old})$$

$$\text{NA}(\text{new}) = \text{NA}(\text{old}) - 1/3\left(\check{S}\acute{U} + \text{NA}\right)(\text{old})$$

The Saros is a very special time interval: 1 Saros = 223 synodic months ≈ 239 anomalistic months ≈ 241 sidereal months ≈ 242 draconic months ≈ 6585 1/3 day ≈ 18 years

All the important periods describing the movement of the Moon repeat after a Saros: the lunar phase and velocity, the position of the Moon within the ecliptic, as well as the lunar latitude. It has turned out that the Goal-Year method is excellent, and the reason why it works is that ŠÚ+NA and most of the variables determining ŠÚ and NA repeat after one Saros. Only the time between opposition and sunrise will be shifted by 8 hours = 1/3 day.[10] The observation of ŠÚ will take place 8 hours later with respect to opposition than one Saros earlier, while NA is observed 8 hours earlier with respect to opposition. These 8 hours = 1/3 day are incorporated through the addition and subtraction, respectively, of 1/3 (ŠÚ+NA), i.e., 1/3 of the daily retardation of the Moon. Figure 1.9 shows how accurately the daily retardation of the Moon ŠÚ+NA repeats after one Saros.

We have seen two different predicting methods: one based on computations within the schematic astronomy and the other, the Goal-Year method, based on observational data. They seem somehow to be connected, and both methods occur on one and the same cuneiform tablet. This leads to some questions: What is the connection between the two? Which method came first, TU 11 §19 or TU 11 §16? Who got the idea to use the daily delay of the Moon for predictions? In addition, there seems to have been a strange crossover between the schematic and the predicting astronomy. In §18 of TU 11, ideas from the Goal-Year method are applied to schematic lunar values. The text tells us to **go 18 years back and take NA of your old year and use the schematic night length** to find NA and the length of the month. The text does, however, not use 1/3, but 1/2 of the daily delay of the Moon. Table TU 11, written by the scribe Anu-uballiṭ, shows that he was familiar with both

[10] See Brack-Bernsen 1997, 115–21 and Brack-Bernsen 1999.

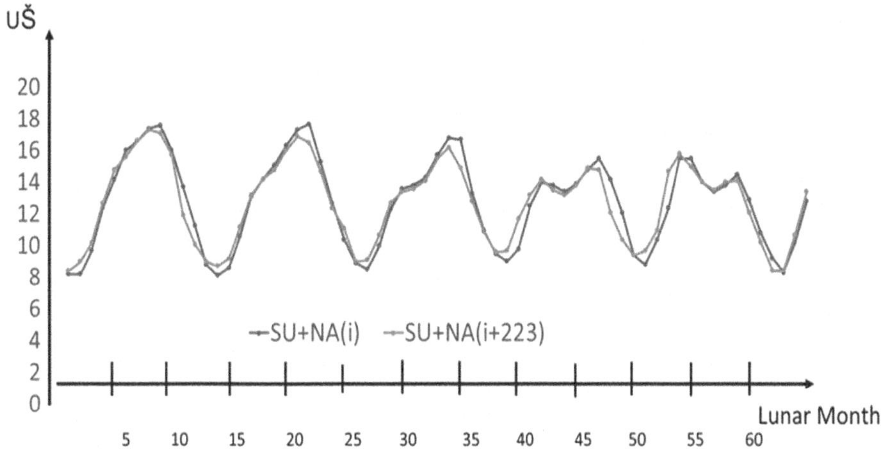

Fig. 1.9 The values of ŠÚ+NA over 5 years (from −700 to −695) are printed as a function of the lunation number i, and compared to those measured 1 Saros later (from −682 to −677). Note. We see that the daily retardation of the setting full Moon ŠÚ+NA repeats rather precisely after 1 Saros = 223 months

types of astronomy, the schematic and the predicting one. It seems that the astronomers in Mesopotamia knew and exercised both types of astronomy. (Some mathematical table texts are written by astronomers with the title "scribes of EAE.") This would explain the crossover between the schematic, the predicting, and the mathematical astronomy. It is typical for scientists (also in our time) to test different theories or approaches against each other, in order to combine them and make predictions based on the different models. I think that this is what TU11 witnesses. Evidently, the Goal-Year method is heavily based on observations, namely (1) that the situations at the horizon are similar at full Moons situated 1 Saros apart, and (2) that the sums ŠÚ+NA and ME+GE_6 repeat after 223 synodic months. However, the calculations from TU 11 §19, presented in Sect. 1.3, may just have been invented by scribes who used the schematic values of the Moon's daily retardation to survey its movement relatively to the Sun. Or the calculations may have been inspired by the Goal-Year method—or §19 may have been a pedagogical way of explaining the ideas behind the Goal-Year method by means of the schematic astronomy, which probably was the first to be taught to young scribes.

How well or how badly does the schematic astronomy describe nature? We have seen how the schematic astronomy was used for astronomical descriptions and predictions. Figure 1.4 showed that the schematic lengths of day and night was quite inaccurate, and Fig. 1.6 indicated that the method for predicting the day of the new crescent as presented in TU 11 §19 does not work. However, we got important insight into astronomical considerations. We shall now check two other schematic approximations against nature:

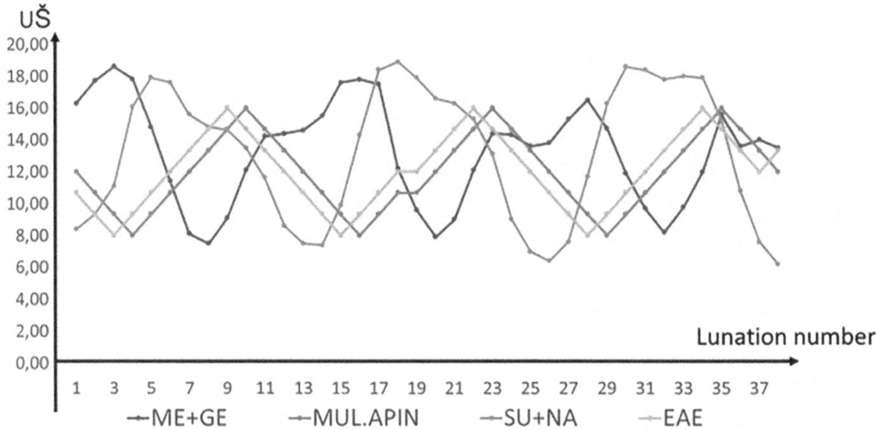

Fig. 1.10 The daily retardation of the full Moon, plotted as a function of the lunar month. Note. The blue curve depicts the retardation of the rising full Moon, while the gray curve depicts the daily retardation of the setting full Moon. The orange and red zigzag function depict the retardation according to EAE and MUL.APIN, respectively. We see that these functions, as compared to "nature," are out of phase by three months

The daily delay of the (full or new) Moon was found as the night length times 1/15

In Fig. 1.10 for 37 lunar months, the daily retardation of the rising and setting full Moon ME+GE$_6$ and ŠÚ+NA, respectively, is printed and compared to 1/15 night length, given as a function of the month according to EAE and MUL.APIN. We must conclude that this is a bad fit to nature. The curves are three months out of phase and, also, the variation is too small. We also see that the blue curve is at its minimum when the gray curve is large (in the fall), and large (in the spring) when the gray curve is at its minimum. But, surprisingly, another schematic prediction rule works very well.

The times of eclipses were found from eclipses observed 1 Saros earlier. Model calculations in TU 11 § 9–12 illustrate how the times of expected eclipses were found from eclipses that had been observed one Saros earlier. This method turns out to be surprisingly good.[11] The method is based on the fact that eclipses repeat after 1 Saros. But the time of the eclipse is shifted by 1/3 day (1 Saros = 6585 + 1/3 day). This was known by the Babylonians who arranged dates of possible lunar eclipses in huge Saros schemes, starting back from 747 BCE. Let T_0 be the time of an "old eclipse." Then the time T of the eclipse occurring 1 Saros later was found as:

$$T = T_0 + 1 + 1/3\,\text{night}.$$

This means that, with respect to sunset, the delay after 1 Saros of lunar eclipses was approximated by 1 + 1/3 night. This was a quite good approximation, as we can see in Fig. 1.11.

[11] Brack-Bernsen and Steele 2005; Steele 2000; and Brack-Bernsen and Hunger 2002, 79–85.

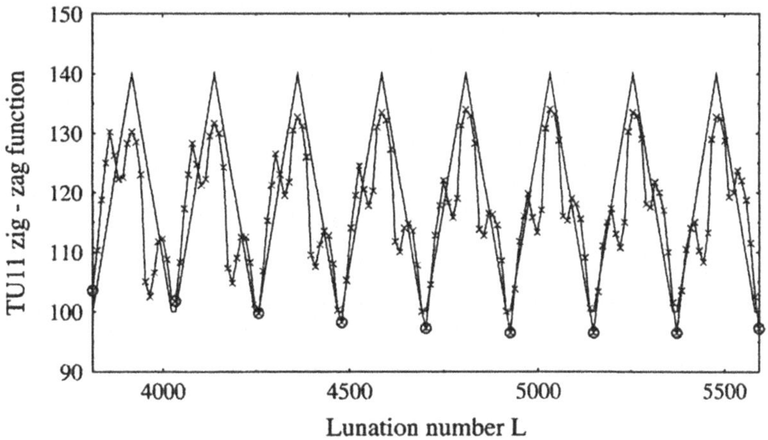

Fig. 1.11 For a series of consecutive lunar eclipse possibilities near the descending node (693–549 BCE), the length of the Saros is found by computation and compared to the TU 11 zigzag function

In Fig. 1.11, the accuracy of the TU 11 procedure for finding Saros's length is checked. For a series of consecutive lunar eclipse possibilities taking place near the descending node, the length of the Saros is found by computation and compared to the TU 11 method. The delay of eclipses was in TU 11 calculated as 1 + 1/3 night, which defines a linear zigzag function varying between 100° = 1,40 UŠ and 140° = 2,20 UŠ. We compare this zigzag function with the computer-simulated "nature": The comparison takes place over eight consecutive Saroi, starting with the first eclipse possibility, 10 March 693 BCE, and ending with the eclipse possibility on 4 June 549 BCE. The Saros length, measured from consecutive eclipse possibilities at the descending node, is marked by crosses and connected by a thin line; the TU 11 zigzag function is drawn with a thick line. As we see in Fig. 1.11, this zigzag function has the correct mean value of 2,00 UŠ = 1/3 day, and it approaches the real duration of the Saros very well. It has the same variation and phase as the computer-simulated times and gives thus a perfect fit to "nature." We do, however, not know if this nice predicting method is based on theoretical considerations within the schematic astronomy, or if it is an empirical fit, or a mixture of both!

1.5 Rising Times of Ecliptic Arcs and Day Length

In the lunar ACT table texts, the durations of day and night were recorded properly: here we find the correct values for the longest and shortest day, and the day-length schemes fit nature quite well. The question is how may the Babylonians have constructed their day-length schemes? We know, that the lengths of day and night are functions of the actual position of the Sun in the ecliptic: Let the Sun be at the

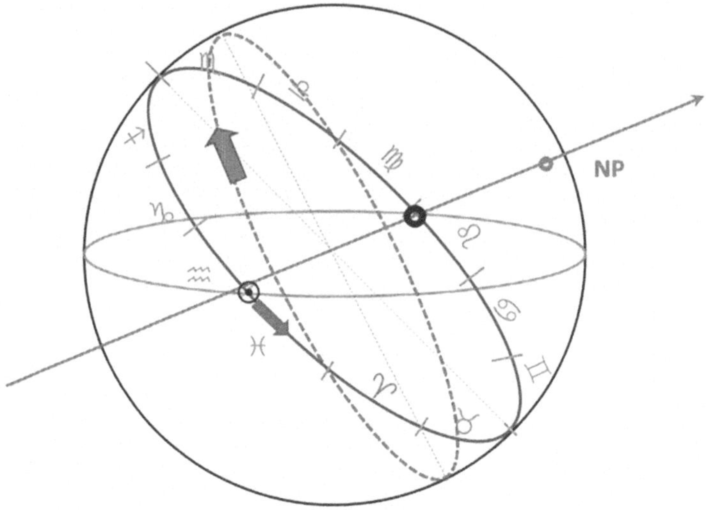

Fig. 1.12 The celestial sphere at the latitude of Babylon one month before spring equinox. Note. The Sun ⊙ is at the beginning of Pisces and about to rise, while the "anti-Sun" ◉ is setting. The direction of the daily revolution of the sky is shown by the red arrow on the red equator. The Sun is moving slowly (1° per day) along the blue ecliptic in the direction given by the blue arrow

position $\lambda\odot$ in the ecliptic. The day begins when the point $\lambda\odot$ of the ecliptic rises, and it ends when it sets. The duration of the day equals the time it takes the ecliptic arc, ($\lambda\odot$, $\lambda\odot$+180°) to rise. Figure 1.12 shows the situation on the celestial sphere when the Sun, at the beginning of Pisces, is about to rise. From spherical geometry we know about symmetries and connections between rising times of ecliptic arcs; but what did the Babylonians know?

We see that the six zodiacal signs ♓♈♉♊♋♌ (= Signs XII, I, II,… V) will rise during daytime, and the signs ♍♎♏♐♑♒ will set. The length of this day equals the rising time of the signs ♓♈♉♊♋♌, and the night length equals their setting time. Note that the rising time of sign n equals the setting time of sign n + 6. The symmetries in Fig. 1.12 result in the following equalities. The rising time of sign I =, ρ(I) equals the rising time of sign XII: ρ(I) = ρ(XII), and so on: all other rising times of the zodiacal signs are pairwise equal ρ(II) = ρ(XI), ρ(III) = ρ(X), ρ(IV) = ρ(IX), ρ(V) = ρ(VIII), and ρ(VI) = ρ(VII). It is also evident that the rising time of sign N equals the setting time of sign N + 6: ρ(N) = σ(N + 6), so that the setting times also will be pairwise equal. Such knowledge of rising and setting times of the zodiacal signs enables us to find the day length for all positions of the Sun. When the Sun is at the beginning of ♓, i.e., at position 0° in sign XII, then the day length is ρ(XII) + ρ(I) + ρ(II) + ρ(III) + ρ(IV) + ρ(V). The question is how the Babylonian astronomers found the day length, without having the spherical model of heavens.

Otto Neugebauer has treated Babylonian day length in ACT I (47, 72) and in HAMA (368; 712–717). He mentions that column C gives the length of daylight as a function of the seasons, and proposes that the day-length schemes are constructed on the following assumptions: The equinox is at position ♈ 10° in System A and at ♈ 8° in System B, and the duration of the longest and shortest days are known to be:

$$\text{Longest day M} = 3{,}36\ \text{U}\check{\text{S}} = 216^\circ \quad \text{Shortest day m} = 2{,}24\ \text{U}\check{\text{S}} = 144^\circ$$

With these extreme values of day length as starting point, Neugebauer proposes that the day-length schemes are the result of arithmetical skill alone and do not origin in observations of rising ecliptic arcs. Thereafter, he uses the day-length scheme to reconstruct the rising times for ecliptic arcs of 30°. In HAMA p. 368 he writes (the bold marking is mine):

> **The arithmetical, non-observational, basis of the Babylonian description of the variable length of daylight** becomes still more evident when one looks in the ACT material at the methods which lead from a given solar longitude λ⨀ to be the corresponding length of daylight (therefore denoted by C). As we have seen before C = d(λ⨀) must be the rising time or oblique ascension of the semicircle of the ecliptic from λ⨀ to λ⨀ + 180°. Making use of this relationship one can use the Babylonian values of C to calculate the corresponding rising times for consecutive 30° sections of the ecliptic, beginning with the vernal equinox. The resulting values turn out to belong to one of the following two arithmetical progressions which we call "System A" and "System B," respectively.

Calling the rising arc of sign N: ρ(N), Neugebauer constructs the following values for the 12 signs (See Table 1.2):

We call the setting arc of sign N: σ(N), and we know that for all N σ(N) = ρ(N + 6) (modulo 12); and the Babylonians knew this, too. From applications made by the Babylonians, we know that they used the same values for nighttime as for day time, however in inverse order.

The values in columns 3 and 4 of Table 1.2 are well chosen, and the symmetries and connections between ρ and σ, i.e., between rising and setting arcs, are all correct. Is that just a "lucky result" of arithmetical skill? I don't think so. It is fair to ask: from where came the knowledge of these symmetries and the numerical values? We know it from spherical geometry, but the Babylonians?

The Babylonians had observed ME+GE_6 and ŠÚ+NA systematically and regularly since 700 BCE, and such observations deliver basic knowledge of rising and setting arcs. I claim therefore that there is an observational basis for the schemes,

Table 1.2 Columns 3 and 4 give Neugebauer's values for the rising times (in arcs) as defined in column 2. In the first column, I have added the setting times σ(N) of the signs

Setting time	Rising time	System A	System B
σ 7 = σ 6	ρ 1 = ρ 12	20°	21°
σ 8 = σ 5	ρ 2 = ρ 11	24°	24°
σ 9 = σ 4	ρ 3 = ρ 10	28°	27°
σ 10 = σ 3	ρ 4 = ρ 9	32°	33°
σ 11 = σ 2	ρ 5 = ρ 8	36°	36°
σ 12 = σ 1	ρ 6 = ρ 7	40°	39°

Fig. 1.13 The obverse side of Table VAT 7847, published as Tafel 6 by Ernst Weidner (1967). Note. We have here a representation of the zodiacal sign Leo, sign V, where only a small part is missing at the end. Leo is subdivided into 12 microsigns through the little squares below the hydra. Ten of the microsigns are fully visible and their names are written in cuneiform; I have added the corresponding number of the microsigns in Roman numbers. Each of the microsigns measures 2½°

and I propose that observations of the daily retardation of the rising and setting Moon, ME+GE$_6$ and ŠÚ+NA, are the basis of the Babylonian description of the variable length of daylight. Other texts document that the Babylonians, indeed, were concerned with rising and setting of ecliptic arcs. From Rochberg (2004) "A Babylonian Rising-Times Scheme" we know that the Babylonians treated rising times, however in this case for microsigns: Microsigns are known from astrological/ medical texts, and they are part of the schematic astronomy, but they have no connection to observational astronomy (Fig. 1.13).

In a rough approximation, the Sun moves 1° each day, while the Moon moves 13° per day, i.e., it moves 12° farther than the Sun. During the 30 days of the schematic fifth month, the Sun will move all way through Leo, and the Moon will move 360° plus the 30° of Leo. Each time the Sun has moved 2½°, the Moon has moved 2½° plus 12 × 2½° = 2½° + 30°, according to this schematic description. When the Sun has moved N microsigns, the Moon will have moved N signs plus N microsigns. This means that the Sun will be at the Nth microsign in sign V while the Moon will be at the Nth microsign of sign V + N. This is, of course, a quite primitive schematic model, but it was important within the astrological and medical practices. But obviously, it has nothing to do with observation, how and where would you identify an ecliptic arc of 2½ degrees? It would have been absolutely impossible to find the rising arc of a microsign by observations; but the schemes show us that rising arcs of ecliptic arcs were of interest. Francesca Rochberg (2004) proposes that the rising-time schemes were forerunners to, or eventually an alternative version of, the rising times underlying column C. In his book "Rising Time Schemes in Babylonian Astronomy," John Steele (2017) presents all known sources connected

Table 1.3 A summary of the numbers behind the rising-time schemes. Note. Column 3 gives the distances between those *ziqpu* stars which were supposed to culminate in the middle of the corresponding months or zodiacal signs

Date	Zodiacal position	Distance between culminating points	Rising time of micro sign	Day length according to MUL.APIN
I 15th	♈ 15°	20	1;40	3,00 UŠ = 12 h
II 15th	♉ 15°	20	1;40	3,20 UŠ = 13;20h
III 15th	♊ 15°	20	1;40	3,40 UŠ = 14;40h
IV 15th	♋ 15°	40	3;20	4,00 UŠ = 16 h
V 15th	♌ 15°	40	3;20	3,40 UŠ = 14;40h
VI 15th	♍ 15°	40	3;20	3,20 UŠ = 13;20h
VII 15th	♎ 15°	40	3;20	3,00 UŠ = 12 h
VIII 15th	♏ 15°	40	3;20	2,40 UŠ = 10;40h
IX 15th	♐ 15°	40	3;20	2,20 UŠ = 9;20h
X 15th	♑ 15°	20	1;40	2,00 UŠ = 8 h
XI 15th	♒ 15°	20	1;40	2,20 UŠ = 9;20h
XII 15th	♓ 15°	20	1;40	2,40 UŠ = 10;40h

to rising-time schemes. He analyses the texts and positions them within the schematic astronomy, and he explains their connection to *ziqpu* stars, i.e., culminating stars, which were used for timing. He summarizes: "I would argue that the purpose of the microzodiac rising-time scheme is not to provide a method to calculate the length of daylight but instead to present a description of the relationship between culminating positions of (parts) of the zodiacal signs in a way that is consistent with the schematic tradition embodied by MUL.APIN" (107).

The rising-time schemes are all based on or agreeing with, the traditional ratio of 2/4 for the length of the shortest to the longest day; but most important, they document a theoretical interest in rising arcs of zodiacal arcs (Table 1.3).

We see that the rising times of the 12 microsigns of each zodiacal sign are fixed to either 1;40° for one-half of the year/zodiac, or to the double 3;20° for the other half of the schematic year/zodiac. The constructed times of culminating *ziqpu* stars agreed to the fraction of 2/4 between the shortest and longest day, where real observations would have delivered the fraction 2/3. Evidently, this scheme is **irreconcilable with astronomical observations.** It is impossible to observe rising times of microsigns, and the times of culminating *ziqpu* stars are constructed so that the astronomers got the result they wanted! The rising-time schemes of the microzodiac is certainly not a forerunner of the day-length schemes behind column C. But it might have been constructed inspired by the day-length schemes based on observations. One may ask: Where did the interest in rising times come from? What was the basis for finding day lengths by rising times of ecliptic arcs? Where did the knowledge of zodiacal symmetries shown in Table 1.2 originate? I propose that the answer to these questions can be found in the Babylonian practice of observing the Lunar Six and their utilization for predictions. The rising-time schemes for microsigns may be the alternative, schematic version of time schemes derived from observations of rising arcs (the second surmise of Rochberg 2004).

We know that rising times were, indeed, observed by the Babylonians. Through centuries, the daily retardation of the full Moon, ME+GE$_6$ and ŠÚ+NA, was observed. This daily retardation of the rising and setting Moon measures the time it takes the elongation arc to pass the horizon. The daily elongation arc is the arc of the ecliptic which measures the movement of the Moon relatively to the Sun during one day, in this case on the day of full Moon. Section 19 in TU 11 and the Goal-Year method in section 16 show us that the Babylonians understood the significance of ME+GE$_6$ and ŠÚ+NA: they measured time differences which originate in the movement of the Moon with respect to the Sun. Therefore, I propose that the rising-time schemes for Systems A and B, as presented in Table 1.2, were constructed from observations of ŠÚ+NA and ME+GE$_6$.

1.6 Day-Length Schemes Derived from Observations of ME+GE$_6$ and ŠÚ+NA

We know that ME+GE$_6$ and ŠÚ+NA measure the time it takes the daily elongation arc to rise and set, respectively. Remember that month M was often identified with sign ZM. Therefore, when measured in month M, ME+GE$_6$ and ŠÚ+NA contain information on the rising- and setting arc of sign ZM. As we shall see, a **simple approximation** can lead to the rising times we are looking for. Let us simply use the standard mean value of 12°/day for the daily movement of the Moon with respect to the Sun. Then it follows that, approximately:

$$\text{ME} + \text{GE}_6 \text{ measures the time it takes } 12^\circ \text{ to rise}$$

$$\check{\text{S}}\acute{\text{U}} + \text{NA measures the time it takes } 12^\circ \text{ to set}$$

All the zodiacal signs ZM measure 30°. Therefore, the rising and setting time for each sign ZM, can be found from the rising and setting times of the elongation arc of 12° measured in month M just by a multiplication with the factor 2.5 = 30/12.

$$\text{Rising time of Sign ZM} = \text{ME} + \text{GE}_6(\text{M}) \times (30/12) = \text{ME} + \text{GE}_6(\text{M}) \times 2.5$$

$$\text{Setting time of Sign ZM} = \check{\text{S}}\acute{\text{U}} + \text{NA}(\text{M}) \times (30/12) = \check{\text{S}}\acute{\text{U}} + \text{NA}(\text{M}) \times 2.5$$

We shall now investigate how computer-simulated setting times fit the numerical values used by the Babylonians. Thanks to Peter Huber, who has computed the circumstances of all new, full, and old Moons for the Lunations from −758 to –68, we have a very useful tool which enable us to follow the behavior of the Moon, as seen from Babylon, during almost 700 years. From Huber's "Creslong.dat," the "observed" values of ME, GE_6, ŠÚ and NA can be found. I have done this and added them up to find the partial sums ME+GE_6 and ŠÚ+NA over longer time periods. Starting with the Babylonian month I = Nissan (Kandalanu 1A I), which began by the first visibility of the new crescent on 26 February in the year 647 BCE. and during 18 years, I have found the "observed" values of ME+GE_6 for all months over 1 Saros. Based on these data, for each month I, II, III, …, XI and XII, the mean values of ME+GE_6 over 1 Saros have been listed in Table 1.4. For each month and thus also for each zodiacal sign, we find the approximate time it takes 12° to rise.

Appropriate values for the rising time of each sign ZM can be found from these values by a multiplication with 30/12 = 2½. Therefore, the rising time of sign $ZM = <ME+GE_6(ZM)> \times 2.5$. These rising times are in Table 1.5 (column 3) compared to the rising times ρ 1 to ρ 12 of the zodiacal signs as reconstructed by Neugebauer in Table 1.2.

With this evidence, I conclude that the time schemes in the ACT texts origin in observations of the Lunar Four in combination with calculations or considerations similar to the arguments above. As a further argument, the practice of the Babylonians also indicated that they knew the identity $\sigma(N) = \rho(N + 6)$, namely that the setting time of a sign equals the rising time of the sign 180° apart. In addition, this identity

Table 1.4 Utilizing the identification month M ~ Sign ZM we have here found the approximate time $< ME+GE_6 >$ it will take 12° of each zodiacal sign to rise

Month	I	II	III	IV	V	VI	VII	VIII	IX	X	XI	XII
Z. Sign	♎	♏	♐	♑	♒	♓	♈	♉	♊	♋	♌	♍
<ME+GE6>	**15,4**	**15,7**	**15,6**	**13,8**	**11,2**	**9,2**	**8,4**	**9,6**	**12**	**14,7**	**15,8**	**15,6**

Table 1.5 The "Babylonian" rising times ρ(M) of the 12 signs are reproduced in column 4 (System A) and 5 (System B)

ρ (sign)	<ME+GE>	<ME+GE > 30/12	Syst A	Syst B
I	15,4	**38,5°**	40°	39°
II	15,7	**39,25°**	36°	36°
III	15,6	**39°**	32°	33°
IV	13,8	**34,5°**	28°	27°
V	11,2	**28°**	24°	24°
VI	9,2	**23°**	20°	21°
VII	8,4	**21°**	20°	21°
VIII	9,6	**24°**	24°	24°
IX	12	**30°**	28°	27°
X	14,8	**37°**	32°	33°
XI	15,9	**39,75°**	36°	36°
XII	15,7	**39,25°**	40°	39°

Note. These numbers can very well be seen as numerical approximations to "Observed" data in column 3. We also see that rising times found by observations do, indeed, possess the correct symmetries

Table 1.6 The symmetries of the rising and setting times are clearly visible in the "Observed" values of ŠÚ+NA and ME+GE$_6$

Month	ŠÚ+NA	ME+GE6
I	**8,6**	**15,4**
II	**9,7**	**15,7**
III	**11,8**	**15,6**
IV	**14,3**	**13,8**
V	**15,7**	**11,2**
VI	**15,7**	**9,2**
VII	**15,3**	**8,5**
VIII	**15,7**	**9,7**
IX	**15,7**	**11,8**
X	**14,2**	**14,7**
XI	**11,5**	**15,9**
XII	**9,1**	**15,7**

Note. For the data in columns 2 and 3, we see two symmetry lines: between month/sign VI and VII, and between month/sign XII and I. In addition are the values in column 2 for month/sign I, II, … VI almost equal to the values in column 3 for month/sign VII, VIII, …, XII

is visible directly from the observed retardations of the Moon, ŠÚ+NA and ME+GE$_6$, as we can see in Table 1.6.

That the Babylonians understood the connection between the lunar position with respect to the Sun and the rising arcs is evident from their Goal-Year method for the prediction of the Lunar Six. The sums ŠÚ+NA and ME+GE$_6$ were heavily used, not only to find the Lunar Four, but also to predict KUR and NA$_1$ (see Sect. 1.4). Around conjunction, where the daily retardation of the Moon could not be observed, the Babylonians used the daily retardations of the full Moon observed six months earlier. They obviously knew that the situation of the old and new Moon at the eastern and western horizon, respectively, was similar to the situation of the rising and setting full Moon taking place half a year, i.e., six months earlier (see Brack-Bernsen 1999, esp. 168–172).

Until now I have argued by means of mean values of Babylonian observations. The system which we have found for the mean values, and the good agreement with Systems A and B, do also show up for values found directly from monthly observations, as Figs. 1.14 and 1.15 and table 1.7 show.

In a simple first approximation, for all months, ME+GE$_6$ gives a fair estimate of the rising time of those 12° which the Moon in average has traveled with respect to the Sun on the day of full Moon. We know, however, that **month M** was identified with **sign ZM.** Therefore**,** we have here also found the time interval ME+GE$_6$ it will take 12° of each zodiacal sign to rise.

The rising time of each sign of 30° can be found by multiplying these values by 30/12 = 2.5. The results for these calculations are in Fig. 1.14 compared to the values in Systems A and B.

In Fig. 1.14 we see a good agreement. The values determined through observations are given by the blue curve, while the values behind the times in Systems A

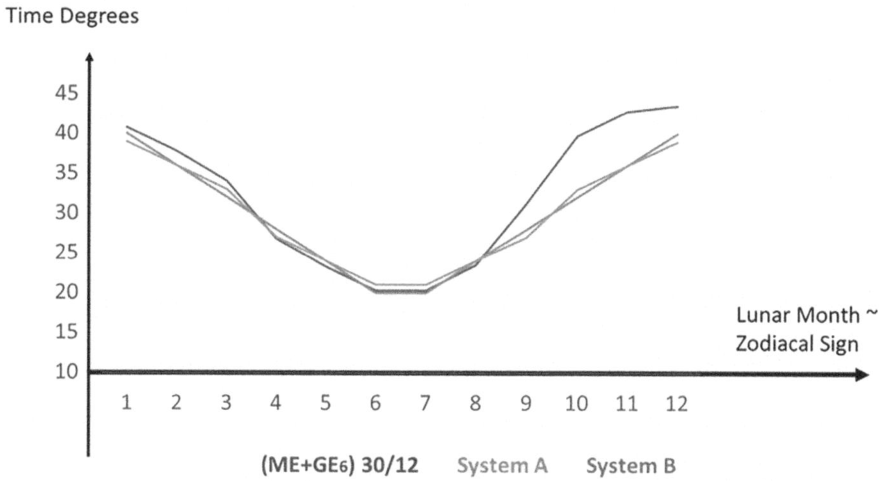

Fig. 1.14 The rising times of zodiacal signs, determined by "observed ME+GE$_6$" during the Babylonian year Artaxerxes 6 (starting on 17 April 459 BCE). Note. The blue curve shows these rising times which are compared to the values behind Systems A and B (orange and gray curves, respectively)

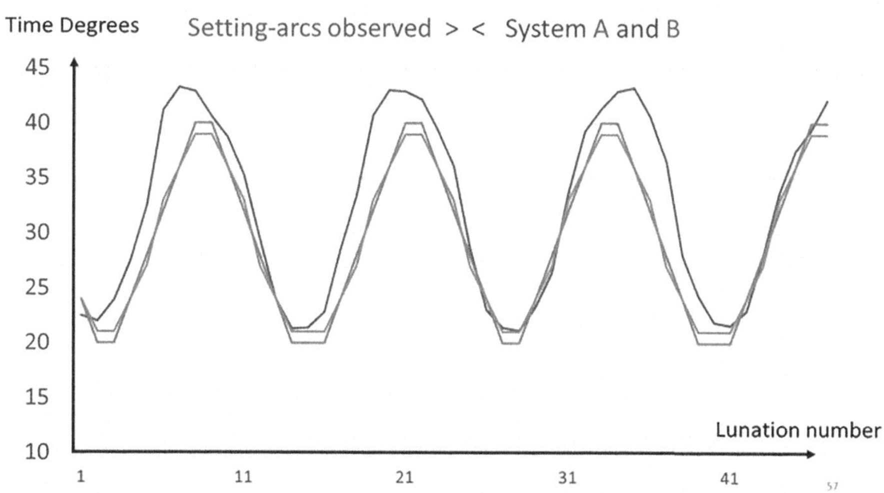

Fig. 1.15 Setting arcs found from ŠÚ+NA, "observed" over 3½ years are compared to the values behind systems A and B. Again, we have a fine fit "to nature."

and B are represented by the orange and gray curves, respectively. The Babylonian rising times fit "nature" well. The same is true for the setting times of zodiacal signs. In Fig. 1.15 the setting arcs found from ŠÚ+NA, "observed" over 3½ years starting on 21 January 500 BCE (year 21 month XI of Darius I) are visually compared to the values behind Systems A and B. The setting arc of sign M is found from ŠÚ+NA observed in month M multiplied by 30/12.

Table 1.7 The rising times ME+GE$_6$ for all months of the Babylonian Year Artaxerxes 6 (Starting on 17 April 459 BCE)

Month	I	II	III	IV	V	VI	VII	VIII	IX	X	XI	XII
Z. Sign	♎	♏	♐	♑	♒	♓	♈	♉	♊	♋	♌	♍
ME+GE6	**16,5**	**15,4**	**14,4**	**11,9**	**10,2**	**8,3**	**8**	**8,6**	**11,2**	**15**	**16,7**	**17,5**

Note: Here the lunar data are taken from the time where the zodiac was invented

I conclude: the observable daily retardation of the Moon (i.e., ŠÚ+NA and ME+GE$_6$) enabled the Babylonians to find the symmetries and connections between rising and setting times of the Moon, and thus to determine the rising times of zodiacal signs. The time schemes in Systems A and B were developed from observations, and the Babylonians, indeed, understood more about the situation of the ecliptic at the eastern and western horizon than has been supposed until now. How and why the equinox was fixed at position ♈ 10° in System A and at ♈ 8° in System B is another question, which is not part of this investigation.

Appendix: How Was Time Measured in Old Babylonian Time?

When the two schemes for day lengths in EAE, Tables A and B, were worked out, time was obviously measured in two types of units, in UŠ and in minas which was subdivided into 60 shekel. In Sect. 1.2 we have seen that the units UŠ and shekel were equal, and that EAE as well as MUL.APIN list the day lengths to be 3 mina at the equinoxes, 2 mina at winter solstice and 4 mina at summer solstice. These (not too precise) values of day lengths were, however, introduced much earlier. They go back to Old Babylonian times, as a table published by Hunger and Pingree (1989, 163) shows. On this table, the daylengths on the four cardinal points of the year were also given to be 3, 4, 3, and 2—here only the numbers were written without the unit mina. We know that in Old Babylonian times, mathematics and geometric figures were used for all kinds of calculations and administrative tasks. The scribes were, e.g., trained in solving quadratic equations by means of geometric figures. One of the geometric figures treated in Old Babylonian times provides a suitable tool for measuring time in early Mesopotamia: the regular hexagon. At the latitude of Babylon, the angle between the points of sunrise at summer and winter solstice is close to 60°. Therefore, I propose that at early times, the day length was measured along the horizon from sunrise to sunset. The northern point, the rising and setting points of the Sun at the solstices, and the southern point of the horizon build a regular hexagon with side 1 (mina) = 60 shekel = 60 UŠ. In Fig. 1.16, we can see what we gain if the hexagon was used as means of time measuring. The standard hexagon with side 1 mina, not only leads naturally to the ratio 4/2 between the longest and shortest day, but it also explains the identity 1 mina ≈ 60 UŠ. An identity which really tells us that we should not understand mina for measuring time in units of

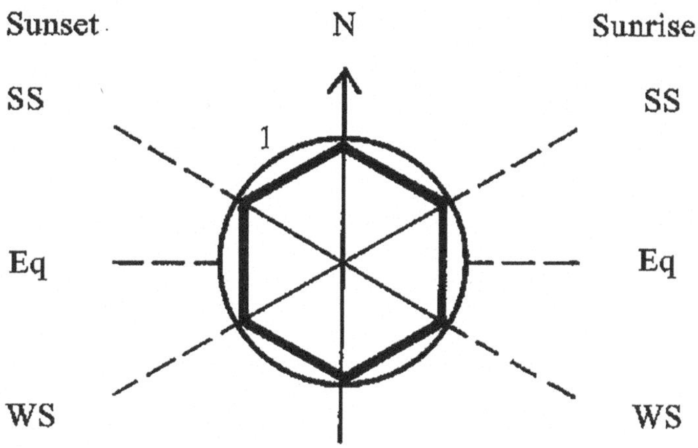

Fig. 1.16 Circle representing the Horizon at Babylon. At summer solstice the Sun rises at SS in the East and sets at SS in the West having moved along four sides of the hexagon, at winter solstice the Sun moves only along two sides

weight, but simply as the standard unit used for the sides of all kinds of geometrical figures in Old Babylonian times. Figures 1.16 and 1.17 undermine this approach.

Measuring time by means of the unit mina, the side of the standard hexagon, we get:

SS: From rising to setting point 4 mina
EQ: From rising to setting point 3 mina
WS: From rising to setting point 2 mina

It makes sense to read mina for measuring times, as in the Old Babylonian mathematical texts, where standard figures had the side length of 1 (mina). The standard figure could be a square, an equilateral triangle, a circle, a pentagon, or a hexagon. Area, diagonal and other quantities for the standard figures were given in coefficient lists. The important quantities for figures with the same shape but with sizes different from the standard figures were found by proportionality. As an illustration, we see in Fig. 1.17 an Old Babylonian mathematical text treating a hexagon with side length ½, i.e., half the side of the standard hexagon. The side of this smaller hexagon is given as 30, and from the numbers in the figure we see that 30 UŠ is meant. Evidently, the side length 1 mina of the standard hexagon was reckoned as 60 UŠ.

In this old Babylonian Tablet, we see a regular hexagon subdivided into six equilateral triangles of length 0; 30 = ½ mina.

The area of such an equilateral triangle is written as 0; 06, 33, 45, which equals ¼ A_6.

$$0;06,33,45 = (0;30)^2 \times 0;26,15.$$

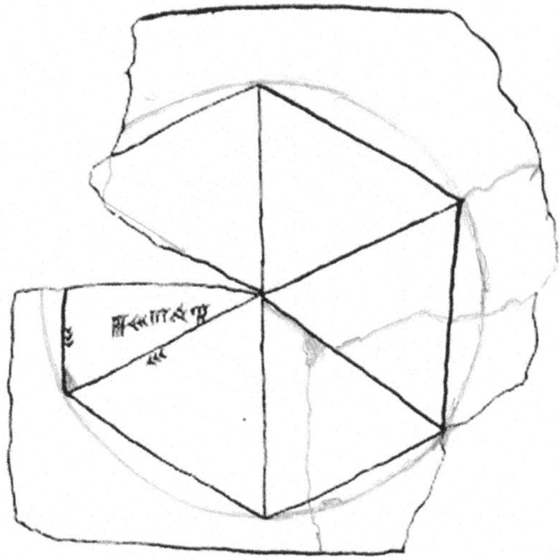

Fig. 1.17 The hexagon drawn on TMS II, an Old Babylonian diagram from Susa. Note. Figure 1.17 shows some cracks and the circle circumscribing the hexagon. A first drawing of the tablet was published by Bruins and Rutten in *Textes mathématiques de Suse* (1961) on p. 24, but without the circumscribed circle. However, the photo of the tablet, published on Plate III, shows some cracks and the circumscribed circle, which now are included in the drawing. Friberg (2007, 218) has also detected the circle around the hexagon

Here $A_6 = 0;26,15$ is the area of the equilateral triangle of length 1. The "standard" hexagon being composed of six such triangles (see Robson 1999, 48–49). The value of A_6 is given in the Babylonian coefficient list J (p. 204).

Note that one mina of water equals half a liter. That the Mesopotamians in Old Babylonian times should have measured time by means of a water clock seems very implausible to me. How could they measure 4 hours = 1 mina by scarcely ½ l of water? I am convinced that the solution is to be found in earlier mathematical texts, where the standard unit was called mana, which was subdivided into 60 shekel or 60 UŠ. As support for this proposal, I quote M. A. Powell (1976) from his "Antecedents of Old Babylonian Place Notation":

> ½ *gin* and 1/3 *gin do not* mean "one-half" or one-third *shekel*" but "one-half" or "one-third *mana*" i.e., 30 and 20 shekels, respectively, just as I have indicated in my translation. The *mana* is the prime unit and *gin* is used to express fractions. I stress this distinction because many Assyriologists do not appear to be aware of it. (Powell 1976, 422)

> Here the term "10 shekels" is used to express "one-sixth." The use of "shekel" here signifies "one-sixtieth" (0;1) and has nothing to do with shekels used in weighing per se, for the number in question refers to bundles of reeds. (Powell 1976, 421)

Powell explicitly points at the fact that the unit mina in Old Babylonian times was used as the prime unit, and that we should not read shekel as a weight, but as 1/60th of the standard unit mina. For further arguments for this hypothesis, see Brack-Bernsen (2005, 7–21).

If I am correct in this proposal, then both ways of measuring time were rooted in observations: in the very early times, day and night were measured along the horizon, while at later times other means of measuring time were introduced: by angles on the celestial sphere, measured in time degrees, by culminating *ziqpu* stars, or by water clocks.

The Bacon conference was held to honor the memory of Noel Swerdlow. Let me tell about the first time I met him; it was in Aarhus. At the occasion of the retirement and 70th birthday of Professor Olaf Pedersen, a whole group of historians of science were invited to a conference at the University of Aarhus 1990. Olaf Pedersen and my teacher Professor Olaf Schmidt knew that I would present a paper on how the period of the lunar velocity could be found from Babylonian horizontal observations. They liked my approach but were skeptical if anyone would be quick enough to realize what it meant! Then, as I showed the interesting curve, based on computer-simulated lunar data, a young man (whom I did not know) jumped up from his bench and ran to me at the blackboard, asking enthusiastically: "does that mean that you have found the period of the lunar velocity?" It was Noel Swerdlow. Since then we have been in contact: he was one of those promoting me to get a habilitation stipend so that I could get back to full-time research. Later we met and could discuss several times at the Dibner Institute. I was very impressed by his competence and wide interests, ranging from the early Babylonians through antiquity and middle ages up to the time of Kepler. In all disciplines Noel showed a fundamental understanding and deep competence.

In *The Babylonian Theory of the Planets* (1998), Noel Swerdlow has not only presented his reconstruction of the planetary theory from observations, he has also—and for the first time—taken a wider view on the Babylonian society and their scholars. Swerdlow also looked at social position and institutional background of the Babylonian scholars, and he examined what the astronomical calculations were used for and looked at the connection between divination and celestial observations. Swerdlow's reconstruction of the Babylonian procedures for calculating planetary phases is based on observed times. Asger Aaboe had pointed at the fact that the positions of successive heliacal risings of a planet are placed closer together on some part of the ecliptic and further apart on others. Therefore, he had proposed that the Babylonian step functions were constructed from positions of the planets observed at their specific phases. Both approaches may be correct. The mathematical astronomy was developed by many different Babylonian scholars over a long time.

References

Al-Rawi, Farouk N. H. and Andrew R. George. 1991–1992. Enūma Anu Enlil XIV and Other Early Astronomical Tables. *Archiv für Orientforschung* 38/39: 52–73.

Brack-Bernsen, Lis. 1997. *Zur Entstehung der Babylonischen Mondtheorie: Beobachtung und theoretische Berechnung von Mondphasen.* Boethius 40. Stuttgart: Franz Steiner Verlag.

Brack-Bernsen, Lis. 1999. Goal-Year Tablets: Lunar Data and Predictions. In *Ancient Astronomy and Celestial Divination*, ed. Noel M. Swerdlow, 149–177. Cambridge, MA: MIT Press.

Brack-Bernsen, Lis. 2005. The 'Days in Excess' from MUL.APIN. On the 'First Intercalation' and 'Water Clock' Schemes from MUL.APIN. *Centaurus* 47:1–29.

Brack-Bernsen, Lis, and Hermann Hunger. 2002. TU 11, A Collection of Rules for the Prediction of Lunar Phases and of Month Lengths. *SCIAMVS* 3:3–90.

Brack-Bernsen, Lis, and John M. Steele. 2005. Eclipse Prediction and the Length of the Saros in Babylonian Astronomy. *Centaurus* 47:181–206.

Bruins, Evert M., and Marguerite Rutten. 1961. *Textes mathématiques de Suse*. Paris: Librairie Orientale Paul Geuthner.

Chapront-Touzé, M., and J. Chapront. 1988. ELP2000-85: A Semi-Analytical Lunar Ephemeris Adequate for Historical Times. *Astronomy and Astrophysics* 190:342–352.

Friberg, Jöran. 2007. *A Remarkable Collection of Babylonian Mathematical Texts*. New York: Springer.

Huber, Peter. 1997. Creslong.dat [computer file]. Huber computed the file 1997 with his PC-program from 1981.

Huber, Peter, and John M. Steele. 2007. Babylonian Lunar Six Tablets. *SCIAMVS* 8:3–36.

Hunger, Hermann, and David Pingree. 1989. *MUL.APIN: An Astronomical Compendium in Cuneiform*, Archiv für Orientforschung 24. Horn: Berger & Söhne.

Hunger, Hermann, and John M. Steele. 2019. *The Babylonian Astronomical Compendium MUL. APIN*. Abingdon: Routledge.

Moshier, S. L. 1996. *Computer Code "AA" (v. 5.4, Public Domain) for Ephemerides of the Solar System, Using Algorithms Published in the "Astronomical Almanac"*. Washington, DC: US Government Printing Office.

Neugebauer, Otto. 1947. Studies in Ancient Astronomy VIII. The Water Clock in Babylonian Astronomy. *Isis* 37 (1–2): 37–43.

Neugebauer, Otto. 1955. *Astronomical Cuneiform Texts, Neugebauer, Otto. 1955. Astronomical Cuneiform Texts, Vols. I–III (ACT I–III)*. London: Lund Humphries.

Neugebauer, Otto. 1975. *A History of Ancient Mathematical Astronomy*. 3 vols. Berlin/New York: Springer.

Ossendrijver, M. 2012. *Babylonian Mathematical Astronomy: Procedure Texts*. New York: Springer.

Powell, Marvin A. 1976. The Antecedents of Old Babylonian Place Notation and the Early History of Babylonian Mathematics. *Historia Mathematica* 3:417–439.

Robson, Eleanor. 1999. *Mesopotamian Mathematics, 2100–1600 BC. Technical Constants in Bureaucracy and Education*. Oxford: Clarendon Press.

Rochberg, Francesca. 2004. A Babylonian Rising Time Scheme in Non-tabular Astronomical Texts. In *Studies in the History of the Exact Sciences in Honour of David Pingree*, ed. C. Burnett, J. P. Hogendijk, K. Plofker, and M. Yano, 56–94. Leiden: Brill.

Sachs, A. J., and H. Hunger. 1988. *Astronomical Diaries and Related Texts from Babylonia*, Diaries from 652 B.C. to 61 B.C. Vol. I. Vienna: Österreichische Akademie der Wissenschaften.

Sachs, A. J., and H. Hunger. 1996. *Astronomical Diaries and Related Texts from Babylonia*, Diaries from 652 B.C. to 61 B.C. Vol. III. Vienna: Österreichische Akademie der Wissenschaften.

Steele, John M. 2000. Eclipse Prediction in Mesopotamia. *Archive for History of Exact Sciences* 54:421–454.

Steele, John M. 2017. *Rising Time Schemes in Babylonian Astronomy*, Springer Briefs in History of Science and Technology. New York: Springer.

Swerdlow, Noel. 1998. *The Babylonian Theory of the Planets*. Princeton: Princeton University Press.

Thureau-Dangin, F. 1922. *Tablettes d'Uruk*." *Textes cunéiformes* 6. Paris: Musée du Louvre.

Weidner, Ernst. 1967. *Gestirn-Darstellungen auf Babylonischen Tontafeln*. Graz: Herman Böhlaus Nachf.

Chapter 2
Neo-Assyrian Investigations into the Prediction of Eclipses

John M. Steele

2.1 Introduction

In the conclusion to his book *The Babylonian Theory of the Planets*, Noel Swerdlow remarked that "from the earliest omen texts and Diaries to the latest almanacs and ephemerides, for century after century, generations of teachers and students observed, calculated, and also pondered deeply how these irregular but periodic phenomena were to be described, reduced to precise rules, and computed for the past, present, and future." But the Babylonian scholars, he continued, "have left no record of their theoretical analyses and discussions."[1] One of the tasks of the historian of Babylonian astronomy, therefore, is to try to reconstruct its development from texts which are the direct result of its practice—in particular texts containing systematic records of observations and texts which contain predicted phenomena calculated either using past observations (goal-year astronomy) or arithmetical systems (mathematical astronomy)—aided by a smaller number of texts which provide instructions for making calculations. A very small number of texts—I know of only two or three—seem to bear witness to a Babylonian scholar actively working on the development and/or testing of methods of predicting astronomical phenomena,[2] but they provide no more than occasional glimpses into the process of the development of methods of computing astronomical phenomena.

[1] Swerdlow 1998, 181.

[2] Steele and de Jong 2023; Steele forthcoming.

J. M. Steele (✉)
Department of Egyptology and Assyriology, Brown University, Providence, RI, USA
e-mail: john_steele@brown.edu

F. Rochberg (ed.), *Studies in Premodern Sciences in Memory of Noel M. Swerdlow*, Archimedes 74, https://doi.org/10.1007/978-3-032-04040-4_2

In his book, Swerdlow presented a detailed attempt at reconstructing the development of the various systems of mathematical astronomy for calculating planetary phenomena from the kinds of planetary observations recorded in the so-called Astronomical Diaries and related texts;[3] other scholars have published proposals for the development of the lunar systems of mathematical astronomy.[4] Similarly, I have attempted to reconstruct the development of Babylonian observational practice and of goal-year methods of prediction.[5]

My aim in the present chapter is to investigate the early stages in the development of methods for predicting eclipses. Although eclipses were a major subject within celestial divination since at least the Old Babylonian period,[6] they are notably absent as a topic in the late second or early first millennium astronomical compendium MUL.APIN.[7] It appears that the author of MUL.APIN tried to include discussions of all astronomical phenomena which could be represented by numerical schemes; the absence of eclipses from this text suggests that at the time of its composition, schemes for mathematically describing or predicting eclipses had not yet been developed. Later texts from Babylon show that a system for predicting eclipse possibilities by tethering the eclipse possibilities in repeating Saros cycles to the observational record, was in operation by about 600 BCE. This system also allowed the time and, for lunar eclipses only, the magnitude of the eclipse to be predicted in advance.[8] Thus, between the composition of MUL.APIN and about 600 BCE, Babylonian scholars had identified the Saros cycle and developed a method for using it to predict eclipses. These developments very likely took place in the century or so before the system was in full operation. Plausible routes for this development can be reconstructed from what we know of astronomy in Babylon during this period, but in the present study I will examine evidence from seventh century Assyria which suggests that scholars in Assyria were also actively trying to develop such methods. In particular, I investigate Assyrian evidence for the concept of eclipse possibilities which are separated by five- and six-month intervals and attempts to identify eclipse periods from observational evidence. This evidence comes from two types of sources: Letters and Reports sent by scholars to the Assyrian king and a small number of (mostly previous unpublished) Neo-Assyrian astronomical texts which concern eclipses.

I begin this chapter with a short overview of Babylonian methods of eclipse prediction using the Saros as known from ca. 600 BCE onward before turning to the evidence for eclipse prediction in the Neo-Assyrian Letters and Reports. I then

[3] Swerdlow 1998. Alternative proposals for the development of the planetary systems have been published by Neugebauer 1968, Aaboe 1980, and de Jong 2019a, b, 2021.

[4] See, for example, Brack-Bernsen 1997, 2020, Britton 2007, 2009, and de Jong 2017.

[5] Steele 2000, 2019, and Steele forthcoming.

[6] Rochberg-Halton 1988.

[7] For an edition, translation, and study of MUL.APIN, see Hunger and Steele 2019.

[8] Steele 2000, 2002, and forthcoming; Brack-Bernsen and Steele 2005.

discuss a few Assyrian tablets which concern the intervals between eclipse possibilities before presenting the previously unpublished tablet Rm 121 and its duplicates which show evidence of Assyrian interest in studying patterns in the record of observed lunar and solar eclipses. Finally, I will consider the question of whether Assyrian investigation into the prediction of eclipses were complementary to or the inspiration for Babylonian investigations.

2.2 Babylonian Use of the Saros Cycle to Predict Eclipses

Babylonian methods of eclipse prediction were formulated around four basic principles:[9]

1. There are months when eclipses are possible and months when eclipses are impossible. We call the former "eclipse possibilities." In modern terms, an eclipse possibility occurs when the moon is sufficiently close to a node at opposition (for lunar eclipses) or conjunction (for solar eclipses).
2. Lunar eclipse possibilities and solar eclipse possibilities occur at neighboring syzygies. I.e., a lunar eclipse possibility occurs at the opposition that either precedes or follows a conjunction which is solar eclipse possibility.
3. Eclipse possibilities of the same kind (i.e., lunar eclipse possibilities or solar eclipse possibilities) occur at intervals of either five or six months. Six-month intervals occur much more frequently than five-month intervals.
4. There exist periods, in particular the Saros period of 223 lunar months, after which eclipses recur. In modern terms, this is because in an eclipse period there are whole numbers of both synodic months and draconitic months. For example, in the Saros period of 223 synodic months there are almost exactly 242 draconitic months. More precisely, 223 mean synodic months = 6585.322 days and 242 draconitic months = 6585.357 days. This small discrepancy (0.035 days) means that the period is not perfect and will gradually break down after many cycles.

By combining knowledge of the five- and six-month intervals between eclipse possibilities and the 223-month Saros cycle, the Babylonian scholars were able to deduce that there are 38 eclipse possibilities within a Saros, 33 of which are at a six-month interval and five of which are at a five-month interval.[10] It is then a matter of distributing the six- and five-month intervals within the 223 months of the Saros.

---

[9] Aaboe 1972; Britton 1989; Steele 2000.

[10] This fact can easily be determined by simple mathematics. In modern terms, if there are a six-month intervals and b five-month intervals within a Saros of 223 months the following must hold: $223 = 6a + 5b$. Knowing that six-month intervals occur significantly more often that five-month intervals quickly leads to the realisation that a must be 33 and b must be 5.

The obvious (and correct) way to do this distribution is to spread the five-month intervals as evenly as possible among the six-month intervals. This results in five groups of eclipse possibilities, the first eclipse possibility of which is at a five-month interval and the remaining are at six-month intervals as in the following pattern:

56666666 5666666 56666666 5666666 56666666

The groups contain respectively eight, seven, eight, seven, and eight eclipse possibilities. Because the Saros is an eclipse period, all eclipse possibilities within a Saros occur one Saros later. Thus, the pattern of five- and six-month intervals repeats again and again:

56666666 5666666 56666666 5666666 56666666
56666666 5666666 56666666 5666666 56666666
56666666 5666666 56666666 5666666 56666666
56666666 5666666 56666666 5666666 56666666
...

Several Babylonian texts reflect this arrangement including the theoretical "Saros Canon,"[11] which presents the months of eclipse possibilities arranged in Saros cycles with horizontal rulings highlighting the five-month intervals, and texts which compile eclipse observations and predictions in Saros cycles.[12] Furthermore, in Steele (2000) I showed that this system for identifying eclipse possibilities was used in practice to predict all of the eclipses found in Late Babylonian astronomical texts (excluding texts of mathematical astronomy). Implementing this system for predicting eclipses requires the pattern of eclipse possibilities to be aligned with the observational record. Essentially, this comes down to deciding which eclipse is at the first five-month interval in the first Saros cycle. The Babylonians (retrospectively) chose the lunar eclipse in Month XII of the accession year Nabonassar (5 February 747 BCE).

Our understanding of the history of the development of this system for predicting eclipses is hampered by a paucity of sources from the early period. What we can say is that it was apparently in full operation by early in the sixth century BCE.[13] Thus, it was probably developed sometime in the seventh or perhaps very early sixth century BCE.

---

[11] Aaboe et al. 1991.

[12] Hunger 2001, texts 2, 3, 4, 9, 10, and 11. See Walker 1997; Huber and de Meis 2004, 1–6; and my appendix to Hunger 2001 for a discussion of these texts.

[13] Steele 2000, and Steele forthcoming.

2.3 Eclipses in the Neo-Assyrian Letters and Reports

The corpus of Letters and Reports sent by scholars to the Assyrian kings Esarhaddon and Assurbanipal provide a detailed window into the practice of the observational and interpretation of ominous celestial (and other) phenomena during the Neo-Assyrian period.[14] In broad terms, the Reports were sent by the scholars to notify the king of the occurrence of ominous celestial phenomena. Many Reports simply quote one or more omens that relate to the observation; some also include details of the observation itself and related remarks. A small number of Reports discuss astronomical phenomena that had been predicted in advance but which had not been observed, either because they did not occur or because bad weather prevented the phenomenon being seen. The Letters were usually written in response to a question from the king, although sometimes scholars wrote to initiate conversations or to provide unsolicited advice or recommendations. It is important to note that neither the Letters nor the Reports contain dates beyond the day of the month and/or the month of the year (and even these are sometimes only implied); the year in which they were written and in which the phenomena occurred is not given. Some Letters and Reports can be dated by modern scholars using a combination of comparing the details of the astronomical observations they contain with modern computation and contextual information such as references to historical events or things discussed in other Letters or Reports and the known period of activity of the scholar writing the letter.

Thirty-seven Reports and forty-two Letters include references to lunar or solar eclipses.[15] Because the king employed many scholars who each wrote to him, in many cases the same eclipse is discussed in several Letters and Reports. Both observed and predicted eclipses are recorded in these texts. Unfortunately, it is not always possible to determine whether the record refers to an observed or predicted event: both observed and predicted eclipsed can be indicated by the phrase "it makes an eclipse."[16] Only if further information is given is it possible to determine whether the eclipse was observed or predicted. In the case of an observation, this additional

[14]The Reports are edited and translated by Hunger 1992 and are cited here as SAA 8. For a study of the dates of the Reports, see de Meis and Hunger 1998. The Letters are edited and translated by Parpola 1993 and are cited here as SAA 10. For a detailed study of many of the Letters and their dating, see Parpola 1983. See also Oppenheim 1969 and Brown 2000.

[15]I omit from these counts a few Letters which discuss ritual activity which took place in response to an eclipse but which do not discuss the eclipse itself and a handful of Letters and Reports which include an eclipse in the apodosis of an omen. For example, the Report SAA 8 308 includes the omen "If the sun at its rising carries(?) red radiance: an eclipse will take place, and Adad will devastate" (Hunger 1992, 172); this omen attests to a *divinatory* rather than an *astronomical* prediction of an eclipse.

[16]The verb is usually written GAR-*an* in the Reports and syllabically (usually either *iš-ša-kan* or *i-šak-kan*) in the Letters. The logographic writing in the reports reflects the usual way that eclipses are indicated in the lunar eclipse omens of *Enūma Anu Enlil*, for which see Rochberg-Halton 1988.

information may be either a description of the appearance of the eclipse, either directly or by quoting an omen which draws on observed details, or other contextual information in the Report such as a reference to ritual activity. Predictions may be identified by the term "pass(ed) by," by the term "it will not occur," or by contextual information indicating that the eclipse had not yet taken place. "Passed by" (written syllabically as *ú-še-taq*, *ú-še-ti-iq*, etc.; in Late Babylonian eclipse reports, this term is always written using the logogram DIB) is used to indicate that the eclipse took place when the luminary was below the horizon or that the Moon was too far from the node for an (umbral) eclipse to take place at that eclipse possibility. In the case of lunar eclipses, it is sometimes coupled with the expression "the god will be seen with the god" or "the moon will be seen with the sun." This expression refers to the Moon and Sun being seen together in the morning sky, which takes place after opposition. Thus, once the two luminaries have been seen together there is no possibility of a lunar eclipse that month.

Most of the Reports of predicted eclipses in the Letters and Reports are brief and seem to reflect the situation where an eclipse possibility had been identified but no eclipse had been visible. The account in the Letter SAA 10 132 is typical:

> To the king, my lord: your servant Nergal-šumu-iddina. Good health to the king, my lord! May Nabû and Marduk bless the king, my lord!
> Concerning the watch about which the king, my lord, wrote to me, the moon let the eclipse pass by, [it did not occ]ur.
> May [Sin and Š]amaš appoint a gu[ardian of he]alth and life for the king, my lord.[17]

The method used to predict these eclipses has been the matter of considerable debate among modern scholars.[18] Parpola (1983, 51) argued that in addition to knowing about the six-month interval between eclipse possibilities, which he calls a "rule of thumb," the Assyrian scholars "of this period certainly had recognized the 47-month period and probably also the 18-year 'Saros.'" Parpola seems to be suggesting that eclipses were predicted by applying an eclipse period to a previously observed eclipse. However, there is no evidence for Mesopotamian knowledge of a 47-month eclipse cycle,[19] and, whether or not the Saros was known to these scholars, for several of the predicted eclipses recorded in the Letters and Reports no eclipse was observable either 47 months or one Saros earlier. Thus, the eclipse predictions cannot be made by the direct application of either of these eclipse periods.

In fact, several Letters and Reports contain clear evidence that the five- and six-month intervals between eclipse possibilities played the central role in predicting eclipses. For example, consider the Report SAA 8 502. The Report seems to have been written near the beginning of Month IX. It begins with several sections concerning the planets before discussing eclipses in Rev. 7–11:

---

[17] Translation by Parpola 1993, 107.

[18] See, in particular, Schnabel 1924; Schaumberger 1935, 251–53; van der Waerden 1974, 115–20; Parpola 1983, 51–52 and 68; Brown 2000, 200–6; and Steele 2000.

[19] See Brown 2000, 206 n. 565.

GISKIM.MEŠ *ma-la il-li-ka-ni šá* KUR URU.KI *ù* LÚ.NUN.MEŠ-*šú ši-na mim-ma ina* ŠÀ *lum-ni-ši-na ana* LUGAL *be-lí-ía ul* TE-*ḫi* AN.KU$_{10}$ ^{d}30 *u* ^{d}UTU *šá ina* ITI.SIG$_{4}$ GAR-*an* GISKIM.MEŠ *an-na-tu šá* ḪUL-*nu šá* KUR URI.KI *ú* LUGAL.MEŠ *šá* KUR MAR.TU.KI *šá* KUR URI.KI *ši-na ù a-du-ú ina* ITI.GAN *a-ga-a* AN.KU$_{10}$ *iš-šak-kan* ITI.⸢ŠE?⸣ ITI *ki me ú* MUL.SAG.ME.GAR *ina* AN.KU$_{10}$-*šá* GUB-*az ana* LUGAL *be-lí-ia šul-mu*

All the signs which have come concern the land of Akkad and its nobles; any evil in them will not come close to the king my lord. An eclipse of the moon and the sun in Month III took place.[20] These signs are of bad fortune for Akkad, for the kings of the Westland and of Akkad; and now, in this Month IX, an eclipse will take place. Month XII(? error for Month IX?) is the month of… and Jupiter stands[21] in its eclipse; well-being for the king, my lord.[22]

According to this passage, lunar and solar eclipses had both been seen in Month III. Six months later, in Month IX, the Moon will again be eclipsed. The Letters SAA 10 45 and SAA 10 358 indicate that sometimes it was necessary to watch for an eclipse in two successive months. Balasî, the author of SAA 10 45, explains:

ina UGU *ma-ṣar-ti šá* ^{d}UTU *šá* LUGAL *be-lí iš-pur-an-ni* ITI EN.NUN *šá* ^{d}UTU *šú-u* 2-*šú* EN.NUN-*šú* UD.28!.KÁM *šá* ITI.APIN UD.28.KÁM *šá* ITI.GAN *ni-na-aṣ-ṣar ki-i ḫa-an-ni-i-e* EN.NUN *šá* ^{d}UTU 2 ITI.MEŠ *ni-na-aṣ-ṣar*

Concerning the watch of the sun which the king, my lord, wrote to me, it is (indeed) the month for a watch of the sun. We will keep watch twice, on the 28th of Month VIII and the 28th of Month IX. Thus, we will keep the watch of the sun for 2 months.[23]

Keeping watch for two months must refer to uncertainty over whether the upcoming eclipse possibility occurs five or six months after the previous one. The fact that eclipses cannot occur at other intervals is discussed in the Letter SAA 10 71 written by Nabû-aḫḫe-eriba in answer to the king's request for information about the watch for a lunar eclipse. Rev. 4–12 reads:

[AN.KU$_{10}$ *ina* Š]À UD.DUG$_{4}$.GA.MEŠ *a-da-an-na-a-te* [*us-se*]-*et-iq* [*ur-ki*?] ⸢5?⸣ ITI.MEŠ ITI. ⸢APIN⸣ *ma-ṣar-tu-šú ú-ma-a* ITI.GAN *ni-na-ṣar* ⸢*a*⸣-[*ki*] *ma-ṣar-*⸢*a*⸣-[*ti an-na*]-*ti* ⸢ITI⸣ [...] ⸢30⸣ [...]

[Eclipses with]in the intervals [pa]ss b[y. After] 5? months in Month VIII, there was a watch and now in this Month IX we will (again) keep watch. [Accord]ing to [the]se wat[ches] the month [...] moon [...]

The passage is both terse and quite damaged. Nabû-aḫḫe-eriba begins by stating that eclipses within "intervals" pass by. The term I translate as "intervals" is a plural form of the Akkadian word *adānu* (written logographically UD.DUG$_{4}$.GA.MEŠ,

[20] Hunger 1992, 279 translates "will take place" but the context suggests that the eclipse has already occurred. This interpretation is supported by the scribe writing the verb GAR-*an* here in contrast to the phonetic *iš-šak-kan* when referring to the eclipse in Month IX below. According to modern computation (Huber and de Meis 2004), lunar and solar eclipses were indeed visible in Month III of this year (1 June 679 BCE and 17 June 679 BCE, respectively).

[21] Hunger 1992, 279 translates "will stand," but the text is ambiguous as to whether this is prediction of Jupiter's visibility during an expected eclipse or a report of Jupiter having been seen during an observed eclipse.

[22] Translation adapted from Hunger 1992, 279–80.

[23] Translation adapted from Parpola 1993, 34.

which is glossed with the syllabic writing *a-da-an-na-a-te*). This word refers to an interval of time or in some contexts, more specifically, the end of fixed interval of time. Eclipses "within" this time interval pass by, meaning that eclipses can only occur at the beginning and end of the time interval. Nabû-aḫḫe-eriba then states that a watch for an eclipse should be made after five months (Month VIII) and then in the following month (Month IX). Unfortunately, the number 5 is badly damaged on the tablet. Only a single vertical wedge in what would be the bottom register of a two-register number sign is clearly preserved. Parpola therefore read the number as 4. However, there appear to be very slight traces of another vertical wedge just to the left of the well-preserved wedge, which would suggest that the number is 5.[24] It seems clear, therefore, that the intervals referred to in this passage are the five- and six-month intervals between eclipse possibilities.[25]

Knowledge of the five- and six-month intervals between eclipse possibilities, especially when coupled with the fact that five-month intervals are the exception to the norm of much more common six-month intervals, would allow the months of eclipse possibilities to be predicted some years ahead. The only uncertainty would be when a five-month interval would occur. If a five-month interval had occurred fairly recently, then it is certain that the eclipse possibility would occur at a six-month interval. Only when it has been three years or so since the last five-month interval would it be necessary to watch for an eclipse in two successive months.

A few of the eclipse predictions in the Letters and Reports contain additional information. Some show greater advance certainty of whether an eclipse will occur. For example, in the Report SAA 8 447, Ṭab-ṣilli-Marduk is firm that a lunar eclipse will not take place that month:

[*ina* ITI.G]U$_4^?$ UD-*mu ú-šal-lam* [UD.14$^?$].KÁM ^{d}30 *u* ^{d}UTU *it-ti a-ḫa-meš* IGI.MEŠ
[UD].13.KÁM *mu-ši šá* UD.14.KAM [UD]-*mu* EN.NUN *u* AN.KU$_{10}$ NU GAR
pa-t[*ú-a*]*-di* 7-*šú na-šá-ka* AN.KU$_{10}$ *ul iš-šak-kan a-mat pa-ri-is-tum a-na* LUGAL *al-tap-ra*
šá ^{I}DÙG.GA-GIŠ.MI-^{d}AMAR-UTU DUMU ŠEŠ-*šú šá* ^{Id}EN-*na-ṣir*

[In Mon]th II (the moon will complete the day; [on the 14th(?)] day the moon and sun will be seen together. The 13th [day], the night of the 14th day, is the [da]y of the watch (to be held), and there will be no eclipse.
I guarantee it seven times, an eclipse will not take place. I am writing a definitive word to the king.

From Ṭab-ṣilli-Marduk, nephew of Bel-naṣir.[26]

Given the Report's focus on the visibility of the Sun and Moon in the morning sky, it seems likely that Ṭab-ṣilli-Marduk's certainty is because he had estimated the

[24] I have examined photographs of the tablet on the British Museum's website and the Electronic Babylonian Library website. In addition, Christopher Walker collated the tablet for me in 2000 and noted that reading the number as 5 was perfectly possible.

[25] It is for this reason that I translate the term as "intervals" not "periods" (as Parpola and others have done); I reserve this later term for periods such as the Saros period of 223 months which is itself made up of a series of five- and six-month intervals.

[26] Translation adapted from Hunger 1992, 251.

expected time of the eclipse and found that it would occur when the Moon is below the horizon.

Predictions of the time of the eclipse are also stated or implied in a few other Letters and Reports. The partially preserved Letter SAA 10 78 written by Nabû-aḫḫe-eriba includes the prediction that a lunar eclipse will occur during the morning watch (i.e., the third and final watch of the night):

> [...] *ša* LUGAL *b*[*e-lí*] *iš-pur-a-*[*ni*] *mu-šú an-n*[*i-ú*] *ina* EN.NUN UD.ZA[L.LA]$^{e\text{-}nu\text{-}un\ ú\text{-}⌈za\text{-}al⌉\text{-}la}$ *ma-ṣar-tu-šú* AN.KU$_{10}$ EN.NUN UD.ZAL.LA *i-šá-kan* [... *šá*] LUGAL [EN]-*ia ma-ṣar-tú* LUGAL *be-lí šul-mu*
>
> [Concerning the watch for a lunar eclipse] about which the king, [my lo]rd, wrote to me, its watch will be (kept) tonight, in the morning watch. The eclipse will occur during the morning watch.
> The watch is (kept) [for the safety] of the king, my [lord]; (all) will be well with the king, my lord.[27]

Another letter written by Nabû-aḫḫe-eriba, SAA 10 71, unfortunately badly damaged at this part, seems to say that the scholar was uncertain about whether the lunar eclipse would take place before or after sunset. This case, at least, indicates an interest in predicting the time of the eclipse. Finally, the unidentified Report SAA 8 502 quoted above says that Jupiter will be visible during the eclipse. If this is indeed a prediction of Jupiter's visibility during the eclipse, it seems to require both knowledge of when Jupiter would be visible that evening and of the approximate time of the eclipse (but see my comments below).

Two possible methods have been put forward for how the times of these eclipse were predicted. Parpola (1983, 68) has suggested that by observing the time interval that the Moon was visible before sunset on the days running up to the eclipse the scholars could extrapolate from these intervals to produce an estimate of the time of full Moon. Such a method would indeed work, and there is later evidence from Babylon that the lunar six time interval ME (moonrise to sunset) could be used to predict the time of the eclipse.[28] Any method which relied upon observing the Moon on the nights before the expected eclipse could only yield short-term predictions. A variation of this method could have been used if the scholars knew how to predict the visibility time of the Moon. Using such a method, the time of the eclipse could be predicted well in advance. A method for predicting the lunar six intervals using involving past observations and the Saros cycle is known to have been used in Babylon by about 600 BCE,[29] but we have no evidence that the Assyrian scholars either knew the method or had the systematic observations of the lunar six that are needed to made these calculations.

A different approach is taken by Brown (2000, 205), who has suggested that the Assyrian scholars made use of the Saros cycle. Knowing that the Saros lasts for

[27] Translation by Parpola 1993, 58.

[28] This evidence is found in BM 42127+46243, which will soon be published by the present author.

[29] For the method, see Brack-Bernsen 1999. For evidence of its use in the early sixth century BCE, see Huber and Britton 2007 and Huber and Steele 2007.

approximately a third of a day more than a complete number of days, if an eclipse had been observed one Saros earlier, then the time of the upcoming eclipse could be estimated simply by adding a third of the day on to that time. The method relies, of course, on Assyrian knowledge of the Saros cycle as well as dated records of observations from one Saros earlier than the predictions that are being made. As discussed earlier, the Letters and Reports themselves do not record the year in which an eclipse occurred; thus, we would need to suppose the existence of another source of eclipse observations which include their dates. Brown argues that the Report SAA 8 502 provides "the clearest piece of evidence that the scholars used the Saros interval,"[30] because at least one of the eclipse predictions it contains was made several months in advance and it contains the statement about Jupiter being visible during an eclipse, which Brown takes to imply advance knowledge of what time of night the eclipse would take place. If the date assigned to this Report by Parpola (1983) and accepted by Hunger (1992) and Brown (2000) is correct, however, then this eclipse was in Month IX of Esarhaddon's second year, corresponding to 26 November 679 BCE. That eclipse took place during daylight, ending well before sunset. Jupiter could therefore not have been seen during the eclipse. The eclipse one Saros earlier, on 15 November 697, began before a little less than two hours before sunrise in Nineveh, the Moon setting during the eclipse. Adding one-third of a day on to this time, would bring us to the middle of the afternoon, clearly before the Moon rose. Thus, it is unlikely that a scholar using the Saros, even given possible uncertainty in the exact time of the eclipse observed one Saros earlier, would predict the eclipse of 26 November 679 BCE as being visible. I suspect that the scholar did not try to predict the time of the eclipse but simply knew that Jupiter was in its retrograde arc (which is mentioned earlier in the report), and that this meant that the planet is visible for most of the night, and therefore assumed that if the eclipse was indeed seen then Jupiter would probably also be above the horizon for at least some of the eclipse. The Report SAA 8 502, therefore, cannot be taken as evidence that its author knew the Saros or used it to predict the time of an eclipse.

In summary, the evidence from the Letters and Reports show that the scholars knew that:

1. There are syzygies at which eclipses are possible—whether or not they actually occur—as opposed to those at which they are not possible. Eclipses which do not occur at an eclipse possibility, either because the luminary was below the horizon or because (in modern terms) the Moon was too far from the node, as said to "pass by."
2. Lunar and solar eclipses possibilities occur at neighboring syzygies: i.e., a solar eclipse possibility occurs either the conjunction before or the conjunction after the opposition at which a lunar eclipse possibility occurs.
3. Eclipse possibilities of the same kind (i.e., either lunar eclipse possibilities or solar eclipse possibilities) occur at intervals of six or sometimes five months. These

[30] Brown 2000, 205.

intervals are termed *adānu* (usually written logographically as UD.DUG$_4$.GA). Eclipse possibilities could be determined well in advance with the exception in some cases where there was uncertainty whether the eclipse possibility was at a five- or a six-month interval. In such cases, a watch would be kept in both months.

4. At least one scholar, Nabû-aḫḫe-eriba, and possibly others, attempted to predict the approximate time of day or night when an eclipse would occur. It is likely that these predictions were made in the days running up to the eclipse by observing the visibility time of the Moon.

The Letters and Report provide no firm evidence that the scholars had identified the Saros cycle or used it either to determine the months of eclipse possibilities or to predict the time of an eclipse.

2.4 Additional Evidence for Neo-Assyrian Knowledge of the Five- and Six-Month Intervals Between Eclipse Possibilities

A few other texts provide evidence for Neo-Assyrian knowledge of eclipse intervals:

2.4.1 82–5–22, 501

82–5–22, 501 preserves the top left corner of what was originally a sizable tablet. The tablet was probably written in two columns of which only parts of the left columns on the obverse and reverse remain. It was previously edited by Rochberg-Halton (1988, 383–84), who recognized that it deals with six- and multiples of six-month intervals between eclipses.[31] I edit the text below (I use indentations to mimic the layout of the tablet).[32]

[31] Rochberg-Halton 1988, 42 states that the tablet has no known duplicate. A Late Babylonian parallel, BM 39779, has recently been identified. This source differs from 82–5–22, 501 in using the logographic writing UD.DUG$_4$.GA for "interval" as well as in which months are mentioned at the beginning of the line. The preserved part reads:

1′ [DIŠ] *ina* ITI.⸢DU$_6$ UD.14.KAM AN⸣.[KU$_{10}$ GAR …]
2′   BE-*ma* TU$_4$: ITI.DU$_6$ x […]
3′ [DIŠ] *ina* ITI.APIN UD.14.KAM AN.K[U$_{10}$ GAR …]
4′   BE-*ma* 2 UD.DUG$_4$.GA.MEŠ […]
5′ [DIŠ *ina* ⸢ITI⸣.ŠE UD.14.KAM AN.K[U$_{10}$ …]
6′   […] ITI.ŠE AN.K[U$_{10}$ GAR …]

[32] My edition is based upon Rochberg-Halton's but I have incorporated a few corrections based upon collation of the tablet on 7 June 2022.

Obv. I

1 [DIŠ *ina* IT]I.⌜SIG$_4$⌝ UD.[14.KAM A]N.KU$_{10}$ GAR-*in* ⌜*ina* ITI⌝.[GAN(?)...]
2 BE-*ma ina* ITI.SIG$_4$ AN.KU$_{10}$ ^{d}UTU GAR-*an* MIN ⌜x⌝ [...]
3 *ina* 3 *a-dan-ni ina* ITI.GAN UD.14.KAM A[N.KU$_{10}$ GAR...]
4 DIŠ *ina* ITI.ŠU UD.14.KAM GAR-*in ina* ITI.A[B...]
5 DIŠ *ina* ITI.IZI UD.14.KAM AN.MI GAR-*in ina* ITI.ZÍZ [...]
6 BE-*ma* ITI DIRI *ina* ITI.AB 1 *a-dan-na* DIB-*iq-ma ina* 2 *a-da*[*n-na-ti*...]
7 DIŠ *ina* ITI.KIN UD.14.KAM AN.KU$_{10}$ GAR-*in ina* ITI.ŠE U[D$^?$...]
8 BE-*ma* 2 *a-dan-na-ti la-am* ITI.KIN AN.KU$_{10}$ NU GAR-*ma* ⌜x⌝ [...]
9 2 *a-dan-na-ti* DIB-*iq-ma ina* ITI.KIN *ina* 3 x [*a-dan-na-ti*...]
10 BE-*ma ina* ITI.KIN AN.KU$_{10}$ ^{d}UTU GAR-*in* 1 *a-dan-n*[*a*...]
11 [DIŠ *ina*] ⌜ITI⌝.DU$_6$ UD.14.KAM AN.KU$_{10}$ GAR-*i*[*n*] ⌜2⌝ *a-dan-*[*na-ti*...]
12 [x x] ⌜x⌝ *a-dan-ni* AN.KU$_{10}$ GA[R...]
13 [... AN].KU$_{10}$ GAR-*an* [...]
14 [...] ⌜x⌝ AN.KU$_{10}$ GAR-[...]
15 [...] ⌜x⌝ GABA IT[I...]
16 [...] ⌜^{d}UTU⌝ [...]

Obv. II and Rev. I lost

Rev. II

1′ [DIŠ$^?$ *m*]*i-i*[*š-ḫu*...]
2′ [DIŠ$^?$ *m*]*i-iš-ḫ*[*u*...]
3′ [DIŠ$^?$ *m*]*i-iš-ḫu* ⌜x GIM⌝ [...]

Obv. I

1 [¶ In Mon]th III, the [14th] day, it makes an eclipse. In Month [IX(?)...]
2 If in Month III the sun makes an eclipse, ditto [...]
3 In 3 intervals, in Month IX, the 14th day, [it makes] an ec[iplse... [
4 ¶ In Month IV, the 14th day, it makes an eclipse. In Month X [...]
5 ¶ In Month V, the 14th day, it makes an eclipse. In Month XI [...]
6 If there was an intercalary month (i.e. a Month VI$_2$), in Month X (after) 1 interval has passed. In 2 int[ervals...]
7 ¶ In Month VI, the 14th day, it makes an eclipse. In Month XII [...]
8 If 2 intervals before Month VI it does not make an eclipse [...]
9 2 intervals have passed and in Month VI (ditto). In 3... [intervals...]
10 If in Month VI the sun makes an eclipse, 1 interval [...]
11 [¶ In] Month VII, the 14th day, it makes an eclipse. 2 inter[vals...]
12 [...] intervals it makes an eclipse [...]
14 [...] it makes an [ecl]ipse [...]
14 [...] it makes an eclipse [...]
15 [...] all month[s...]
16 [...] the sun [...]

Obv. II and Rev. I lost

Rev. II

1′ [¶$^?$ lum]ino[sity$^?$...]
2′ [¶$^?$ lum]ino[sity$^?$...]
3′ [¶$^?$ lum]inosity$^?$... like [...]

We are here concerned only with the obverse.[33] The text presents a series of statements beginning with a DIŠ sign. Following the DIŠ sign, which is used as an item marker (and not to be read as a logogram for *šumma* "if"), a month of the year is given in which a lunar eclipse is said to take place on the 14th day. This is followed by months that are 6, 12, and 18 months later at which an eclipse is either said to occur or not occur. The dates at 12 months are said to be at two "intervals" and those at 18 months are at three "intervals." The term "interval" is written syllabically as *a-dan-na* or *a-dan-na-ti*. The example for Month V is particularly interesting because it provides two options: the next eclipse will be in Month XI or, if there has been an intercalary Month VI, then in Month X. Thus, in both cases, one interval corresponds to six months. For several entries, the text repeats the same pattern for solar eclipses.

2.4.2 K.12404

This small fragment was also edited by Rochberg-Halton (1988, 42–43), who states it is a parallel to 82–5–22, 501. In fact, the text seems to be doing something slightly different. The text reads:[34]

1′ [...]
2′ [...]
3′ [...] ⸢x⸣

4′ [...] *ša* ITI DIRI GAR-*un* [...]
5′ [... U]D.DUG$_4$.GA *ina* ITI.ŠE *ina* 5 I[TI...]
6′ [... A]N.KU$_{10}$.MEŠ EGIR *a-ḫa-meš* GAR-*u*[*n*...]
7′ [...] ⸢x⸣ *a-ḫa-meš ina* AN.KU$_{10}$ ^{d}30 GAR [...]

8′ [... UD.DU]G$_4$.GA ITI.ZÍZ DIB-*iq* [...]
9′ [... IT]I.AB *u* ^{d}30 *ina* ITI.[ZÍZ...]
10′ [...] ITI.IZI GAR-*u*[*n*...]
11′ [... ITI.B]ÁR GAR-[*un*...]

1′ [...]
2′ [...]
3′ [...]...

----

[33] Too little of the reverse is preserved to analyse. It may very well be a copy of a completely unrelated text.

[34] My edition is based upon Rochberg-Halton 1988, 42–43 with a few minor corrections taken from collation using a photograph. Most important, I read the number toward the end of line 5′ as a 5 rather than Rochberg-Halton's 6.

4′ [...] of the intercalary month, it makes (an eclipse) [...]
5′ [... i]nterval in Month XII at 5 m[onths...]
6′ [... e]clipses behind eachother it make[s...]
7′ [...] ⌜x⌝ eachother, it makes an eclipse of the moon [...]

8′ [... inter]val Month XI passes. [...]
9′ [... Mon]th X and the moon in Month [XI?...]
10′ [...] Month V it makes (an eclipse) [...]
11′ [... Mon]th I it makes (an eclipse) [...]

Although the text is badly damaged, enough remains to see that like 82–5–22, 501 it concerns the intervals between eclipses. However, whereas 82–5–22, 501 dealt only with six-month intervals K.12404 also apparently refers to the existence of five-month intervals. Furthermore, K.12404 refers to the fact that a solar eclipse can either precede or follow a lunar eclipse; thus, a solar eclipse may be in one month and the lunar eclipse in the next or they month be in the same month.

2.4.3 Sîn ina tāmartīšu 1

Schaumberger first noted that a commentary to the omen "If the moon makes an eclipse at an unaccounted time" offers as one interpretation that six months have not passed since the previous eclipse.[35] The entry appears in the first tablet of the commentary series *Sîn ina tāmartīšu* and is within a series of omens about the Moon appearance at an unaccounted time.[36] The entry reads:[37]

DIŠ 30 *ina* NU ŠID.MEŠ-*šú* AN.KU$_{10}$ GAR-*un-ma ša* 6 ITI.MEŠ NU DIR-*ú* KI.MIN UD.12.KAM UD.13.KAM AN.KU$_{10}$ GAR-*ma*

¶ "(If) the moon makes an eclipse at an unaccounted time"—six months have not passed. Secondly, it makes an eclipse on the 12th day or the 13th day.

The commentary offers two interpretations of "at an unaccounted time": that the eclipse occurred before six months have passed, or that the eclipse occurred on the 12th or 13th day of the month. In *Enūma Anu Enlil*, lunar eclipse omens are given for days 14, 15, 16, 20, and 21 (and occasionally days 1, 17, 18, and/or 19).[38] The second interpretation of "at an unaccounted time" given in this commentary therefore refers to eclipses occurring earlier than the norm. Crucially, note that eclipses

[35] Schaumberger 1935, 251. See also Rochberg-Halton 1988, 41; van der Waerden 1974, 116; Brown 2000, 200; Steele 2000; and Wainer 2016, 87. The omen itself is quoted in the Report SAA 8 4 without explanation.

[36] I translate phrase NU ŠID.MEŠ-*šú* (Akkadian: *lā minâtišu*) as "unaccounted time." Literally, ŠID.MEŠ refers to count, both in the sense of making a count and making a calculation. Rochberg-Halton translates the phrase as "uncalculated"; Koch translates "unexpected time."

[37] *Sîn ina tāmartīšu* I 25; transliteration taken from Koch-Westenholz (1999, 154), my translation.

[38] Rochberg-Halton 1988, 38–39.

on such days are possible and eclipses on the 13th day happen quite often. Indeed, omens for eclipses on days 12 and 13 are included in the so-called 29th *aḫû* tablet of containing additional omens which compliment those of *Enūma Anu Enlil* proper.[39] The first interpretation explicitly refers to the six-month interval between eclipse possibilities. By analogy with the second interpretation, and in line with the phrase "have not passed," it is highly likely that an eclipse at an "unaccounted time" can refer to an eclipse after a five-month interval. Thus, this commentary entry implicitly refers to both the six-month and the five-month intervals between eclipse possibilities.

The omen "(If) the moon makes an eclipse at an unaccounted time" is quoted in the Report SAA 8 4 written by Issar-šumu-ereš. That eclipse is said to take place on the 14th day of Month III. If we assume that Issar-šumu-ereš knew and used this interpretation of the omen from *Sîn ina tāmartīšu*,[40] then he must here be saying that the eclipse took place five rather than six months after the preceding eclipse possibility.

2.5 A Neo-Assyrian Eclipse Compilation

Three previously unpublished tablets contain copies of a Neo-Assyrian compilation of eclipse reports: Rm 121, K.12595+20083, and K.13633.[41] This text is remarkable because (i) it provides firm evidence that scholars in the Neo-Assyrian period had access to dated records of eclipse observations, evidence that we did not have before (recall that the Letters and Reports are not dated to the year and so cannot have been the source), and (ii) it demonstrates that these scholars were interested in analyzing observations of particular astronomical phenomena over time, probably with the intention of better understanding the regularity of these phenomena and developing or improving methods for their prediction.

The three tablets have the same layout and content.[42] Rm 121 is a substantial fragment preserving roughly the lower half and almost all the width of the original

[39] Rochberg-Halton 1987.

[40] Wainer 2016 has demonstrated that the Neo-Assyrian scholars very often used *Sîn ina tāmartīšu* in the interpretation of omens.

[41] I learnt of Rm 121 and K.12595 from the notebooks of Asger Aaboe, who had himself been alerted to these tablets by Erica Reiner. Aaboe made a preliminary transcription of the two tablets on 5 September 1974 and jotted a few notes alongside the transcription but apparently did not investigate their content further. More recently, scholars under the auspices of the Electronic Babylonian Library project have expanded the corpus: Aino Hätinen identified another fragmentary copy, K.13633, and Zsombor Földi identified a join between K.12595 and K.20083; I thank Jon Taylor for checking this join for me at the British Museum.

[42] Photographs of the tablets are available in the British Museum's online collection database (https://www.britishmuseum.org/collection). Note that at the time of writing the join between K.12595 and K.20083 is not reflected in the database. Note, there are minor orthographic differences between the tablets. For example, Rm 121 uses KÁM whereas K.12595+20083 uses KAM.

tablet; its left and bottom edges are partially preserved. The tablet is set out as a four-column table which extends from the obverse onto the reverse. Horizontal and vertical rulings mark the rows and columns. Often, the scribe writes on top of the horizontal ruling, almost as if it is a guideline rather than a row divider. Furthermore, the scribe often extends the text across the columns when there is insufficient space. K.12595+20083 is a one-sided flake which duplicates part of the rightmost column of the reverse of Rm 121; the right edge is preserved. K.13633 is a flake from the left edge. Its placement is uncertain.

Full editions of the three tablets are presented in the appendix. Figs. 2.1 and 2.2 present a composite text and a translation following the line numbering of the best preserved source, Rm 121. Many cells of the table are empty; these are represented in my edition by "(blank)." Cells where the text is lost or unreadable are indicated by [...]; it should be noted that these cells could also be empty. For legibility, in my transliteration and translation, I have kept the text within the columns divisions even where in the sources it stretches into the next column—see the critical apparatus to the editions of the individual tablets in the appendix for details of when this occurs.

I begin my analysis by considering the fourth column. This column contains the names of one or more months and a year number. No king's names are given, although the transition between one king and another is indicated through the use of the term "accession year" (MU SAG) in Rev. 9. Unfortunately, none of the preceding lines preserve the year number, which would have allowed the length of the reign of a king to be determined and thus the dating of the entries trivial to establish. It is worth noting that years are indicated by regnal year rather than by eponym, which was the everyday practice in the Neo-Assyrian Empire. Counting years by eponyms, however, does not make it easy to compare data over several years and would thus not be suitable for the type of analysis of past observations that seems to be the purpose of this text.

Some entries in the fourth column also include the term "intercalary month" (ITI DIRI), which is used to indicate only that the year in question contained an intercalary month; it does not mean that Month XII_2 is also the subject of that row of the table. Two entries preserve the phrase "passed" (DIB-*iq-ma*) at the end; in both cases, the first column contains the term "interval(s)" (UD.DUG$_4$.GA(.MEŠ)). Most entries give only one month name. In such cases, the first column usually contains the phrase "1 interval" (1 UD.DUG$_4$.GA) or one of the first three columns contains a DIŠ sign and/or some remark (in a few cases, all columns are blank; see the discussion below). If two month names are given, the first column contains the phrase "2 intervals" (2 UD.DUG$_4$.GA.MEŠ). No cases with three or more months are preserved but context suggests that in one or two lines three or four month names need to be restored and we can assume that the first column contains the phrase "3/4 intervals." Using the preserved entries in the first three columns and the presence or absence of the plural marker on the term "interval" allows a significant number of month names and year numbers to be restored alongside those that are still preserved. These restored entries have been included in Figs. 2.1 and 2.2.[43]

[43] Note, in Figs. 2.1 and 2.2 only month names that can be restored by internal content are included.

Obv.	I	II	III	IV
1′	[…]	(blank)	[…]	[…]
2′	[…]	(blank)	[…]	[…]
3′	[…]	DIŠ *i-na* 5 IT[I]	[…]	[…]
4′	[…]	(blank)	[…]	[…]
5′	[…]	(blank)	[…]	[…]
6′	[x UD.DUG$_4$.G]A.MEŠ	(blank)	[…]	[…]
7′	[…] TIL-*tim*	(blank)	(blank?)	[…]
8′	[…] (blank?)	(blank)	(blank)	[…]
9′	[…] (blank?)	(blank)	(blank)	[…]
10′	[x UD.D]UG$_4$.GA.MEŠ	(blank)	(blank)	[…]
11′	(blank)	*i-na* 6 ITI SILIM-*ma*	(blank)	[…]
12′	[x] UD.DUG$_4$.GA.MEŠ	(blank)	(blank)	[…]
13′	(blank)	DIŠ *i-na* 5 ITI	(blank)	[…]
14′	(blank)	(blank)	(blank)	[…]
15′	1 UD.DUG$_4$.GA	(blank)	(blank)	ITI.A[B$^?$ …]
16′	(blank)	DIŠ TIL-*tim*	(blank)	ITI.ŠU […]
17′	(blank)	DIŠ TIL-*tim*	(blank)	ITI.A[B? …]
18′	(blank)	DIŠ ^{d}UTU *ina* 5 ITI	(blank)	ITI.SI[G$_4$ …]
19′	⸢2$^?$⸣ UD.DUG$_4$.GA.MEŠ	(blank)	(blank)	ITI.APIN x […]
20′	(blank)	(blank)	DIŠ ^{d}UTU *li-la-tum* *ád-riš* ŠÚ-*bi*	ITI.APIN […]
21′	1 UD.DUG$_4$.GA	(blank)	(blank)	ITI.GU$_4$ M[U …]
22′	(blank)	(blank)	(blank)	(blank) […]
23′	(blank)	(blank)	DIŠ ^{d}UTU *ka-ṣa-a-tum ina* 5 ITI	ITI.ŠE […]
24′	(blank)	1 ŠU.SI *kíd-dum šá* ^{d}UTU *ina* 2 ITI	(blank)	ITI.x […]
25′	[…]	(blank)	DIŠ TIL-*tim*	I[TI.x …]
26′	[x UD.DUG$_4$.GA].MEŠ	(blank)	(blank)	[…]
27′	[…]	[…]	DIŠ	[…]

Fig. 2.1 Composite edition of Rm 121 and K.12595+20083 Obv. and Rev.

Rev.	I	II	III	IV
1	[…]	[…]	(blank)	[…]
2	[…]	(blank?)	(blank)	[…]
3	[…]	(blank)	DIŠ *ád-riš* […]	[…]
4	[…]	(blank)	(blank)	[…]
5	[…]	(blank)	(blank)	⸢ITI⸣ […]
6	[…]	⸢DIŠ⸣ TIL-*tim*	(blank)	ITI.GAN […]
7	[…]	⸢DIŠ⸣ TIL-*tim*	(blank)	ITI.SIG$_4$ […]
8	[…] ⸢x x⸣ *li-la-a-ti i-na* 5 ITI	(blank)	(blank	ITI.APIN M[U …]
9	[*i-n*]*a* 2-*i* ITI EGIR ^{d}UTU	(blank)	(blank)	ITI.GAN MU SAG ^{d}30 *u* ^{d}UTU x […]
10	(blank)	(blank)	(blank)	ITI.SIG$_4$ MU.1.KAM …]
11	⸢1 UD⸣.DUG$_4$.GA	(blank)	(blank)	ITI.⸢APIN⸣ MU.1.KÁM ITI.DIRI DIB-*iq-m*[*a*]
12	DIŠ *i-na* 5 ITU	(blank)	(blank)	ITI.BÁR MU.2.KAM
13	(blank)	DIŠ	(blank)	ITI.DU$_6$ MU.2.KAM
14	DIŠ	(blank)	(blank)	ITI.BÁR MU.3.KAM
15	DIŠ TIL-*tim*	(blank)	(blank)	ITI.KIN.2.KÁM MU.3 ITI.DIRI
16	(blank)	DIŠ	(blank)	ITI.ŠE MU. 3.KAM
17	⸢2$^{?}$⸣ UD.DUG$_4$.GA.MEŠ	(blank)	(blank)	ITI.KIN ITI.ŠE DIB-*iq-ma*
18	⸢^{d}⸣UTU *ka-ṣa-a-tum i-na* 5 ITI	(blank)	(blank)	ITI.⸢IZI$^{?}$⸣ MU.5$^{!}$.KAM ITI.DIRI
19	(blank)	*i-na* 1 ITI DIŠ ^{d}30 *u* ^{d}UTU	(blank)	⸢ITI⸣.[…] MU.5$^{!}$.KAM ITI.DIRI
20	(blank)	(blank)	DIŠ 20 *u* 30 *ád-*⸢*riš*⸣	[…] MU.6.KAM
21	(blank?)	DIŠ TIL-*tim* 20 5 ITI	(blank)	[…] MU.6.KAM
22	[x UD].⸢DUG$_4$.GA.MEŠ⸣	(blank)	(blank)	[…] x x x […]
23	[…]	⸢x x x⸣	(blank)	[…]
24	[…]	[…]	[…]	[…]

Fig. 2.1 (continued)

Obv.	I	II	III	IV
1′	[…]	(blank)	[…]	[…]
2′	[…]	(blank)	[…]	[…]
3′	[…]	¶ at 5 mon[ths]	[…]	[…]
4′	[…]	(blank)	[…]	[…]
5′	[…]	(blank)	[…]	[…]
6′	[x interval]ls	(blank)	[…]	[…]
7′	[…] complete	(blank)	(blank?)	[…]
8′	[…] (blank?)	(blank)	(blank)	[…]
9′	[…] (blank?)	(blank)	(blank)	[…]
10′	[x in]tervals	(blank)	(blank)	[…]
11′	(blank)	at 6 months, intact	(blank)	[…]
12′	[x] intervals	(blank)	(blank)	[…]
13′	(blank)	¶ at 5 months	(blank)	[…]
14′	(blank)	(blank)	(blank)	[…]
15′	1 interval	(blank)	(blank)	Month X? […]
16′	(blank)	¶ complete	(blank)	Month IV […]
17′	(blank)	¶ complete	(blank)	Month X? […]
18′	(blank)	¶ The sun at 5 months	(blank)	Month II[I …]
19′	⌜2?⌝ intervals	(blank)	(blank)	Month VIII x […]
20′	(blank)	(blank)	¶ The sun morning it set dark	Month VIII […]
21′	1 interval	(blank)	(blank)	Month II Ye[…]
22′	(blank)	(blank)	(blank)	(blank) […]
23′	(blank)	(blank)	¶ the sun morning at 5 months	Month XII […]
24′	(blank)	1 finger on the edge of the sun at 5 months	(blank)	Month […]
25′	[…]	(blank)	¶ complete	M[onth …]
26′	[x interval]s	(blank)	(blank)	[…]
27′	[…]	[…]	¶	[…]

Fig. 2.2 Composite translation of Rm 121 and K.12595+20083 Obv. and Rev.

Rev.	I	II	III	IV
1	[...]	[...]	(blank)	[...]
2	[...]	(blank?)	(blank)	[...]
3	[...]	(blank)	¶ dark [...]	[...]
4	[...]	(blank)	(blank)	[...]
5	[...]	(blank)	(blank)	⸢Month⸣ [...]
6	[...]	⸢¶⸣ complete	(blank)	Month IX [...]
7	[...]	⸢¶⸣ complete	(blank)	Month III [...]
8	[...] ... evening at 5 months	(blank)	(blank)	Month VIII Ye[ar ...]
9	[a]t the 2nd month after the sun	(blank)	(blank)	Month IX Accession Year moon and sun ... [...]
10	(blank)	(blank)	(blank)	Month III Year 1 [...]
11	1 interval	(blank)	(blank)	Month VIII Year 1 intercalary passe[d]
12	¶ at 5 months	(blank)	(blank)	Month I Year 2
13	(blank)	¶	(blank)	Month VII Year 2
14	¶	(blank)	(blank)	Month I Year 3
15	¶ complete	(blank)	(blank)	Month VI_2 Year 3 intercalary
16	(blank)	¶	(blank)	Momth XII Year 3
17	⸢2$^?$⸣ intervals	(blank)	(blank)	Month VI Month XII passed
18	The sun morning at 5 months	(blank)	(blank)	Month $V^?$ Year $5^!$ intercalary
19	(blank)	In 1 month ¶ moon and sun	(blank)	⸢Month⸣.[...] Year $5^!$ intercalary
20	(blank)	(blank)	¶ sun and moon dark	[...] Year 6
21	(blank?)	¶ complete the Sun 5 months	(blank)	[...] Year 6
22	[x in]tervals	(blank)	(blank)	[...] x x x [...]
23	[...]	⸢x x x⸣	(blank)	[...]
24	[...]	[...]	[...]	[...]

Fig. 2.2 (continued)

The pattern of months found in the fourth column immediately suggests a connection with eclipses: many entries are at an interval of six months from the preceding entry. Thus, we can interpret the remarks in the first three columns as referring to eclipses. The fact that the Sun and the Moon are both mentioned in these remarks suggests that both lunar and solar eclipse are represented in the text. First, let us assume that entries where the first column contains a reference to one or more "intervals" represent eclipse possibilities where no eclipse was seen and that those where there is a DIŠ sign and/or other text represent observed eclipses. This interpretation is supported by the two preserved cases of "passed" at the end of the line, both of which occur in lines where "interval(s)" appear in the first column. Indeed, of those rows of the table where the end of a line is preserved, the term "passed" appears whenever the first column has the word "interval(s)" and does not appear where any of the first three columns contain anything else. The term "passed" is used in the Letters and Reports as well as in later texts from Babylon to indicate eclipse possibilities where no eclipse is seen, either because the Moon was too far from the node to produce an umbral eclipse, because the eclipsed luminary was below the horizon at the time of the eclipse, or, in the case of a solar eclipse, the path of the eclipse was fully to the north or south of the place of observation.[44] Knowing that rows with the term "interval(s)" indicate eclipse possibilities with no visible eclipse and assuming that rows with other text in the first three columns refer to visible lunar and/or solar eclipses, we can attempt to date the tablet using the months and year numbers preserved in the fourth column. Tables 2.1 and 2.2 contain all lunar and solar eclipses that were theoretically visible in Nineveh according to modern computation between 690 and 660 BCE.[45] Comparing the dates of these eclipses with those in the text we find a good agreement if we take the "Year 1" in Rev. 10 to correspond to the first year of the reign of Assurbanipal. It should be noted that this fit also demonstrates that the dates given in the text follow the Assyrian, rather than the Babylonian, calendar. In the relevant period, the two calendars differed both in their numbering of the years of the king and in when intercalation was performed.

Figure 2.3 presents a comparison of the contents of the text with modern eclipse data as given in Tables 2.1 and 2.2. I omit from this analysis Obv. 1–9 and Rev. 24, which are too broken for analysis. The first column of the table gives the line

[44] Babylonian astronomers were able to predict the first two circumstances in advance, but not whether a solar eclipse would be seen at their latitude.

[45] I have considered all eclipses from ca. 750 BCE to 600 BCE but, for reasons of space, omit them from these tables and the following discussion. The data in Tables 2.1 and 2.2 is derived from the canons of lunar and solar eclipses computed for Babylon by Huber and De Meis 2004, 177–213. I have applied a correction of 0.08 hours to all the times to take into account the difference in terrestrial longitude of Babylon and Nineveh. In principle, the solar eclipses should be recomputed for the longitude and latitude of Nineveh because their time, magnitude and duration are dependent upon both latitude and longitude. For the precision necessary for this analysis, however, the discrepancy is trivial. Assyrian dates during the reign of Esarhaddon and Assurbanipal have been found using the tables of Parpola 1983, 382. Dates during the reign of Sennacherib are derived from Huber and De Meis 2004 on the basis of attested and extrapolated Babylonian intercalations and may differ by one month from the actual Assyrian calendar.

Table 2.1 Lunar eclipses visible at Nineveh between 690 and 660 BC (a * after the time indicates that the luminary was below the horizon at that time)

Julian year	Assyrian date	Magnitude	Time	Comments
–688 Dec 16	Sennacherib 16 X$^?$	Total	15.20*–19.05	Rose eclipsed
–687 Jun 10	Sennacherib 17 IV$^?$	Total	0.25–3.81	
–687 Dec 5	Sennacherib 17 X$^?$	0.14	19.17–21.06	
–685 Apr 22	Sennacherib 19 II$^?$	0.55	1.61–4.44	
–685 Oct 15	Sennacherib 19 VII$^?$	0.59	6.03–8.94*	Set eclipsed
–684 Oct 3	Sennacherib 20 VII$^?$	Total	18.89–22.59	
–680 Jan 24	Sennacherib 23 XI$^?$	Total	22.80–2.44	
–680 Jul 22	Sennacherib 24 V$^?$	Total	17.85*–21.91	Rose eclipsed
–679 Jan 16	Sennacherib 24 XI$^?$	0.32	15.37–17.62	
–679 Jul 11	Esarhaddon 1 IV	0.20	22.12–0.18	
–678 Jun 1	Esarhaddon 2 III	0.13	1.22–2.91	
–677 May 22	Esarhaddon 3 III	Total	17.42*–21.04	Rose eclipsed
–676 Nov 3	Esarhaddon 4 VIII	0.51	17.24*–20.03	Rose eclipsed
–673 Sep 3	Esarhaddon 7 VI	Total	22.52–2.78	
–672 Feb 27	Esarhaddon 7 XII	0.41	17.70–20.23	
–670 Jul 2	Esarhaddon 10 IV	0.94	19.21–22.66	
–670 Dec 27	Esarhaddon 10 X	Total	23.52–3.38	
–669 Dec 17	Esarhaddon 11 IX	0.15	3.21–5.13	
–688 Jun 10	Esarhaddon 12 III	0.06	1.08–2.91	
–666 Apr 21	Assurbanipal 2 I	Total	16.49*–20.52	Rose eclipsed
–666 Oct 15	Assurbanipal 2 VII	Total	3.66–7.35*	Set eclipsed
–665 Apr 10	Assurbanipal 3 I	0.48	18.65–21.57	
–665 Oct 4	Assurbanipal 3 VII	0.62	19.57–22.80	
–663 Feb 17	Assurbanipal 4 XII	0.81	16.13–19.30	
–663 Aug 13	Assurbanipal 5 V	0.64	1.00–4.20	
–662 Aug 2	Assurbanipal 6 V	Total	0.66–4.75	
–661 Jan 27	Assurbanipal 6 XI	0.35	23.84–2.16	
–660 Dec 6	Assurbanipal 8 IX	0.64	21.05–0.25	
–659 Jun 1	Assurbanipal 9 III	Total	0.54–4.51	
–659 Nov 25	Assurbanipal 9 IX	Total	20.08–0.12	

number of the text. The next four columns reproduce the translation given in Fig. 2.2. In the sixth column of Fig. 2.3 I give proposed Assyrian dates (year and month) for the entries in each row. In columns seven to ten I reproduce the modern eclipse data for the eclipses in those months.

The data in Fig. 2.3 confirms the assumption that whenever the term "interval(s)" appears in the first column, there is no eclipse. In most cases, where a DIŠ sign and/or other text appears in one of the first three columns, there was either a lunar or a solar eclipse (or both) that month. The only exceptions are:

Table 2.2 Solar eclipses visible at Nineveh between 690 and 660 BC (a * after the time indicates that the luminary was below the horizon at that time)

Julian year	Assyrian date	Magnitude	Time	Comments
–689 Jan 22	Sennacherib 14 XI?	0.18	12.31–14.37	
–689 Jul 18	Sennacherib 15 IV?	0.42	8.39–10.46	
–688 Jan 11	Sennacherib 15 X?	0.78	10.36–13.92	
–686 Nov 10	Sennacherib 18 VIII?	0.29	7.41–9.01	
–682 Aug 28	Sennacherib 22 VI	0.20	12.84–14.67	
–678 Jun 17	Esarhaddon 2 III	0.74	6.17–8.50	
–675 Apr 15	Esarhaddon 5 I	0.07	11.01–11.57	
–668 May 27	Esarhaddon 12 II	0.28	3.92*–5.27	Rose eclipsed
–663 Aug 28	Assurbanipal 5 V	0.13	7.75–8.55	
–661 Jan 12	Assurbanipal 6 X	0.86	15.62–18.02*	Set eclipsed
–660 Jun 27	Assurbanipal 8 III	0.96	15.52–18.08	

Obv. 20: The third column has a reference to the Sun setting while "dark" in Month VIII of Esarhaddon's third year. Assuming I have correctly restored the date of this row there was no solar eclipse visible in Nineveh that month. A solar eclipse did occur, but was only visible in northeast Asia.

Obv. 23: The third column refers to a solar eclipse in the morning at a five-month interval in Month XII of Esarhaddon's fourth year. No eclipse took place that month. However, an eclipse did take place the following month (Month I of Esarhaddon's fifth year). This eclipse is reported in the next row of the table (Obv. 25) and is at a six-month interval from the previous eclipse possibility. It seems likely, therefore, that the text reflects uncertainty in whether an eclipse took place at a five- or six-month interval until the eclipse was seen in the later month.

Obv. 25: The third column reads "¶ complete" but, if I have identified this eclipse possibility correctly as Month VIII of Esarhaddon's fifth year, there was no eclipse of either the sun or moon visible that month.

Rev. 8: The first column refers to an eclipse in the evening at a five-month interval in Month VIII of Esarhaddon's twelfth year. No lunar or solar eclipses were visible in Nineveh that month; a solar eclipse did occur but was only visible across the Americas and parts of western Africa.

Rev. 9: The first column refers to a lunar eclipse in Month IX of the same year. The entry states "at the 2nd month after the sun," suggesting that a lunar eclipse possibility was considered at the opposition both before and after the solar eclipse possibility at the end of Month VIII.[46]

[46] It is possible that the broken sign at the end of the fourth column of this line is DIB ("passed"). The end of the line would then read ^{d}30 *u* ^{d}UTU D[IB-*iq-ma*] "the moon and the sun, passed," confirming that no eclipses were observed at the lunar and solar eclipse possibilities that month.

Line	Col. I	Col. II	Col. III	Col. IV	Date	Lunar eclipse		Solar eclipse	
						Magnitude	Time	Magnitude	Time
Obv. 10	[2 int]ervals			[...]	Se 21 VII 22 I				
11		at 6 months, intact		[...]	Se 22 VII				
12	[2] intervals			[...]	Se 22 XII 23 V				
13		¶ at 5 months		[...]	Se 23 XI	total	22.80–2.44		
14				[...]	Se 24 V	total	17.85*–21.91		
15	1 interval			Month [...]	Se 24 XI				
16		¶ complete		Month IV [...]	Es 1 IV	0.2	22.14–0.18		
17		¶ complete		Month [III ...]	Es 2 III	0.13	1.22–2.91		
18		¶ sun at 5 months		Month III [...]	Es 2 III			0.74	6.17 – 8.50
19	[2] intervals			Month VIII [Month II]	Es 2 VIII 3 II				
20			¶ sun evening dark set	Month VIII [...]	Es 3 VIII				
21	1 interval			Month II Ye[ar...]	Es 4 II				
22				(VIII)	Es 4 VIII	0.51	17.24*–20.03		
23			¶ sun morning at 5 months	Month XII [...]	Es 4 XII				
24		1 finger on the edge of the sun in 2 months		Month [...]	Es 5 I			0.08	11.01–11.57
25	[...]		¶ complete	Month [...]	Es 5 VIII				
26	[3] intervals			[...]	Es 6 I VII 7 I				
27	[...]		¶	[...]	Es 7 VI	total	22.52–2.78		
Rev. 1	[...]	[...]		[...]	Es 7 XII	0.41	17.70–20.23		
2	[...]			[...]	8 V 8 XI 9 IV 9 X				
3	[...]		¶ dark [...]	[Month IV Year 10]	Es 10 IV	0.94	19.21–22.66		
4	[...]			[Month X Year 10]	Es 10 X	total	23.52–3.38		
5	[...]			Month [IV? Year 11]]	Es 11 IV				
6	[...]	¶ complete		Month IX [Year 11]	Es 11 IX	0.15	3.21–5.13		
7	[...]	¶ complete		Month III [Year 12]]	Es 12 III	0.06	1.07–2.91	0.28	3.92*–5.27
8	[...] x x evening at 5 months			Month VIII Year [12]	Es 12 VIII				
9	[at] the 2nd month after the sun			Month IX Accession year moon and sun	As 0 IX				
10				Month III Year [1]	As 1 III				
11	1 interval			Month VIII Year 1 Intercalary, passed	As 1 VIII				
12	¶ at 5 months			Month I year 2	As 2 I	total	16.49*–20.52		
13		¶		Month VII year 2	As 2 VII	total	3.66–7.35*		
14	¶			Month I Year 3	As 3 I	0.48	18.65–21.57		
15	¶ complete			Month VI2 Year 3 intercalary	As 3 VII	0.62	19.57–22.80		
16		¶		Month XII Year 3	As 3 XII				
17	2 intervals			Month VI Month XII passed	AS 4 VI XII				
18	Sun, morning at 5 months			Month V Year 5! Intercalary	As 5 V			0.13	7.75–8.55
19		in 1 month, ¶ moon and sun		Month [...] year 5! Intercalary	As 5 V	0.64	1.00–4.20		
20			¶ sun and moon dark	[Month ...] Year 6	As 6 V	total	0.66–4.75		
21		¶ complete sun 5 months		[Month ...] Year 6	As 6 X			0.86	15.62–18.02*
22	[2] intervals			[...]	As 7 IV X	0.35	23.842.16		
23	[...]	x x x		[...]	As 8 III			0.96	15.52–18.08

Fig. 2.3 Analysis of Rm 121 and K.12595+20083

Rev. 16: a DIŠ sign is written in the second column in Month IX of the third year of Assurbanipal. Although no lunar or solar eclipse was visible in Nineveh that month, a fairly large penumbral lunar eclipse took place when the moon was above the horizon. Perhaps that was mistakenly identified as a regular eclipse. A solar eclipse did occur that month but was only visible across South America and parts of western Africa.

A common feature of three of these cases is that they directly or indirectly concern solar eclipses at a five-month interval. Thus, it may be that predictions of these eclipses were noted to indicate when the five-month interval occurred (which would not be obvious from the term "interval").

All but three of the lunar and solar eclipses that were visible in Nineveh correspond to entries where either a DIŠ sign or some other comment (or both) appear in one of the first three columns. The three exceptions are:

Obv. 14: In Month V of Sennacherib's 24th year the moon rose eclipsed and became total. The first three columns are all blank.

Obv. 22: In Month VIII of Esarhaddon's eighth year the moon rose eclipsed. All of the columns (including the date column) are blank.

Rev. 21: The second column seems to record a solar eclipse at a five-month interval. This eclipse would correspond to Month X of Assurbanipal's sixth year. However, in Month XI the moon rose partially eclipsed. There is apparently no row for Month XI. Perhaps we are to understand the entry in this line as referring to both eclipses in two successive months.

A plausible explanation of the first two cases is that no records for that month were available to the scribe who made this compilation and so he simply left all the entries blank.

Although the basic structure and content of this text are clear, many problems of interpretation remain. In particular, the meaning of certain terminology does not seem to correspond to uses elsewhere.

Four Reports include the term "morning" (*ka-ṣa-a-tum*) or "evening" (*li-la-tum/li-la-a-tum*). It is unclear to me whether these are to be understood as the period close to sunrise and sunset or more broadly to the first and last daytime watch. All of the four extries where these terms occur concern solar eclipse possibilities, and in three cases they state that they are at a five-month interval. At one of these (Rev. 18) an eclipse was observable and the eclipse did indeed occur in the morning as indicated. However, it is puzzling that the other three cases are all predicted eclipses. Perhaps the explanation lies in the fact that in these cases, the scribe has copied this detail along with the statement of the five-month interval—detail that would normally be subsumed under the broad concept of an "interval" which passed.

More confusing is the use of the term TIL-*tim* "complete." In Late Babylonian eclipse reports, this term always refers to a total eclipse. It is not attested in the *Enūma Anu Enlil* lunar eclipse omens, nor in the Letters and Reports, however. Comparison of the cases where it is used here with the computed magnitudes of the eclipses shows that it cannot refer to a total eclipse in this text. In fact, it appears in none of the entries where the eclipse was total. I can think of two other possible interpretations of the term: (i) It could be used to indicate eclipses that were visible throughout the eclipse. This interpretation would fit most, but not all, cases where "complete" appears but there should be several other eclipses which were visible from start to finish; perhaps in these cases, the whole eclipse was not seen because of clouds. (ii) The term "complete" could refer to the eclipse possibility being at a normal six-month interval rather than at a five-month interval.

Three entries (Obv. 20′, Rev. 2, Rev. 20) in the third column use the term "dark" (*ád-riš*). This term is used in *Enūma Anu Enlil* to refer either to eclipses which rose or set while eclipsed or to the "darkening" of the Moon or Sun caused by things other than an eclipse, such as clouds or other atmospheric factors. In Obv. 20′, the sun is said to set, pointing to the former meaning. But the eclipses in Rev. 2 and Rev. 20 would be visible all night.

I cannot offer a confident explanation for the arrangement of entries within the three columns. It is possible that the columns indicate the time at which the eclipse took place. For the lunar eclipses, entries in the first column seem to be for eclipses which took place before midnight and those in the second column after midnight. For the solar eclipses, only entries in the second column are preserved which relate to observable eclipses and these are mostly in the morning. But two solar eclipses are also mentioned in the second column which took place in the late afternoon. Only two cases of observable eclipses are given in the third column: one took place well before midnight, the other after midnight. Therefore, it is possible, but far from certain, that the three columns refer to the three watches of the day/night in which an eclipse took place.

Despite these problems of interpretation, the text on Rm 121 and its duplicates is of considerable importance for our understanding of eclipse prediction in the Neo-Assyrian Empire. At a basic level, the text shows that Neo-Assyrian scholars were interested in exploring the patterns of eclipse possibilities and the link between the months of lunar and solar eclipse possibilities. Such an analysis was likely the first step toward identifying longer eclipse periods, such as the Saros, within which there is a fixed sequence of eclipse possibilities at six- and five-month intervals. The text also demonstrates that Assyrian scholars had access to dated Reports of eclipses. A little over half the height of the main source, Rm 121, is preserved. The preserved entries cover the period ca. 690–660 BCE. Thus, when complete, the compilation covered ca. 700–650 BCE. We do not have the source of these observations. They are unlikely to have been drawn from the Letters and Reports because, as discussed above, these texts were of immediate relevance and so usually do not include any indication of the year. Thus, we must postulate that (i) the Assyrian scholars kept separate records of dated observations, which have not been preserved to us, (ii) that they had access to the observations that were being recorded in Babylon, in which case they would have had to convert the Babylonian dates into Assyrian dates, which is not trivial given the evidence for different intercalation practices in the two regions, or (iii) that the text is a copy of an original composition from Babylon, in which the dates have been converted to the Assyrian calendar, which seems very unlikely. I suspect that the first postulate is correct, which would imply a much more organized and systematic tradition of astronomical observation in Assyria than has previously been recognized.[47]

2.6 Final Remarks

The evidence discussed in the preceding sections shows that Assyrian scholars during the first half of the seventh century BCE kept systematic records of eclipse observations, compiled and analyzed these records, had developed the concept of eclipse possibilities and identified the five- and six-month intervals between them, and, almost certainly, were attempting to identify patterns in these five- and six-month intervals and to identify eclipse periods. We have no clear evidence that they were successful in this last endeavor—if the evidence of records of eclipse observations over a 50-year period provided by Rm 121 is taken to be the sum total of what was available to these scholars, there was probably insufficient data to identify the Saros, for example. But this activity shows clearly that the practice of observation and interpretation of astronomical phenomena for the royal court that we see in the Letters and Reports was only part of Neo-Assyrian interest in the astral science. The observation and analysis of astronomical phenomena *as astronomical phenomena* and attempts to develop effective methods of prediction of these phenomena—even if within the broader context of the needs of divination and ritual practice—is closer in spirit to astronomical activity from Babylonia than has previously been recognized.

This evidence for Assyrian interest in the astronomical analysis of celestial phenomena and in the development of methods of prediction raises the question of its interaction with developments in Babylonia at the same time. Were these two completely separate endeavors? Or did they interact, with one influencing—perhaps even triggering—the other? Chronologically, we can be certain that in Babylon by about 600 BCE, roughly 50 years after the eclipse compilation on Rm 121 and duplicates was put together, the Saros cycle had been identified and methods of using it to predict eclipses and other astronomical phenomena (in particular, the lunar six). It is possible that these Babylonian scholars build upon the work of the Assyrian counterparts. But it should also be recalled that we have firm evidence for systematic eclipse observation and recording keeping from Babylonia during the late eighth and seventh century BCE. Ultimately, in the absence of additional

[47] Additional support for this conclusion may come from a few Assyrian fragments, which Pingree and Reiner 1975 interpret as containing observations of Mercury.

evidence, we cannot answer this question. But it seems likely that there was at least some amount of communication between the Assyrian and Babylonian scholars and that this communication would have extended to their shared interest in understanding patterns of eclipses and developing methods for their prediction.

Acknowledgments I thank the Trustees of the British Museum for permission to study the tablets discussed in this chapter. I also wish to thank the Electronic Babylonian Library and its contributors for identifying joins and duplicates to Rm 121 and K.12595, and Hermann Hunger for useful discussions. Finally, I remember the late Asger Aaboe whose notes on Rm 121 and K.12595 prompted my work on this topic.

Appendix: Editions of Rm 121, K.12595+20083, and K.13633

Rm 121

Edition: Fig. 2.4

Critical Apparatus:

Obv. 20′ This row is double width to accommodate two lines of text in column III.

Obv. 24′ The text in column I extends into column II.

Rev. 8 The text in column II extends into column III.

Rev. 9 The text in column I extends into column II.

Rev. 17 In column IV, ITI.KIN.2.KÁM is a scribal error for ITI.KIN ITI.ŠE probably caused by mistakenly copying line 15.

K.12595+20083

Edition: Fig. 2.5

Critical Apparatus

1′ The final sign might be the beginning of DIB, although the first wedge looks like a winkelhaken rather than a horizontal.

10′ 7 must be a mistake for 5.

11′ 6 must be a mistake for 5.

K.13633

Edition: Fig. 2.6

Critical Apparatus:

8′ The text in column I extends into column II.

References

Aaboe, Asger. 1972. Remarks on the Theoretical Treatment of Eclipses in Antiquity. *Journal for the History of Astronomy* 3:105–118.

Aaboe, Asger. 1980. Observation and Theory in Babylonian Astronomy. *Centaurus* 24:14–35.

Aaboe, Asger, John P. Britton, Janice A. Henderson, Otto Neugebauer, and Abraham J. Sachs. 1991. *Saros Cycle Dates and Related Babylonian Astronomical Texts*. Philadelphia: American Philosophical Society.

Obv.	I	II	III	IV
1′	[…]	(blank)	[…]	[…]
2′	[…]	(blank)	[…]	[…]
3′	[…]	DIŠ *i-na* 5 IT[I]	[…]	[…]
4′	[…]	(blank)	[…]	[…]
5′	[…]	(blank)	[…]	[…]
6′	[x UD.DUG$_4$.G]A.MEŠ	(blank)	[…]	[…]
7′	[…] TIL-*tim*	(blank)	(blank?)	[…]
8′	[…] (blank?)	(blank)	(blank)	[…]
9′	[…] (blank?)	(blank)	(blank)	[…]
10′	[x UD.D]UG$_4$.GA.MEŠ	(blank)	(blank)	[…]
11′	(blank)	*i-na* 6 ITI SILIM-*ma*	(blank)	[…]
12′	[x] UD.DUG$_4$.GA.MEŠ	(blank)	(blank)	[…]
13′	(blank)	DIŠ *i-na* 5 ITI	(blank)	[…]
14′	(blank)	(blank)	(blank)	[…]
15′	1 UD.DUG$_4$.GA	(blank)	(blank)	ITI.A[B$^?$ …]
16′	(blank)	DIŠ TIL-*tim*	(blank)	ITI.ŠU […]
17′	(blank)	DIŠ TIL-*tim*	(blank)	ITI.A[B? …]
18′	(blank)	DIŠ ^{d}UTU *ina* 5 ITI	(blank)	ITI.SI[G$_4$ …]
19′	⸢2$^?$⸣ UD.DUG$_4$.GA.MEŠ	(blank)	(blank)	ITI.APIN x […]
20′	(blank)	(blank)	DIŠ ^{d}UTU *li-la-tum* *ád-riš* ŠÚ-*bi*	ITI.APIN […]
21′	1 UD.DUG$_4$.GA	(blank)	(blank)	ITI.GU$_4$ M[U …]
22′	(blank)	(blank)	(blank)	(blank) […]
23′	(blank)	(blank)	DIŠ ^{d}UTU *ka-ṣa-a-tum ina* 5 ITI	ITI.ŠE […]
24′	(blank)	1 ŠU.SI *kíd-dum šá* ^{d}UTU *ina* 2 ITI	(blank)	ITI.x […]
25′	[…]	(blank)	DIŠ TIL-*tim*	I[TI.x …]
26′	[x UD.DUG$_4$.GA].MEŠ	(blank)	(blank)	[…]
27′	[…]	[…]	DIŠ	[…]

Fig. 2.4 Edition of Rm 121 Obv. and Rev.

Rev.	I	II	III	IV
1	[…]	[…]	(blank)	[…]
2	[…]	(blank?)	(blank)	[…]
3	[…]	(blank)	DIŠ *ád-riš* […]	[…]
4	[…]	(blank)	(blank)	[…]
5	[…]	(blank)	(blank)	⸢ITI⸣ […]
6	[…]	⸢DIŠ⸣ TIL-*tim*	(blank)	ITI.GAN […]
7	[…]	⸢DIŠ⸣ TIL-*tim*	(blank)	ITI.SIG$_4$ […]
8	[…] ⸢x x⸣ *li-la-a-ti i-na* 5 ITI	(blank)	ITI.APIN M[U …]	
9	[*i-n*]*a* 2-*i* ITI EGIR ^{d}UTU	(blank)	(blank)	ITI.GAN MU S[AG …]
10	(blank)	(blank)	(blank)	ITI.SIG$_4$ MU […]
11	⸢1 UD⸣.DUG$_4$.GA	(blank)	(blank)	ITI.⸢APIN⸣ MU.1.KÁM IT[I …]
12	DIŠ *i-na* 5 ITU	(blank)	(blank)	ITI.BÁR MU.2 […]
13	(blank)	DIŠ	(blank)	ITI.DU$_6$ MU.2 […]
14	DIŠ	(blank)	(blank)	ITI.BÁR MU.3 […]
15	DIŠ TIL-*tim*	(blank)	(blank)	ITI.KIN.2.KÁM MU.3 […]
16	(blank)	DIŠ	(blank)	ITI.ŠE MU. ⸢3⸣ […]
17	⸢2?⸣ UD.DUG$_4$.GA.MEŠ	(blank)	(blank)	ITI.KIN.2.KAM MU […]
18	⸢^{d}⸣UTU *ka-ṣa-a-tum*	*i-na* 5 ITI	(blank)	ITI.⸢IZI?⸣
19	(blank)	*i-na* 1 ITI DIŠ ^{d}30 *u* ^{d}UTU	(blank)	⸢ITI⸣.[…]
20	(blank)	(blank)	DIŠ 20 *u* 30 *ád-*⸢*riš*⸣	[…]
21	(blank?)	DIŠ TIL-*tim* 20 5 ITI	(blank)	[…]
22	[x UD].⸢DUG$_4$.GA.MEŠ⸣	(blank)	(blank)	[…]
23	[…]	⸢x x x⸣	(blank)	[…]
24	[…]	[…]	[…]	[…]

Fig. 2.4 (continued)

1	[… M]U SAG ^{d}30 *u* ^{d}UTU x […]
2	[…] MU.1.KAM
3	[… IT]I.APIN MU.1.KAM ITI.DIRI DIB-*iq-m*[*a*]
4	[… IT]I.BÁR MU.2.KAM
5	[… IT]I.DU$_6$ MU.2.KAM
6	[… ITI.B]ÁR MU.3.KAM
7	[… KIN].2.KAM MU.3.KAM ITI.DIRI
8	[…] MU.3.KAM
9	[…] KIN$^{?}$ ITI.ŠE DIB$^{?}$-*iq-ma*
10	[…] MU.7.KAM ITI.DIRI
11	[…] MU.6 KAM ITI.DIRI
12	[…] MU.6.KAM
13	[…] MU.6.KAM
14	[…] x x x […]

Fig. 2.5 Edition of K.12595+20083

1′	⸢1 UD.DUG$_4$⸣.[GA …]	[…]
2′	(blank)	[…]
3′	2 UD.DUG$_4$.G[A.MEŠ]	[…]
4′	(blank)	DIŠ TIL-*t*[*im* …]
5′	1 UD.DUG$_4$.GA	(blank)
6′	(blank)	DIŠ TIL-*ti*[*m* …]
7′	(blank)	DIŠ TIL-*ti*[*m* …]
8′	DIŠ ^{d}UTU TIL-*tim* ⸢*li-la-a-ti*$^{?}$⸣ […]	

Fig. 2.6 Edition of K.13633

Brack-Bernsen, Lis. 1997. *Zur Entstehung der Babylonischen Mondtheorie: Beobachtung und theoretische Berechnung von Mondphasen. Boethius 40*. Stuttgart: Franz Steiner Verlag.

Brack-Bernsen, Lis. 1999. Goal-Year Tablets: Lunar Data and Predictions. In *Ancient Astronomy and Celestial Divination*, ed. Noel M. Swerdlow, 149–177. Cambridge, MA: MIT Press.

Brack-Bernsen, Lis. 2020. Babylonian Astronomy: A New Understanding of Column Φ. Schematic Astronomy, Old Prediction Rules, Riddles, Loose Ends, and New Ideas. *Archive for History of Exact Sciences* 74:605–640.

Brack-Bernsen, Lis, and John M. Steele. 2005. Eclipse Prediction and the Length of the Saros in Babylonian Astronomy. *Centaurus* 47:181–206.

Britton, John P. 1989. An Early Function for Eclipse Magnitudes in Babylonian Astronomy. *Centaurus* 32:1–52.

Britton, John P. 2007. Studies in Babylonian Lunar Theory: Part I. Empirical Elements for Modeling Lunar and Solar Anomalies. *Archive for History of Exact Sciences* 61:83–145.

Britton, John P. 2009. Studies in Babylonian Lunar Theory: Part II. Treatments of Lunar Anomaly. *Archive for History of Exact Sciences* 63:357–431.

Brown, David. 2000. *Mesopotamian Planetary Astronomy-Astrology*. Groningen: Styx.

de Jong, Teije. 2017. On the Origin of the Lunar and Solar Periods in Babylonian Lunar Theory. In *Studies on the Ancient Exact Sciences in Honor of Lis Brack-Bernsen*, ed. John M. Steele and Mathieu Ossendrijver, 105–126. Berlin: Edition Topoi.

de Jong, Teije. 2019a. A Study of Babylonian Planetary Theory I. The Outer Planets. *Archive for History of Exact Sciences* 73:1–37.
de Jong, Teije. 2019b. A Study of Babylonian Planetary Theory II. The Planet Venus. *Archive for History of Exact Sciences* 73:309–333.
de Jong, Teije. 2021. A Study of Babylonian Planetary Theory III. The Planet Mercury. *Archive for History of Exact Sciences* 75:491–522.
de Meis, Salvo, and Hermann Hunger. 1998. *Astronomical Dating of Assyrian and Babylonian Reports*. Rome: Istituto italiano per l'Africa e l'Oriente.
Huber, Peter J., and John P. Britton. 2007. A Lunar Six Text from 591 B.C. *Wiener Zeitschrift für die Kunde des Morgenlandes* 97:213–217.
Huber, Peter J., and Salvo de Meis. 2004. *Babylonian Eclipse Observations from 750 BC to 1 BC*. Milan: Mimesis.
Huber, Peter, and John M. Steele. 2007. Babylonian Lunar Six Tablets. *SCIAMVS* 8:3–36.
Hunger, Hermann. 1992. *Astrological Reports to Assyrian Kings. State Archives of Assyria 8*. Helsinki: Helsinki University Press.
Hunger, Hermann. 2001. Astronomical Diaries and Related Texts from Babylonia. 5: *Lunar and Planetary Texts*. Vienna: Österreichischen Akademie der Wissenschaften.
Hunger, Hermann, and John M. Steele. 2019. *The Babylonian Astronomical Compendium MUL. APIN*. Abingdon: Routledge.
Koch-Westenholz, Ulla. 1999. The Astrological Commentary *Šumma sîn ina tāmartīšu* Tablet 1. *Res Orientales* 12:149–165.
Neugebauer, Otto. 1968. The Origin of 'SystemB' of Babylonian Astronomy. *Centaurus* 12:209–214.
Oppenheim, A. Leo. 1969. Divination and Celestial Observation in the Last Assyrian Empire. *Centaurus* 14:97–135.
Parpola, Simo. 1983. *Letters from Assyrian Scholars to the Kings Esarhaddon and Assurbanipal. Part II: Commentary and Appendices*. Neukirchen-Vluyn: Butzon & Bercker Kavelaer.
Parpola, Simo. 1993, *Letters from Assyrian and Babylonian Scholars, State Archives of Assyria 10. Helsinki*: Helsinki University Press.
Pingree, David, and Erica Reiner. 1975. Observation Texts Concerning the Planet Mercury. *Revue d'Assyriologie* 69:175–180.
Rochberg-Halton, Francesca. 1987. The Assumed 29th *aḫu* Tablet of *Enūma Anu Enlil*. In *Language, Literature and History: Philological and Historical Studies Presented to Erica Reiner*, ed. Rochberg-Halton, 327–350. New Haven: American Oriental Society.
Rochberg-Halton, Francesca. 1988. *Aspects of Babylonian Celestial Divination: The Lunar Eclipse Tablets of Enūma Anu Enlil*. Horn: Berger & Söhne.
Schaumberger, Johann. 1935. *Sternkunde und Sterndienst in Babel. 3. Ergänzunscheft zum ersten und zweiten Buch*. Münster in Westfalen: Aschendorffsche Verlagsbuchhandlung.
Schnabel, Paul. 1924. Die Sarosperiode der Finsternisse schon in der Sargonidenzeit bejannt. *Zeitschrift für Assyriologie und Vorderasiatische Archäologie* 35:297–318.
Steele, John M. 2000. Eclipse Prediction in Mesopotamia. *Archive for History of Exact Sciences* 54:421–454.
Steele, John M. 2002. A Simple Function for the Length of the Saros in Babylonian Astronomy. In *Under One Sky: Astronomy and Mathematics in the Ancient Near East*, ed. John M. Steele and Annette Imhausen, 405–420. Münster: Ugarit-Verlag.
Steele, John M. 2019. The Early History of the Astronomical Diaries. In *Keeping Watch in Babylon: The Astronomical Diaries in Context*, ed. Johannes Haubold, John M. Steele, and Kathryn Stevens, 19–52. Leiden: Brill.
Steele, John M. Forthcoming. *Astronomy in Babylonia: The Development of the Astral Sciences in the First Millennium BC*.
Steele, John M., and Teije de Jong. 2023. An Early System A-Type Scheme for Saturn from Babylon. *Archive for History of Exact Sciences* 77:501–535.

Swerdlow, Noel. 1998. *The Babylonian Theory of the Planets*. Princeton: Princeton University Press.
van der Waerden, Bartel L. 1974. *Science Awakening II: The Birth of Astronomy*. Leiden: Noordhof.
Wainer, Zackary M. 2016. *The Series If the Moon at Its Appearance and Mesopotamia Scholarship of the First Millennium BCE*. PhD diss: Brown University.
Walker, Christopher B. F. 1997. Achaemenid Chronology and the Babylonian Sources. In *Mesopotamia and Iran in the Persian Period: Conquest and Imperialism 539–331 BC*, ed. John Curtis, 17–25. London: British Museum.

Chapter 3
Babylonian Horoscopy and Genethlialogy

Francesca Rochberg

This chapter raises the question of the nature and roots of Babylonian horoscopy and genethlialogy. It considers the fundamental difference between Babylonian and Greek/Greco-Roman astrological practices, devoting particular attention to the ways the world was thought to work in terms of a causal structure underpinning the idea that human life was indicated or even determined by the heavens. Recognition of this essential metaphysical distinction in causality is, I suggest, essential to our historiography of early astral sciences as well.

By the sixth century BCE, shortly before the fall of Babylon to Cyrus the Persian, astrology was so emblematic of Babylonia that the biblical prophet we call second Isaiah (47:13) commanded "the daughter of the Chaldeans," that is, Babylonia itself, to sit in the dust and the darkness to wish and hope in vain that her "astrologers, stargazers, and monthly prognosticators" might stand up and save her from imminent downfall. The fame of the Babylonians, or Chaldeans, as they were also called by later Hellenistic Greco-Roman writers, persisted in the realms of astronomy and astrology, albeit in a more positive light than that cast by the biblical text. It was with this biblical passage that Noel Swerdlow introduced his *Babylonian Theory of the Planets* (Swerdlow 1998, 1), which was one of the first works to emphasize astrology and omen divination as integral to Babylonian mathematical astronomy. "Whatever the sophistication of ancient mathematical astronomy, both Babylonian and Greek," he said (Swerdlow 1999, 14), "and even though at its highest level it was a science practiced for its own sake and guided by the curiosity and ingenuity of its practitioners, there is no reason to doubt that its origins lie in celestial divination."

F. Rochberg (✉)
Department of Middle Eastern Languages and Cultures, University of California, Berkeley, CA, USA
e-mail: rochberg@berkeley.edu

F. Rochberg (ed.), *Studies in Premodern Sciences in Memory of Noel M. Swerdlow*, Archimedes 74, https://doi.org/10.1007/978-3-032-04040-4_3

Swerdlow was partial to the cuneiform "Scribes of *Enūma Anu Enlil*" (*ṭupšar Enūma Anu Enlil*), as they were designated by title, and who were the compilers and copyists of celestial divinatory texts. He insisted on the importance of celestial divination to both mathematical astronomy and the Late Babylonian astrological and nonmathematical astronomical texts for the history of early science. He was emphatic about the connections between the rigorous mathematical astronomy of the period post-600 BCE and the prior tradition of celestial divination, as in the above-quoted statement. In *Babylonian Theory of the Planets* Swerdlow said of the Scribes of *Enūma Anu Enlil* that:

> The people who wrote and used our texts [by which he meant the astronomical cuneiform texts in general] cannot really be called astronomers, nor can they be called astrologers, although they were both. They were scribes priests, and temple officials ... at times advisors to the king, skilled in various sorts of ritual, divination, exorcism, and magic, in general wise men, scholars learned in the wisdom of their day. (Swerdlow 1998, xiv)

The history of astrology in its most general sense depends upon our recognition of the Babylonian development of the idea that human fortune can be read in the stars, and the history of astronomy similarly depends upon our recognition of the Babylonian development of observational and theoretical knowledge of heavenly phenomena. Not only does our understanding of the histories of astrology and astronomy depend upon our knowledge of these Babylonian manifestations, but how we interpret the early history of science itself also depends upon our understanding of the astral sciences' distant Babylonian forms and how those scholarly disciplines functioned interdependently within an intellectual and textual culture many millennia in the making. Again, Swerdlow was convinced of the scientific nature of Babylonian scholarly divination. He said:

> If our modern understanding of nature may be characterized as above all scientific, the ancient understanding may be characterized as above all ominous; every natural occurrence, regular or irregular, may be an omen. This was true when the compilation of the omen series began early in the second millennium, and perhaps remains true for much of the world even now. But the two are not necessarily mutually exclusive, indeed, in the belief in omens lies the birth of science. For the systematic collection of celestial and meteorological omens as carried out in Mesopotamia required a systematic study of celestial and meteorological phenomena; and thus it is our belief that the collection *and* the interpretation of omens must be regarded as a science, in fact, as the first and the most important science of antiquity—whether we consider it right or wrong, true or false, is irrelevant—and the ultimate origin of science as we understand it today. (Swerdlow 1998, 5)

In the following discussion I am less interested in pinpointing the birth of science than in showing how our modern interpretation of the causal structure of Babylonian astrology, that is celestial divination and horoscopy, played a decisive role in the modern study of its importance for early science.

Genethlialogy, or natal astrology, in the Greek-speaking world was the consequence of widespread influence from the late first millennium centers of cuneiform intellectual culture in Babylon and Uruk to Greek intellectuals in Hellenistic Egypt, mediated in no small measure by Egyptian culture. Otto Neugebauer characterized this in the following way:

> Astrology is in some sense typical of the changes which result from a transmission of Babylonian material to the Greeks. ... Before the fifth century B.C. celestial omina probably did not include predictions for individuals, based on planetary positions in the signs of the zodiac and on their mutual configurations. In this latest and most significant modification astrology became known to the Greeks in the Hellenistic period.... [There] the doctrine was changed in Greek hands to a universal system in which form alone it could spread all over the world. Hence, astrology in the modern sense of the term, with its vastly expanded set of "methods" is a truly Greek creation. (Neugebauer 1975, 613)

In this passage, Neugebauer granted to the Late Babylonian Period (which dates cuneiform texts post-500 BCE that cannot be dated by means of regnal years) the modification of celestial divination to horoscopy and genethlialogy, which he defined as "predictions for individuals based on planetary positions in the signs of the zodiac and their mutual configurations." Documents that fit this description are found in cuneiform and span the fifth to the first centuries BCE (Rochberg 1998). Neugebauer assigned to the Greeks the transformation of this Babylonian form of genethlialogy to the universal system associated with Greek and Greco-Roman theoretical treatises as well as practical handbooks on astrology, exemplified by Ptolemy's *Tetrabiblos* (transl. Robbins 1971) and Vettius Valens's *Anthology* (edition Pingree 1986; transl. of passages in Neugebauer and van Hoesen 1959), respectively.

The mid-twentieth-century characterization of astrology as a universal system, as in the above-quoted passage of Neugebauer, was important for establishing its place in the history of science. Also key to astrology's classification as a science was its physical and mechanical-causal rationale, illustrated in the following statement by Lynn Thorndike on what he called "the true place of astrology in the history of science":

> My present purpose is to emphasize a single point, but one which is the most important and fundamental of all. Briefly stated, it is that, during the long period of scientific development before Sir Isaac Newton promulgated the universal law of gravitation, there had been generally recognized and accepted another and different universal natural law, which his supplanted. And that universal natural law was astrological. (Thorndike 1955, 273)

This distilled and universalist approach to ancient and medieval astrology had already been championed by George Sarton, who, in a review of Thorndike's monumental *A History of Magic and Experimental Science during the First Thirteen Centuries of Our Era* (1923–1958), called astrology "a very remarkable scientific system" (Sarton 1924, 85). He added that "a history of science which did not include an account *of the purer aspects* [my emphasis] of astrology ... would be incomplete" (1924, 85). In spite of his defense of "*the purer aspects*" of astrology, Sarton found much of the contents of Thorndike's magnum opus unscientific and worse, namely, the subjects of divination and magic. Even Neugebauer invoked the *purity* of astrology's mechanical system, saying in 1969:

> To Greek philosophers and astronomers, the universe was a well defined structure of directly related bodies. The concept of predictable influence between these bodies is in principle not at all different from any modern mechanistic theory. And it stands in sharpest contrast to the ideas of either arbitrary rulership of deities or of the possibility of influencing

> events by magical operations. Compared with the background of religion, magic and mysticism, the fundamental doctrines of astrology are *pure science*. (Neugebauer 1969, 171, my emphasis)

That statement was made over 70 years ago. Today we no longer appeal to notions of purity in our efforts to define and characterize the history of science. Moreover, the invidious comparison of (an overly narrow definition of) the category of knowledge we call science to (overly broad definitions of) categories of what we call magic and religion often sets up an immediate distortion for historical analysis of premodern evidence.

The mechanical-causal principle, the universal law to which Thorndike, Sarton, and Neugebauer all referred, was celestial or stellar influence, and this point later became the central element of David Pingree's definition of astrology, which appeared in the opening paragraph of his entry "Astrology" in the *Dictionary of the History of Ideas*. There he said:

> Astrology is the study of the impact of the celestial bodies—Moon, Mercury, Venus, Sun, Mars, Jupiter, Saturn, the fixed stars, and sometimes the lunar nodes—upon the sublunar world. It presupposes a geocentric and finite universe. The influence of the celestial bodies is variously considered to be absolutely determinative of all motions of the four sublunar elements (Aristotelian physics is accepted as the basis for describing this influence ...); ... Astrology so defined could not have existed before the Hellenistic period, and is certainly *not* of Babylonian, Egyptian, or Indian origin. (Pingree 1973, 118)

In addition, Pingree defined genethlialogy as "the science of *interpreting the horoscopic diagram* representing the positions of the planets and the zodiacal signs at the moment of a native's conception or birth" (Pingree 2001, 4; my emphasis). The horoscopic diagram required the use of the *horoscopus* or ascendant (the point where the ecliptic meets the horizon at the moment of birth) and the other centers, or cardines, namely, midheaven, lower midheaven and setting point, the four points the ecliptic will have in common with the horizon and the meridian at a given moment, such as that of a birth.

This is not to say that every horoscope makes explicit reference to the *horoscopus* and cardines. John North pointed out, referring to Neugebauer and van Hoesen's 1959 edition of the Greek horoscopes, that "out of 168 horoscopes, from a period of nearly seven centuries (71 BCE to 621 CE), only 27 horoscopes give an indication of both the ascendant and either mid-heaven or lower mid-heaven" (1986a, 6). He added:

> As for the Greek horoscopes recorded by Neugebauer and van Hoesen, only two out of fifty-one non-literary examples (for A.D. 95 and 137) include a calculated degree for mid-heaven. It seems likely that the average astrologer found the exercise a difficult one. If, as Neugebauer and van Hoesen seem to imply, one of these two (that for A.D. 95) is poorly calculated, our suspicion is reinforced. (North 1986a, 7)

Absent the horoscopic diagram, those methodological problems were not the sort confronted by the scribes who prepared cuneiform horoscopes, and accounts for Pingree's terminology "proto-horoscopes" for the cuneiform counterparts to the Greek texts edited by Neugebauer and van Hoesen. The use of the centers in Greek but *not* Babylonian horoscopes also explains Pingree's further historical claim that genethlialogy "was invented by Greeks, probably in Egypt, in –100" (Pingree 2001, 4).

Pingree's focus on whether or not certain elements were to be found in cuneiform horoscopic texts gave rise to a persistent terminological problem (Brown et al. 2018, 379 and 879 sub proto-horoscopes). In the most cogent response to this problem, Dorian Greenbaum and Micah Ross said, "the varying definitions of 'horoscope' alter the identification of horoscopes in antiquity, the modern publications in which they appear, and the historical conclusions drawn from these pieces of evidence. In fact, depending on which definition is used, the origin of the horoscope could have occurred in Babylon, Egypt or Greece" (Greenbaum and Ross 2010, 146). Historical significance, therefore, rides on the definition of the term, including what is expressed by the broader rubric "genethlialogy" of which the horoscope was its most important tool.

It is my position here, however, that the distinction between Babylonian and Greek horoscopy goes beyond methodology to include questions of ontology, metaphysics and causality. The introduction of the *horoscopus* and cardines required a spherical cosmology and was combined with ideas of physical causes, as Pingree pointed out in his 1973 definition quoted above. His assertion that astrology could not have existed before the Greeks or in any other part of the ancient Middle East (or India) is a direct function of his definition. A broader definition, or at least one that does not set a standard that begins with the Greek material will avoid the circularity of the assertion that astrology defined in Greek terms did not exist in any but Greek terms. It is in the spirit of such a broader definition that I take a different approach to the history of horoscopy and genethlialogy in the following discussion. I begin with description of the various kinds of cuneiform materials for celestial prognostication in chronological order and end with some commentary on the underlying causalities reflected in those texts.

Cuneiform astrological texts fall into three major types: (1) celestial omens whereby individual phenomena of the Sun, Moon, or planets, whether associated with specific dates or not, were thought to indicate future events for the king and the population of his kingdom as a whole and which are attested from the Old Babylonian through to the Seleucid period; (2) natal omens whereby attributes or fortunes of an individual are correlated with a zodiacal sign, and for that reason date only to the period after ca. 500 BCE; and (3) horoscopes, in which the situation of the heavens on or near the date of a birth, such as the positions of the five planets, and other astronomical events of interest (solstice and equinox dates), are combined in order that they may be correlated with the individual born on the date in question. The derivation of data in Babylonian horoscopes was a complex process of extraction and compilation from nonmathematical astronomical texts (NMAT), direct calculation or interpolation, but certainly *not*, as is sometimes still stated, from observations of the planetary positions, solstices, equinoxes, or any other celestial event "observable" (Fincke 2016, 131) on the date of birth (on the derivation of astronomical data for horoscopes, see Rochberg 1998, 8–9 and 2004, 145–63, especially 160–63).

The first example is an eclipse omen from the second millennium BCE followed by a slightly expanded version of the same omen from the later codified series entitled *Enūma Anu Enlil* from the seventh century BCE:

1. Celestial omens

   (If) an eclipse occurs on the 14th of Nisannu: There will be famine and the population will decrease. (BM 22696:18, George and Taniguchi 2024, 143)

   And

   If an eclipse occurs on the 14th of Nisannu in the evening watch and is red; it begins and clears in the north: The king of Akkad will be well on account of that red eclipse; (but) there will be famine in that land (Akkad). (EAE 17 ¶ I.2, Rochberg-Halton 1988, 116).

   Second, and only emerging in the latter part of the fifth century BCE, are natal omens that foretell personal matters on the basis of the appearance of a planet or a fixed star at the birth of the individual concerned (as in the first passage), or indeed in what zodiacal sign the individual was said to have been born (as in the second):

2. Natal omens

   If a child is born and Jupiter comes out (= rises): He will be normal, healthy; he will be rich (and) grow old; long days. (TCL 6 14:29, Sachs 1952, 66 and 68).

   And

   Born in the region of Sagittarius: (he will be expert in) shooting the bow, riding horses. (LBAT 1593 rev.8′, Reiner 2000, 425).

The qualities of the planets signaled certain life fortunes, as in the following passage:

1. (If) Jupiter: favorable; …; wealth (and) long days. (If) Ve[nus: very calm, wherever he goes good fortune and long days.]
2. (If) Mercury: heroic, lordly; great strength. (If) [Mars: ambiguous; (he will be) qui]ck [to anger.]
3. (If) Saturn: dark, disturbed; sick and constrained. 4 (If) the moon: bright, good; true and long(lived) // If the eclipsed moon: dark, disturbed, not bright; [no true omen. If] the eclipsed sun: divided, confused. (Rochberg-Halton 1987, 215)

These planetary qualities were derived from a system long predating natal astrology in which the planets were considered benefic or malefic on the basis of the deities with whom they were associated, as well as whether a planet was bright or dim. Hence the eclipsed Moon or Sun turn the positive benefic indications from the luminaries into negative malefic ones.

In natal omens, the life of the child born in the region of a zodiacal sign was prognosticated by virtue of the symbolism and imagery of the zodiacal signs. Physiognomy was also predicted, as in the following example:

2′ Region of Libra: [(break)] narrow of forehead; …
5′ …. Region of Aquarius: (the child will be) male, His eyes (face) red;
region of Aquarius: (it means(?)) that the lower part(?) of the ears will be large, …
(LBAT 1593 obv. 1′–7′, Reiner 2000, 423)

And

rev. 6′ Born in the region of Libra: true or untrue …
rev. 7′ [Born in?] the region of Scorpius: … she will be a sorceress, she will be joyous? …
rev. 8′ and will die by a scorpion (sting). … Born in the region of Sagittarius: (he will be expert in) shooting the bow, riding horses. (LBAT 1593 rev. 6′–8′, Reiner 2000, 425; Schreiber 2020, 128–29 for lines 2′–5′)

Facial features are associated with certain zodiacal signs and correspondences between body parts and months or zodiacal signs are attested in versions of the *melothesia* employed in astrological medicine (Geller 2014; Schreiber 2020, 124–27). The idea of the correspondence between body parts and the heavens has roots in material prior to the Hellenistic period, as discussed by Schreiber (2020, 125–26). And, he added, "Zodiacal physiognomy was part of a wider worldview in which not only the human body, which was connected to the heavens, but also constellations and zodiac signs had their own anatomy. This idea predates the zodiac: some astronomical texts mention the anatomical features of certain constellations" (Schreiber 2020, 129).

Third are the horoscopes or, following Pingree's terminology, proto-horoscopes (Pingree 1997, 22–24), that depict the situation of the heavens on the date and usually also the time of a birth (*šerru alid*) or conception (*šerru rehi*) by enumerating the zodiacal positions of the seven planets. The following third century example is rare in that it gives the name of the person in question and in this case he, Anu-bēlšunu, was a famous scribe of *Enūma Anu Enlil*, the title used by the Hellenistic period Babylonian astronomers. Anu-bēlšunu's horoscope is a fairly pared down example, but the positions of all the planets in the zodiacal signs at the time of his birth are given:

3. Babylonian horoscope
Year 63 (SE)
Month X, in the evening, day 2
Anu-bēlšunu was born.
That day, Sun 9;30° Capricorn
Moon 12° Aquarius: His days will be long.
Jupiter (in the) beginning of Scorpius: Someone will help the prince.
… in Aquarius; region of Venus, he will have sons.
Mercury in Capricorn; Saturn in Capricorn
Mars in Cancer (BH 9–248 29 December, Rochberg 1998, 79–80)

The contents of this and other cuneiform horoscopes distinguish these texts from the old celestial or even contemporary natal omens in several ways. The characteristic formulation of omens as conditionals—"if this happens that happens," or If P, Q—has been abandoned and the goal is the depiction of the situation of the heavens at the time of birth. In addition to the planetary longitudes, other elements were included, such as the seasonal hour of the birth, the passing of the Moon by an ecliptical fixed star, the date of either the equinox or solstice nearest to the birth, and lunar and solar eclipses, also chosen for proximity to the birth and favoring those separated by a half month between the lunar and the solar eclipse. The only traces of the old divination by omens are the occasional prediction for the native, such as "he will have long days," reminiscent of much older omens from the month of someone's birth.

The most important difference between horoscopes and omens lies in the derivation of the required astronomical data. Celestial phenomena listed in the omen series *Enūma Anu Enlil* were based on centuries of watching the sky for lunar, solar, planetary, stellar, and weather events deemed ominous. By the time these phenomena were written down and serialized, the omen statements compiled in the series do not represent datable observations but rather exemplifications of celestial events either seen at some time or given dates based on calendrical schemata, or indeed of imagined events of a similar kind never in fact "seen" (such as the Sun at midnight), but which nonetheless held potential, if imaginary, mantic significance.

In contrast to the celestial omens, the data in cuneiform horoscopes, such as planetary positions in the zodiac, were derived by extraction and quotation from nonmathematical astronomical texts such as Diaries and Almanacs (NMAT) and possibly also, when planetary longitudes were given in degrees and fractions of degrees, from interpolations that could have been made (at least in theory) from mathematical astronomical tables (Rochberg 2004, 145–63). This unique relationship to astronomical texts sets the horoscopes apart from contemporary natal omens. It also demonstrates how integrated all the parts of Babylonian astral science were, in spite of the fact that the scribes kept the various text types apart in distinct genres for different methods and aspects of their study of the heavens, and of the relation of the heavens to humankind.

Divinatory prognostication not only for kings and whole peoples but also for individuals had roots in older practices of Assyro-Babylonian divination. The interest in prognosticating the life of an individual from the month of his birth is already attested in a hemerological omen series from the seventh century BCE (Labat 1965). This series, titled *Iqqur īpuš*, meaning "He tore down and rebuilt," concerns activities undertaken on different dates in the 12 months of the year. It includes one section (§64) of 12 omens for events in the life of a child born in a certain month, such as "If in the month of Nisannu (Mo I), a child is born: wherever he goes, that child will squander his father's wealth (lit.: "scatter his father's house")." Other forecasts are that "he will die a sudden death," or "will prosper and be content," "will live long days," or "lose his virility" (DIŠ (*ina* ITI.x *šerru alid* ^{d}Ištar *ina lalīšú* UŠ.[MEŠ-šú] lit. "the goddess Ishtar will persecute him in the prime of his life"; Labat 1965, 134:12). These birth omens are, however, not genethlialogical as they are not based on celestial phenomena on the date of a birth; they give only months, no days, and no concern for the time of birth. The natal omens and horoscopes are the first true examples of genethlialogy in cuneiform.

From the preceding description of the contents of the three major types of astrological texts and how they reflect continuities as well as discontinuities from the early to the late period, I turn now to the question of causation reflected in Babylonian horoscopy and genethlialogy. As mentioned at the outset, the widespread principle, or rationale, for ancient astrology in Greek and Greek-derived Western systems, was celestial or stellar influence. Stellar influence, in the Greek astrological system, implies a certain causality, ontology and metaphysics derived from a variety of Greek philosophies (Platonic, Aristotelian, Stoic). Of course there were those, such as Plotinus (third century CE), who questioned the causal principle and favored the notion of the stars as signs rather than causes (e.g., *Enneads* 2.3, Armstrong 1967a, 54–102; and more recently, Gerson 2017, 145–63). Nonetheless, the idea of stellar influence and the causal ramifications of that principle was already widely authoritative, and physical influence was made authoritative by Ptolemy. The development of this idea, as Pingree emphasized, was Greek and followed from concerns with matter and motion rooted in Greek philosophies of nature.

Just as the rationale for Greek astrology followed from Greek ideas about natural phenomena and the physical structure of the cosmos, the Babylonian rationale for celestial prognostication was consistent with their own ideas about phenomena and the structure of their world, including what was conceived as the relationship of the divine to the heavenly bodies. In other words, what it took for these distinct ideas to be valid had to do with the metaphysical underpinnings of their respective systems. On the Greek side, as just mentioned, stellar influence was dependent upon natural philosophy. On the Babylonian side, ideas about how celestial omens function and why astral phenomena are ominous are found in the knowledge corpora of what the scribes called *ṭupšarrūtu*, or "the art of the scribe."

The rubric *ṭupšarrūtu*, used during the seventh century by Neo-Assyrian court scholars, encompassed the scribal scholarly disciplines that organized knowledge of the phenomenal world and the practices that depended upon that organization (Rochberg 2021). While the term itself is not found in Late Babylonian scholarly text colophons, the fact of the continuation of the subjects and canonical texts of *ṭupšarrūtu* after 500 BCE argues for the direct descent of knowledge from the earlier textual traditions to those preserved in the final centuries of the cuneiform writing tradition. A great many texts from the corpora of *ṭupšarrūtu*, which in the seventh century included divination, magic, medicine, astrology and astronomy, persisted in post-500 BCE Babylonia (Rochberg 2010, 237–56). Nowhere within this extensive corpus, in either early or late periods, can we discern an interest on the part of the scribes in the physics of motion or physical-causal influence predicated on a philosophy of nature (Rochberg 2016, 164–90).

In order to consider the ideas that informed Babylonian astrological prognostication, evidence outside of the celestial divinatory and genethlialogical material, astral medical texts are a good place to start, as these have been at the center of the modern attempt to harmonize the causal rationale for Babylonian astrology and astral medicine with that of the Greek in the form of stellar influence. On the basis of pre-Seleucid texts in which medical preparations were set out under the stars and left overnight to be administered to the patient in the morning, Erica Reiner said:

> The stars exert a direct influence and serve as mediators between man and god. Directly, through astral irradiation, they transform ordinary substances into potent ones that will be effective in magic, medicine, or ritual, as *materia medica*, amulets, or cultic appurtenances. (Reiner 1985, 593 and 595; 1995, 15, 43–60)

She was referring to the expression found in medical texts, namely, *ina kakkabi tušbât* lit. "to keep overnight under the stars" (CAD s.v. *bâtu*, mng. 3b, with medical literature cited). Reiner went on to qualify her interpretation for fear that it might suggest that such practices were "religious" in nature:

> I do not claim that every prayer to Sin and Šamaš is an appeal for lunar or solar irradiation, just as I do not hold that invocations of Ištar and Nabû in reality seek the benefic influence of Venus and Mercury; in other words, *I do not intend to lead you astray into astral religion*. Nevertheless, we are bound to consider such prayers as appeals to the planetary aspect of the deity addressed when their concern is with establishing a favorable moment for making magic means more efficacious. (Reiner 1985, 595; my emphasis)

The problem, however, is not so much a question of religion vs. science—which is some form of a category mistake in this historical context—as it is a problem with the analysis of causation.

Reiner's appeal to the idea of stellar influence or irradiation in this context seems to follow from her use of a Latin sense of influence, from *influere* as an "inflowing." In the Assyro-Babylonian case, however, no physics of matter or motion yet existed to account for the same or a similar sense of such influence. John North (1986b, 45–46) discussed the question of celestial influence in a paper concerning medieval astrology and maintained that the sense of "inflowing" was not at stake in that context; rather, celestial influence had to do with the physics of motion as a cause of change. If stellar irradiation by virtue of a physics of motion was not a native Babylonian, or Assyro-Babylonian concept, as indeed it was not, was some form of stellar irradiation, or what Reiner compared to Greek *astronomein* or German *bestrahlen*, that which produced the desired effect for the Assyro-Babylonian learned scribes and medical practitioners?

Akkadian prayers to the celestial deities shed some light on this question. They indicate the beneficial effect of the presence of the heavenly bodies for obtaining prognostications of the future, as in this prayer to the Sun god Shamash and Moon god Sin:

> The lands rejoice at your appearance.
> Day and night they entrust (to you) their ability to see.
> You *stand by* to let loose the omens of heaven and earth.
> I, your servant, who keep watch for you,
> who look upon your faces each day,
> who am attentive to your appearance,
> make my evil omens pass away from me.
> Set for me propitious and favorable omens. (Ebeling 1949, 179 and 181. Quoted here is the translation of Benjamin R. Foster, *Before the Muses: An Anthology of Akkadian Literature* [Bethesda, MD: CDL 1993], 2.684; my emphasis)

Similarly, in a prayer to the gods of the night, namely the constellations, they are asked to be present, to "stand by" for the success of an extispicy:

> The great ones, the gods of the night,
> Bright Girra,
> Warrior Erra,
> The Bow, the Yoke,
> Orion, the Furious Serpent,

> The Wagon, the Goat,
> The Bison, the Horned Serpent,
> May they *stand by* so that,
> In the extispicy I am performing,
> In the lamb I am offering,
> You may place the truth. (Oppenheim 1959, 282–301, lines 14–24; also Cooley 2011, 82; my emphasis)

Here the stars-as-gods were asked to be present, to *stand by* so that they would "place the truth" in the liver by writing the appropriate signs on its surface. The prayer is recited in advance of the inspection of the liver in hopes that the extispicy will yield a good outcome. In yet another prayer, in this case to the Wagon Star (Ursa Major), the constellation is invoked for the purpose of bringing a dream omen:

> O Wagon star, heavenly Wagon! Your yoke is Ninurta, your pole is Marduk, your axles are the two heavenly daughters of Anu. You rise in Assur, you turn toward Babylon. Let a dream bring me a sign whether so-and-so, son of so-and-so, will become healthy and well! (STT 73:71–75 and dupl., cited Reiner 1995, 71, note 272)

The circumpolar Wagon Star was a favorite object of prayer for the well-being of humankind as it was always visible at night. But in each case, the practice of exposure of something earthly to something heavenly, in the same way as exposing medicines or amulets to the stars, was effective because the divine-human relation at stake in maintaining a good human life was derived from the very world order in which these practices functioned, namely, that the heavens above and Earth below were conceived as interconnected counterparts. The seventh century Diviner's Manual explained that "the signs on earth just as those in heaven give us signals. Heaven and earth both produce portents, though appearing separately, they are not separate (because) heaven and earth are interconnected" (*idāt erṣetim itti šamê ṣadda inašša šamê u erṣetim ištēniš giskimma ubbaluni aḫennā ul* BAR.MEŠ *šamê u erṣetim itḫuzu*, Oppenheim 1974, 199: 38–40).

This cosmological relationship enabled myriad interpretations to be made by the scribes in their study of signs above and below, using translation techniques traceable to the Sumero-Akkadian lexical lists in which relationships and associations were found among words, signs (that is, literally, the cuneiform signs) and things (Crisostomo 2019). This principle of correspondence is everywhere clear in omen texts of all kinds, not only celestial omens (Rochberg 2024, 185–92). Whether for a divination requiring the direct inquiry of the divine, such as in the technique of extispicy, or in the request for mantic dreams, or, for the reading of signs that occurred regardless of whether they were impetrated directly from the gods or not, such as in celestial divination, the signs were read, or perhaps we can better say *translated* as a text written in a divine script. The methodology of translation established in the Sumero-Akkadian lexical text tradition of the Old Babylonian period ca. 1800 BCE was extended beyond the linguistic environment of lexicography to the phenomena of heaven and Earth, as well as the signs on the liver of the sacrificed sheep. Each could be viewed, metaphorically, as surfaces upon which the divine might "write."

In extispicy prayers (*ikribu*) to the gods of divination, Shamash and Adad are asked to write their judgment for the client upon the liver, just as a judge would register a written verdict in a legal case. A text concerning the origin of this practice tells of when Shamash and Adad gave the antediluvian sage Enmeduranki "the tablet of the gods, the liver, a secret of heaven and [Earth]" (Lambert 1998, 147, 152). The "heavenly writing" (*šiṭir šamê*, or the more literary form *šiṭir burūmê*), on the other hand, was a poetic metaphor used in Assyrian and Babylonian royal inscriptions to refer to temples made beautiful "like the stars," literally, "like the heavenly writing." However, as the noun *šiṭru* also means "text" or "inscription," the expression can also be taken to mean "heavenly text or inscription."

The heavenly phenomena as a script, or an inscription, implied authorship by the gods and the capacity to be read and interpreted. The idea of the legibility and interpretability of the heavens, a text written upon by the gods with stars on the analogy of clay tablets written upon by the scribes with the cuneiform signs, applies equally to Old Babylonian celestial omens as to Babylonian horoscopes. The metaphor of the liver as the tablet of the gods and the celestial phenomena as a divine script upon the surface of heaven suggest the notion of cuneiform divination as a kind of hermeneutic interpretation of texts in which the scribes employed the methods of philology, etymology, etymography, analogy, and association that were a central feature of their training.

The metaphor of the stars and their configurations as a script did not disappear with cuneiform texts, but reappears in the religious and philosophical literature of late Hellenism, the Greco-Latin, Hebrew and Arabic literature of the Middle Ages and Renaissance. I rely here on secondary scholarly literature in order to make the point that the letters of the alphabet (the *stoicheia*) could be projected onto the heavens in much the same way as in the Akkadian expression "heavenly writing." Thus we find Plotinus in *Enneads* 2.3 expressing his incredulity at the astrological effect of the planets in the zodiac: He said (my emphasis):

> *Let us suppose that the stars are like characters always being written onthe heavens*, or written once for all and moving as they perform their task, a different on: and let us assume that their significance results from this.... For instance, we can come to conclusions about someone's character, and also about the dangers that beset him and the precautions to be taken, by looking at his eyes or some other part of his body.... All things are filled full of signs, and it is a wise man who can learn about one thing from another. (*Enneads* 2.3.7, translation Armstrong 1967a, 69; cf. Gerson 2017, 152)

In another passage, he says:

> Those who know how to read this sort of writing [that is, the celestial bodies] can, by looking at them as if they were letters, read the future from their patterns, discovering what is signified by the systematic use of analogy." (*Enneads* 3.1.6, translation Armstrong 1967b, 29; cf. Gerson 2017, 246)

In a fifth-century treatise on dream divination, Synesius (*De Insomniis* 2, 131C–132A, apud Acevedo 2020, 272) said:

> If all things are symbols appearing through all things, being like siblings within one single living creature, the cosmos, so also they are written in characters of every kind, just as those

in a book, some of which are Phoenician, some Egyptian, and others Assyrian; and the wise person reads them.

As late as the sixteenth century, Anthony Grafton (1999, 6) points out that in 1512, Giovanni Pontano published a treatise entitled *On Celestial Things*. There, as Grafton describes the passage, the "stars and planets … formed the letters of a cosmic alphabet.… Every planet … played the role of a letter with defined qualities. Every astrologically significant configuration of two or more planets … resembled a word or a phrase, the sense of which the astrologer could determine" (Grafton 1999, 6). And for the Renaissance author of *Magia naturalis* (1589) Giambatista della Porta, William Eamon said, "'Reading' nature was like reading a text. The Book of Nature could be expected to contain many of the same delights and surprises he found in Pliny's *Natural History* or in Ovid's *Metamorphoses*" (Eamon 1996, 216). From the heavenly writing to the Book of Nature, there is a spiraling—or meandering—line of sorts, a connection worth mentioning, if difficult to pin down.

Returning to the cuneiform medical and magical literature, we see that in the pursuit of human well-being, the beneficent "hand" of a god was desirable, but the maleficent hand was feared and made the object of magical incantation to calm the angry deity. Illness was also often attributed to a specific demon, sometimes called the "spawn of heaven" or "spawn of the stars," whose "hand" had produced human suffering. The planets themselves, as divine manifestations were as capable of bringing on illness as they were of good fortune, and we have reference to the hand of Venus as an illness, as well as the hand of Jupiter (CAD s.v. *qātu*). Similarly, illness was attributed to the patient having been "touched" (TAG = *lapātu*), that is, in its extended usage, meaning "attacked" or "afflicted" by the offending god or demon, and later even being touched by zodiacal signs, as in this passage:

1′ If a man has been seized by (the toilet-demon) Šulak, pound up Akkad-plant, mix it in river water, rub it on him and he will get better; (put) the remainder together with these (things) on his neck etc.

2′ … when a man suffers ditto (meaning from affliction by Šulak, the toilet-demon), the Great-star (Aquarius) has touched (that is, afflicted) him, his right thigh hurts him, place cypress in catskin on his neck, secondly, rub him with oil and he will improve.

3′ Ditto (= if a man has been afflicted by the toilet-demon Šulak), ditto, a man suffers from ditto, the Field-star (*Ikû*, or Pisces) has afflicted him and his left thigh hurts him, place *barīrātu*-plant in wolfskin on his neck, secondly, rub him with oil and he will improve.

6′ Ditto (= if) ditto, the Bull of Heaven-star (Taurus) has afflicted him and his right hand and scaly skin hurt him, place *kiškanû* in buckskin, secondly rub him with oil and he will improve.

7′ Ditto (= if) the Great Twins (Gemini) has affected him and his head hurts him, place *murrānu*-wood in snakeskin and secondly, rub him with oil and he will improve. (Geller 2014, BM 47755 etc., 85–)

Just as the zodiacal constellations, or signs, could afflict a person, presumably due to the exposure a person had had to the sign or constellation of the zodiac—the conditions for such exposure are not given, at least not in this text, but in a similar way, a planet could further determine the future of an individual depending on the region that planet occupied, where "region" refers to a zodiacal sign. A rare statement that a planet can "decree fate" is found in one of the astrological explanatory texts quoted before that made the analogical association between the child born in the region of Sagittarius and his ability to ride horses and shoot the bow. The passage in question reads:

> ... BAR DIŠ LÚ.TUR *a-lid*
> [d]UDU.IDIM *šá ina* IGI *šá* [d]UDU.IDIM.MEŠ *ana* KI *šú-a-tú* KUR-*du-šú u šim-tum šá-am-šú*
> [...] Month I: If a child is born, (and) the planet which is in front of the planets for that region reaches it (the region), it will decree his fate. (LBAT 1593 obv. 12′–13′, Reiner 2000, 422 and translation on 424)

What I understand from this passage is the idea that the planets were thought to prognosticate an individual's fate, not by physical emanation, irradiation or physical causality, but by virtue of astrological significance conveyed by a planet's assuming, that is "reaching" a particular position. Of course the "reaching" or assuming a certain position may have been the prior requirement for "influence" to occur, but the text refers only to the reaching, not to influencing.

The interest in fixed stars "approaching" or "reaching" other fixed stars goes back to older omen texts where such approachings or reachings represent ominous phenomenon. Colophons identify the nonmathematical astronomical texts we call Almanacs as "measurements of the 'reachings' of the planets" (*mešhi ša kašādī ša* [d]*bibbī*). Although these zodiacal "reachings," as summarized in Almanacs and Diaries, do not make any mention of astrological meaning, the contents of horoscopes surely imply that the arrival of a planet into a sign was astrologically significant. Confirmation of this is found when we see that the horoscope's zodiacal positions of the planets were taken directly from Diaries and/or Almanacs, that is, from the sections in which the planetary "reachings" are enumerated (Rochberg 2004, 145–57).

One significant difference can be drawn between horoscopes on one hand and the prayers for the presence of a heavenly body, or indeed the medical texts that call for leaving medical preparations out overnight on the other. The *exposure*, if we can call it that, of an individual to the planets occupying particular zodiacal signs in a horoscope was effected by virtue of his date and time of birth, not by being somehow physically present, particularly when one considers the time of birth during daylight. The principle of correspondence that underpinned celestial omens, astral medicine, magic and prayer applied in the horoscopes regardless of the physical presence, or visibility of the planets. It was the interconnectedness of the two principle world parts, heaven and Earth, that guaranteed the meaning of the heavens for human life. One's birth, therefore, linked the individual to the situation of the heavens regardless of any particular heavenly body being visible at the time.

In Late Babylonian texts where astrology is merged with extispicy we can see the principle of correspondence operative in the following guide to correlating features of the liver to deities, months, and stars:

The "station" is Enlil; month I; [Aries].
The "path" is Šamaš; month II; Taurus.
The "sweet mouth" is Nusku; month III; Orion.
The "strength" is Uraš; month IV; Cancer, Plow-star.
The "gate of the palace" is Ninegal; month V; Regulus.
The "bubble" is the storm god Adad; month VI; Raven star.
The gall bladder is Anu; month VII; Libra.
The "finger" (identifiable as the processus pyramidalis) is god (broken); month VIII; Goat star. (Reiner 1995, 78)

It seems unlikely that the validity of these correspondences depended upon the visibility of any of the stars, or zodiacal signs named. The scheme simply systematized the correlations of features of the liver with the other elements for a greater scope of translation of the tablet of the gods, that is, for the interpretation of the liver.

Finally, I return to the passage quoted before in which a planet is said to "decree the fate" of the child. The passage suggests how the Babylonian notion of fate is tied to the principle of correspondence underpinning celestial divination, natal astrology and horoscopy. By virtue of its Semitic root, the Akkadian word "fate" (*šīmtu*) means "that which is fixed or determined by (written) decree," as *šīmtu* is a nominal form of the verb *šâmu* "to decree." (Rochberg 2010, 19–30). Such decrees were not exclusive to the gods, but in our context, it was the highest gods who were said to decree fate. The decrees were to be found in the texts written on the heavens or on the liver, that is to say, in the ominous signs. Thus the gods were considered to be responsible for (to cause) the ominous signs. Moreover, the notion of fate and therefore also of causality was tied to the gods in their judicial decision-making and decree-giving role. Indeed, the term for the prediction or consequence of an omen was "verdict." Thus the connections made by associations and analogies between the sign and its consequence (the verdict) lack the dimension of necessity that connects cause to effect according to our modern way of thinking, because magic was available to dispel the effect of the cause. Incantation, prayer and ritual to appease the divine judges could dispel untoward signs and bad decrees, but unless such measures were taken, the consequence or verdict of a sign would surely occur.

The underpinnings of the new sciences of Babylonian horoscopy and genethlialogy, therefore, follow directly from much older ideas of fate and divination, in which correspondences between human life and the stars were made within a system of interpretable signs for which the gods were thought to be responsible. Fundamental differences from the universal system with an underlying mechanical-causal rationale invented by the Greeks could not be more stark, as Babylonian celestial divination, horoscopy and genethlialogy operated within a different world structure, indeed one without concern for matter, motion, or nature (Rochberg 2016). Different as the Babylonian idea of the connection between heavenly phenomena and human life was from the principle of stellar influence, the idea of the stars, planets and luminaries manifesting signs from the gods and legible as a script upon the surface of heaven continued to find further resonance in the Greek context of astrology with similar ideas about the stars as forming letters to be read on the heavens, such as is seen in Plotinus's *Enneads*, quoted above.

Developments in Greek, Greco-Egyptian and Greco-Roman astrology added to the Babylonian astrological inheritance new elements (such as the places, sects, lots and *climata*) and methods (such as using the values of the rising times, according to Babylonian Systems A or B, in accordance with the appropriate *clima* of the place of birth to obtain the length of life, a critical year, or the exaltation of the nativity; Neugebauer 1975, 712–36). Some of these elements, in particular the development of the seven *climata*, are indicative of the ways Greek astrology departed from Babylonian methods to become a system with universal effects within a spherical cosmic framework.

Babylonian horoscopy, as it is clear from the rare inclusion of prognostications in cuneiform horoscope texts (Rochberg 1998, 13), retained ideas consistent with omen divination. This must be what Pingree had in mind when he said the Babylonian horoscopes were "nativity omens" and not "horoscopes" (Pingree 2001, 4, note 3). But in spite of being rooted in celestial divination, the innovative elements of the cuneiform horoscopes, as well as of the natal omens, reflect new forms for the study of births in relation to the heavens. These forms of cuneiform genethlialogy in turn gave rise to astrological medicine and physiognomy.

The ways the new Babylonian genethlialogical sciences of natal omens and horoscopy were modified by the Greeks of the Hellenistic period and integrated within Greek ideas about the geocentric cosmos, the divine heavens, or physical causality, and their particular claims of stellar influence, would be determinative of the rest of the history of astrology. The Babylonian forms of birth divination attested in cuneiform texts from the seventh century, and genethlialogy proper attested in Late Babylonian natal omens and horoscopes, were practices belonging wholly within Babylonian ideas about causality vis-à-vis the gods and the heavenly bodies. The cuneiform scribes, celebrated by Swerdlow in *The Babylonian Theory of the Planets* (1998, 5), were unaware of the universal law of stellar influence as defined in the Greek milieu and were equally unconcerned by matters of physical causality. In view of these differences, the history of astrology does not begin with the manifestation of the astrological law of stellar influence but goes back to the Babylonian hermeneutic interpretation of celestial signs and the eventual application in the second half of the first millennium BCE of that science to the genethlialogical aims of horoscopy.

References

Acevedo, Juan. 2020. Alphanumeric Cosmology from Greek into Arabic: The Idea of Stoicheia, Through the Medieval Mediterranean. In *Forschungen zum Alten Testament 2. Reihe*. Tübingen: Mohr Siebeck.

Armstrong, A. H. 1967a. *Plotinus II: Enneads II.1–9. Loeb Classical Library 441*. Cambridge, MA: Harvard University Press.

Armstrong, A. H. 1967b. *Plotinus III: Enneads III.1–9. Loeb Classical Library 442*. Cambridge, MA: Harvard University Press.

Brown, David, et al. 2018. *The Interactions of Ancient Astral Science*. Bremen: Hempen Verlag.

Cooley, Jeffrey L. 2011. An OB Prayer to the Gods of the Night. In *Akkadian Prayers and Hymns: A Reader*, ed. A. Lenzi, 71–83. Atlanta: Society of Biblical Literature Press.

Crisostomo, Jay. 2019. Translation as Scholarship: Language, Writing, and Bilingual Education in Ancient Babylonia. In *SANER 22*. Berlin: De Gruyter.

Eamon, William. 1996. *Science and the Secrets of Nature: Books of Secrets in Medieval and Early Modern Culture*. Princeton: Princeton University Press.

Ebeling, E. 1949. Beschwörungen gegen den Feind und den bösen Blick aus dem Zweistromlande. *Archiv Orientalni* 17:172–211.

Fincke, Jeanette. 2016. *Divination as Science: A Workshop Conducted during the 60th Rencontre Assyriologique Internationale, Warsaw, 2014*. Winona Lake: Eisenbrauns.

Foster, Benjamin R. 1993. *Before the Muses: An Anthology of Akkadian Literature*. Vol. 2. Bethesda: CDL Press.

Geller, Markham J. 2014. *Melothesia in Babylonia: Medicine, Magic, and Astrology in the Ancient Near East*. Berlin: De Gruyter.

George, Andrew, and Junko Taniguchi. 2024. Old Babylonian Lunar-Eclipse Omen Tablets in the British Museum. *Journal of Cuneiform Studies* 76:127–154.

Gerson, Lloyd P., ed. 2017. *Plotinus: The Enneads*. Translated by George Boys-Stones, John M. Dillon, R. A. H. King, Andrew Smith, and James Wilberding. Cambridge, UK: Cambridge University Press.

Grafton, Anthony. 1999. *Cardano's Cosmos: The Worlds and Works of a Renaissance Astrologer*. Cambridge, MA: Harvard University Press.

Greenbaum, Dorian, and Micah Ross. 2010. The Role of Egypt in the Development of the Horoscope. In *Egypt in Transition. Social and Religious Development of Egypt in the First Millennium BCE*, ed. Ladislav Bares, Filip Coppens, and Květa Smolarikova, 146–182. Prague: Czech Institute of Egyptology.

Labat, René. 1965. *Un calendrier Babylonien des travaux des signes et des mois (séries Iqqur Îpuš)*. Paris: Librairie Honoré Champion.

Lambert, W. G. 1998. The Qualifications of Babylonian Diviners. In *Tikip Santakki Mala Bašmu: Festschrift für Rykle Borger*, ed. Stefan M. Maul, 141–158. Groningen: Styx Publications.

Neugebauer, Otto. 1969. *The Exact Sciences in Antiquity*. New York: Dover.

Neugebauer, Otto. 1975. *A History of Ancient Mathematical Astronomy*. 3 vols. Berlin/New York: Springer.

Neugebauer, Otto, and H. B. van Hoesen. 1959. *Greek Horoscopes*. Philadelphia: American Philosophical Society.

North, John D. 1986a. Celestial Influence: The Major Premise of Astrology. In *Astrologi Hallucinati: Stars and the End of the World in Luther's Time*, ed. Paola Zambelli, 45–100. Berlin: De Gruyter.

North, John D. 1986b. *Horoscopes and History*. Warburg Institute Surveys and Texts 13. London: Warburg Institute, University of London.

Oppenheim, A. Leo. 1959. A New Prayer to the 'Gods of the Night'. *Analecta Biblica* 12:282–301.

Oppenheim, A. Leo. 1974. A Babylonian Diviner's Manual. *Journal of Near Eastern Studies* 33:197–220.

Pingree, David. 1973. Astrology. In *Dictionary of the History of Ideas*, ed. Philip P. Wiener, vol. 1, 118–126. New York: Charles Scribner's Sons.

Pingree, David, ed. 1986. *Vettii Valentis Antiocheni Anthologiarum Libri Novem*. Leipzig: Teubner.

Pingree, David. 1997. *From Astral Omens to Astrology: From Babylon to Bīkāner*. Serie Orientale Roma 78. Rome: Istituto italiano per l'Africa et l'Oriente.

Pingree, David. 2001. From Alexandria to Baghdād to Byzantium. The Transmission of Astrology. *International Journal of the Classical Tradition* 8:3–37.

Plotinus. *Enneads*. See Armstrong 1967.

Ptolemy. *Tetrabiblos*. See Robbins 1971, 1980.

Reiner, Erica. 1985. The Uses of Astrology. *JAOS* 105:589–595.

Reiner, Erica. 1995. *Astral Magic in Babylonia*. Philadelphia: American Philosophical Society.

Reiner, Erica. 2000. Early Zodiologia and Related Matters. In *Wisdom, Gods, and Literature: Studies in Honour of W. G. Lambert*, ed. A. R. George and I. L. Finkel, 421–427. Winona Lake: Eisenbrauns.

Robbins, F. E. 1971. *Ptolemy Tetrabiblos*. Loeb Classical Library 350. Cambridge, MA: Harvard University Press and London: W. Heinemann.

Robbins, F. E., ed. and trans. 1980. *Claudius Ptolemy: Tetrabiblos*. Cambridge, MA: Harvard University Press.

Rochberg, Francesca. 1998. *Babylonian Horoscopes*. Philadelphia: American Philosophical Society.

Rochberg, Francesca. 2004. A Babylonian Rising Time Scheme in Non-tabular Astronomical Texts. In *Studies in the History of the Exact Sciences in Honour of David Pingree*, ed. C. Burnett, J. P. Hogendijk, K. Plofker, and M. Yano, 56–94. Leiden: Brill.

Rochberg, Francesca. 2010. In the Path of the Moon: Babylonian Celestial Divination and Its Legacy. In *Studies in Ancient Magic and Divination 6*. Leiden/Boston: Brill.

Rochberg, Francesca. 2016. *Before Nature: Cuneiform Knowledge and the History of Science*. Chicago: University of Chicago Press, 164–90.

Rochberg, Francesca. 2021. *Ṭupšarrūtu* and the Historiography of Science. *Claroscuro* 2:1–27.

Rochberg, Francesca. 2024. *Worldmaking and Cuneiform Antiquity: An Anthropology of Science*. Cambridge, UK: Cambridge University Press.

Rochberg-Halton, Francesca. 1987. TCL 6 13: Mixed Traditions in Late Babylonian Astrolog. *Zeitschrift für Assyriologie* 77:207– 228. Republished in Rochberg 2010, 113–33.

Rochberg-Halton, Francesca. 1988. *Aspects of Babylonian Celestial Divination: The Lunar Eclipse Tablets of Enūma Anu Enlil*. Horn: Berger & Söhne.

Sachs, A. J. 1952. Babylonian Horoscopes. *Journal of Cuneiform Studies* 6:49–75.

Sarton, George. 1924. Review of Lynn Thorndike. *A History of Magic and Experimental Science during the First Thirteen Centuries of Our Era, 1923*. *Isis* 6:74–89.

Schreiber, Marvin. 2020. Late Babylonian Astrological Physiognomy. In *Visualizing the Invisible with the Human Body: Physiognomy and Ekphrasis in the Ancient World*, ed. J. Cale Johnson and Alessandro Stavru, 119–140. Berlin: De Gruyter.

Swerdlow, Noel. 1998. *The Babylonian Theory of the Planets*. Princeton: Princeton University Press.

Swerdlow, Noel, ed. 1999. *Ancient Astronomy and Celestial Divination*. Cambridge, MA: MIT Press.

Thorndike, Lynn. 1923–1958. *A History of Magic and Experimental Science During the First Thirteen Centuries of Our Era*. 8 vols. New York: Columbia University Press.

Thorndike, Lynn. 1955. The True *Place of Astrology* in the History of Science. *Isis* 46:273–278.

Vettius, Valens. *Anthology*. See Pingree 1986.

Part II
Greece

Chapter 4
What Plato Meant by Saying That God Is Always Doing Geometry: Responses to the Question Posed by Plutarch in *Table-Talk* 8.2

Liba Taub

Abstract In *The Renaissance of Astronomy: Regiomontanus, Copernicus, Tycho, Kepler, Galileo*, Noel Swerdlow opens his chapter on Kepler, "The New Astronomy," with a section on "God's Plan of the Universe." He describes how, in the *Mysterium Cosmographicum*, Kepler discusses God's use of regular polyhedra as the basis of the cosmos: the five regular solids representing the planets are situated between six spheres, which provide motion for the planets. Furthermore, Kepler stated that he agreed with Plato, that God is always a geometer. Kepler's engagement with Plato's ideas is also seen in his translation of and comments on Plutarch's dialogue *On the Face on the Moon*. Elsewhere, in his *Table-Talk* (8.2), Plutarch describes a discussion of what Plato meant by saying that God is always doing geometry. The responses to Plutarch's question leave us with no certain answer. Instead, the *Table-Talk* emphasises the importance of gathering as a community, discussing, debating, and sharing ideas. And, especially, to remember friends, teachers and colleagues, and all that we have learned from them. The *Table-Talk* suggests that we may not be able to have certain knowledge, but we can raise questions, have stimulating conversations, and recall esteemed colleagues.

In his magisterial *The Renaissance of Astronomy: Regiomontanus, Copernicus, Tycho, Kepler, Galileo*, Noel Swerdlow opens his chapter on Kepler, "The New Astronomy," with a section on "God's Plan of the Universe." He describes how, in the *Mysterium Cosmographicum*, Kepler discusses in detail God's use of regular polyhedra as the basis of the cosmos: the five regular solids representing the planets

L. Taub (✉)
Department of History and Philosophy of Science, Newnham College,
University of Cambridge, Cambridge, UK
e-mail: lct1001@cam.ac.uk

F. Rochberg (ed.), *Studies in Premodern Sciences in Memory of Noel M. Swerdlow*, Archimedes 74, https://doi.org/10.1007/978-3-032-04040-4_4

are situated between six spheres, which provide motion for the planets. In the Preface to the *Mysterium Cosmographicum*, Kepler wrote:

> I propose, reader, to demonstrate in this little book that the most Good and Great Creator, in the creation of this moving world, and the arrangement of the heavens, referred to those five regular solids, well known from Pythagoras and Plato to our own time, and that to their nature he fitted the number of the heavens, their proportions, and the plan (*ratio*) of their motions.[1]

He explains, "We know the orbit by the motion and the solid bodies by their number and sizes" and goes on to add:

> What else remains except to say with Plato, "God is always a geometer," and in this structure of moving stars he has inscribed solids within spheres, and spheres within solids, until no further solid was left which was not robed outside and inside with moving spheres.[2]

Swerdlow reminds us that while studying for the clergy, Kepler devoted a good deal of attention to theology. He suggests that Kepler may even have considered himself to be a theologian, with a new revelation of God, understood as reason itself, that is, the mind of God.[3]

Kepler's theological views, and their connection to ideas of the Pythagoreans and Plato, have been commented on by others. Rhonda Martens has explained that:

> The *Mysterium* is primarily about Kepler's notion of archetypes. Kepler believed that God created the world according to a plan. This plan, or pattern, is the archetype for the structure of the world (much like an architectural blueprint). Kepler believed, following the Pythagoreans and Plato, that this pattern was mathematical in nature and aesthetic in character. Because God created the physical world as a representation of this pattern, an archetypal model explains why the universe is the way it is.[4]

As Peter Barker and Bernard Goldstein have noted, in the *Mysterium* Kepler "claimed to have uncovered … the structure of God's providential plan for the cosmos as a whole, and particularly for the arrangement of the planets." They argued, "Rather than an exercise in astronomy or a defense of Copernicanism as a novel cosmology, Kepler's first book [the *Mysterium Cosmographicum*] must be read as essentially theological."[5] Kepler's description of God as Geometer is significant, given that Plutarch's question presupposes that this was Plato's view.

Kepler's theology has links to Plato's *Timaeus*, in which mathematics was crucial to the creation of the world.[6] Kepler presented himself as an intellectual heir

---

[1] Kepler *Mysterium Cosmographicum* (*MC*) 1596, 6; trans. Barker in Barker and Goldstein 2001, 99.

[2] Kepler, *MC*, chap. 2, trans. Duncan 1981, 97.

[3] Swerdlow (2025), chap. 4.

[4] Martens 2000, 39.

[5] Barker and Goldstein 2001, 99.

[6] The harmony present in the motions and arrangement of the heavenly bodies is mathematically structured and analogous to that of heard music. On the harmonic tradition, see also Rothman 2017, 8–11.

both to Pythagoras and Plato.[7] In Book Four of the *Harmonice Mundi* (*The Harmony of the World*), on the essence of harmonic proportions, in a note on a passage in Proclus's commentary on Book One of Euclid's *Elements* that he quoted there, Kepler suggested that the *Timaeus* was a Pythagorean interpretation of the first book of the Hebrew Bible, *Breishit*, or Genesis.[8] Kepler's engagement with Plato's ideas is also seen in his translation of and comments on Plutarch's dialogue *On the Face on the Moon*.[9]

Plutarch, who was born in Chaeronea in Boeotia in central Greece before the year 50 and died after 120 CE, studied philosophy in Athens, visited both Egypt and Italy, and lectured and taught in Rome. Known as a philosopher and biographer, he was also a priest at Delphi, and played an important role in the revival of the shrine. He was a self-avowed Platonist. George Karamanolis has argued that "Plutarch's significance as a philosopher … lies in his attempt to do justice to Plato's work as a whole."[10]

Many of Plutarch's works are dialogues, but very different from those of Plato. Plutarch's dialogues often feature long speeches and the appearance of the author himself. Plutarch was particularly interested in Plato's *Timaeus*. He wrote a work, dedicated to two of his grown-up sons, *On the Generation of the Soul in the* Timaeus. In the final section there, he considers the musical ratios used in the construction of the World-Soul. Christian Tolsa has pointed to some of the parallels there to ideas of Claudius Ptolemy, including analogies between the heavens and musical structures.[11] Kepler also investigated musical structures and their relation to the structure of the cosmos.

Plutarch's dialogue *On the Face on the Moon* (*De facie*) is a work in which Plato's *Timaeus* looms large; we might say it is lurking throughout in the background. The participants in Plutarch's dialogue discuss the nature of the "face" that appears on the Moon, as well as the role of the Moon in the cosmos. Plutarch's dialogue shares with the *Timaeus* a focus on cosmology, and both works incorporate myth as part of their cosmological explanation. As in the *Timaeus*, *On the Face on the Moon* juxtaposes the ideas and arguments of philosophers and mathematicians with mythological accounts. The discussions presented in *On the Face on the Moon* include both the consideration of questions relating to the appearance of the Moon, its size and material constitution, and the recounting of what seem to be rather exotic stories (*mythoi*) about the Moon—whether it is inhabited, its relationship to the soul, and its association with various gods and goddesses.

---

[7] See also Jardine and Segonds 1999, 222–26 regarding Kepler's interest in Pythagoreanism (On the Face on the Moon).

[8] Kepler [1619] 1940, *Gesammelte Werke*, vol. 6, *Harmonices Mundi* IV, chap. 1, 221; Kepler [1619] 1997, 301. See also Martens 2003.

[9] For the translation see Kepler [1634] 1870, 8: 76–123. On Plutarch's influence on Kepler, see Dick 1980, 9; 1982, 22, 70–71, 75; Grafton 1997, 208–13.

[10] Karamanolis 2020, 1.

[11] Tolsa 2014.

Rival views associated with several different philosophical schools (the Academy, the Lyceum, and the Stoa) are aired. The principal participants are well-educated men but not specialists. Nevertheless, the standard of conversation is sophisticated, and the discussants demonstrate that they are familiar with specialist, technical work, for example quoting Aristarchus's *On the Sizes and Distances*. That Plutarch can be read as a well-informed non-specialist is to some extent corroborated by both Thomas Heath and Otto Neugebauer, who cite Plutarch as a source a number of times, even though in certain instances the latter suggests that Plutarch's ideas are rather vague or garbled.[12]

Nevertheless, knowledge of Plutarch's writings contributed to Renaissance debates on whether the Earth or the Sun was the center of the cosmos, in particular, his *On the Face on the Moon* and *Platonic Questions*, as well as the *Opinions of the Philosophers (Placita Philosophorum)* attributed to him. Plutarch's works influenced Renaissance mathematicians and astronomers, including Copernicus, Galileo and Kepler. [13] As a student, in 1593, Kepler wrote a disputation on creatures living on the Moon and astronomical phenomena as seen from there. (The work was clearly Copernican in intention, as it described the appearances seen from a moving body.) The disputation was not delivered at the time, but Kepler saved what he had written.[14] He returns to these topics in his *Discussion with the Sidereal Messenger* of Galileo (1610) and—in their most complete form—in the *Somnium* (*Dream*), along with his Latin translation and commentary on Plutarch's *On the Face on the Moon*. Kepler provided 226 notes to his translation of this work and, as Swerdlow has highlighted, some of these were quite technical.[15] (The *Somnium* and the translation of and commentary on Plutarch were published together posthumously, in 1634.)

Plato's *Timaeus* was a significant work for both Kepler and Plutarch. Karamanolis has argued that Plutarch "focuses primarily and quite strongly on the *Timaeus* for his understanding of Plato's [ideas], and his interpretation of this dialogue shapes his understanding of all of Plato." Plutarch adheres to a literal interpretation of the *Timaeus,* according to which the creator god (the Demiurgos) imposes order on the pre-cosmic disorder.[16] That order is realized through the application of mathematical principles. This takes us to the second question of Book 8 of Plutarch's

[12] Heath 1932, e.g., 108–9, citing *De facie* (*On the Face on the Moon*) and *Platonic Questions*; Neugebauer (1942), in his consideration of some astronomical passages in Plutarch's *On the Generation of the Soul in the* Timaeus, concluded that he is the earliest representative of the appearance in Greek literature of a theory of the unequal lengths of days. Nevertheless, he states (Neugebauer [1975], 2:723, citing his 1942 article) that "a rather garbled parallel to Cleomedes's statement [at 1.6] about the increments in the length of daylight is found in Plutarch (about CE 100) who obviously did not understand what he was talking about."

[13] Harker 2021, 352–53.

[14] Describing himself in the analysis of his horoscope he said, "He preserved even trifling papers he wrote, he tenaciously retained whatever books he brought forth as though they were to be useful at some future time." Quoted by Swerdlow 2025, chap. 4.

[15] Swerdlow 2025, chap. 4.

[16] Karamanolis 2020, 1.

Table-Talk, describing God [*ho theos* (ὁ θεός)] as always doing geometry.[17] But first, let us consider the *Table-Talk* itself, and what Plutarch was aiming to achieve in that work.

In Plutarch's *Table-Talk*, conversations are reported that are said to have taken place at symposia. Both there and in *On the Face on the Moon*, Plutarch's accounts of conversations are in some cases linked to other written texts, sometimes with direct reference to particular works (like Aristarchus, already mentioned), in others by various intertextualities, including allusions to other accounts of symposia. While it may be argued that some of these conversations might have had educational ambitions, the settings and circumstances were intended to foster social and cultural interactions that were not only intellectual.[18]

At the beginning of the *Table-Talk*, Plutarch makes clear his project. The work is addressed to Sosius Senecio (to whom the *Lives* was also dedicated).[19] Plutarch explains that:

> Plato, Xenophon, Aristotle, Speusippus, Epicurus, Prytanis, Hieronymus, and Dio of the Academy, ... all considered the recording of conversations held at table a task worth some effort, and since, moreover, you thought that I ought to collect such talk as suits our purpose [that is, the recording of conversations] from among the learned discussions in which I have often participated in various places both at Rome in your company and among us in Greece, with table and goblet before us, I have applied myself to the task and now send you three of the books, each containing ten questions which we have discussed, and I mean to send you the rest very soon if these seem to you not altogether lacking in charm nor yet irrelevant to Dionysus. (612D-E, trans. Clement)[20]

Calling to mind earlier accounts of symposia, in what follows over nine books presenting discussions of ten questions each, Plutarch's interlocutors range across a vast variety of topics.

Plutarch also explains a bit about his composition of the work, noting that (at 629D–E):

> The conversations which follow have been written in a haphazard manner, not systematically but as each came to mind. Nor must my readers be surprised if ... I have introduced some of your own past conversation also for indeed, if the getting of knowledge does not

---

[17] In the *Timaeus*, the Demiurge crafted the World Soul and divided it into parts, using mathematical proportions (at 35b–36a). Three different mathematical means—the arithmetic, geometric, and harmonic—had been distinguished, we are told, by Archytas of Tarentum (ca. 400–350 BCE), the last prominent member of the early Pythagorean tradition (Archytas, "Fragment 2," in Huffman [2005], 162–63). Plato was well acquainted with Archytas, and in the *Timaeus* arithmetic and harmonic means were crucial in structuring the World Soul. Plutarch, in his work *On the Generation of the Soul in* the Timaeus (1012D), complained that it would take a huge effort to recount all of the disagreements between different interpreters addressing the mathematics contained there. See also Taub 2021.

[18] There were certain conventions adhered to in symposia; these developed and changed over time. Proper (or acceptable) symposium etiquette and activities are discussed at length by Plutarch's interlocutors in *Table-Talk*. Athenaeus also discusses such matters in *Deipnosophistae*.

[19] Quintus Sosius Senecio was also friends with Pliny the Elder, see Campbell 2012.

[20] All translations from *Table-Talk* that follow are from Clement 1969, unless otherwise noted.

insure that one remembers it, frequently the same end is attained by recollection as by learning.

There is an emphasis on recording past conversation; that is one of the points of the exercise. (There are, of course, questions about the historicity of the reports, which I will set aside for now.) And, there is the seemingly Platonic echoing of the importance of recollection, remembering and memorializing.

We can understand the minidialogues as dramatic performances of learned conversations in social gatherings. Plutarch takes pains to identify the "players"; in some cases he gives details of where people were seated (or standing) in relation to one another and what they were doing. The different elements that he includes emphasize the personal and social character of each individual symposium. Participants are named and often described by their interests or expertise; for example, at the beginning of 8.9, Philo is introduced as a physician; at 2.2 we learn that Glaucias is a rhetor.

The conversations cover a wide range of topics. In Book 1, following the discussion of whether philosophy is a fit topic for conversation at a symposium while drinking, there are other questions related to symposia, including the arrangement of the seating. Book 4 considers friendship and various things to do with food, including why Jews do not eat pork (4.5). There are a number of questions that are reported at the end of Book 4 with no discussion, including one concerned with the order of the weekdays compared to the order of the planets (4.7), and why women do not eat heart of lettuce (4.10).

The beginning of Book 8 is concerned with the birthdays of eminent people. Our question—Was God doing geometry all the time?—was posed in honor of Plato's birthday. As the discussants consider the role of geometry in Plato's writings, Plato himself is invited.

In considering the question of what Plato meant by saying that God is always doing geometry, it is crucial to recognize that Plutarch stresses that different answers to this question were offered by each discussant. Five interlocutors participate in the discussion: Diogenianus, Tyndares, Florus, Autobulus (the name of Plutarch's father) and Plutarch himself. At the beginning Diogenianus suggests that since the discussion is taking place on Plato's birthday, we should "take Plato himself as a partner in the conversation."[21] (I will return to this point later.) The friends agree to focus on the question of what Plato had in mind when he asserted that God is always doing geometry.

Significantly, doubt is raised immediately as to whether Plato actually said this anywhere. Plutarch-as-interlocutor remarks that "while this statement is not made explicitly in any of Plato's writings, it is well enough attested and is in harmony with his character" (718C). Plutarch almost presents *himself* here as the "resident expert" on Plato's geometry, declaring that the assertion is consistent with his thinking (718C).[22]

[21] The translations of Book 8.2 are by Minar 1961, unless otherwise noted (718C).

[22] Pelling 2011, 229.

The first to offer a view, Tyndares (a Spartan), asks whether the statement attributed to Plato "conceals a reference to some … difficult doctrine, and not merely to what he himself said and wrote many times, when he sang the praise of geometry for drawing us away from the world of [the senses] … turning us toward the intelligible and eternal level of existence" (at 718C–D).[23]

Tyndares asserts that "in all of the so-called mathematical sciences, as in smooth and undistorted mirrors, there appear traces and ghost-images of the truth about objects of intellectual knowledge; but geometry especially, being, as Philolaus [the Pythagorean] says, the source and mother-city of the rest,[24] leads the understanding upward and turns it in a new direction, as it undergoes, so to speak, a complete purification and a gradual deliverance from sense-perception." He goes on to assert that "it was for this reason that Plato himself reproached Eudoxus and Archytas and Menaechmus for setting out to remove the problem of doubling the cube into the realm of instruments and mechanical devices, as if they were trying to find two mean proportionals not by the use of reason but in whatever way would work." This passage in Plutarch is often cited as definitive regarding Plato's views on the use of instruments. A similar account appears in his *Life of Marcellus* (14), where Plutarch reports that Plato was incensed at Eudoxus and Archytas as corrupters of the pure excellence of geometry, when they used mechanical arrangements to solve the problem of two mean proportional lines. Tyndares argues that Plato thought "the advantage of geometry was dissipated and destroyed" by the use of instruments, because "it slipped back into the realm of sense-perception instead of soaring upward and laying hold of the eternal and immaterial images in the presence of which God is always God" (718E–F).[25] According to Tyndares, the proper practice of geometry conveys theological meaning.

The next speaker, Florus, who we are told is a jokey sort of fellow, suggests that Tyndares has proven that "geometry is necessary not for the gods, but for us." He argues that "surely a god does not need mathematics as an instrument to turn *his* understanding from created things and bring it to bear on what *really* exists [my emphasis]." For, as he notes, it is *in* the god that things exist, with his help and through connection with him (719A).

Instead, Florus focuses on the significance of geometrical proportion for humankind, arguing that it instills social justice, law and order. According to Florus, Plato combined "the spirit of Socrates [with] that of Lycurgus no less than that of Pythagoras." He explains that Lycurgus, the [eighth-century BCE] Spartan lawgiver, had "expelled arithmetical proportion from [Sparta], because of its

[23] Here there is an allusion to Plato *Phaedo* 83d. See Jones 1912, 76–77 on other allusions to Plato in *Table-Talk* 8.2.

[24] See Diels and Kranz [1952] 1972, 44A7a.

[25] This contrasts with the way that the author of the *Letter to King Ptolemy*, attributed to Eratosthenes, presents the history of solving the problem with reference to Plato and his students; the author of the *Letter* does offer an instrumental solution; see Taub (2008b).

democratic and rabble-rousing character."[26] According to Florus, Lycurgus "introduced the geometric proportion, which is appropriate to a moderate oligarchy or a lawful monarchy," whereas "the arithmetical distributes an equal amount to each, measuring by number, whereas the geometric distributes to each an amount corresponding to his worth, measuring by proportion."[27] Florus asserts that "the equality which the mob seeks, which is in reality the greatest injustice of all, God roots out, as far as is feasible; and he maintains distinction by worth, setting the proportionate relation, in geometrical fashion, as the standard of lawfulness" (719B–C). Here, the application of geometry was not abstract, but practical.[28]

As the discussion in Problem 2 of Book 8 develops, pressed by Tyndares to offer a rebuttal to Florus, Autobulus declined, instead giving his own response, turning from matters of the polis to the creation of the cosmos. He asserts that "geometry has no other subject than the proportions and characteristics of limits, and that God in the process of world-making uses no other method than that of imposing limitation on matter, which is by itself unlimited" (719C–D). He explained that "the ancients used the term "unlimited" [*apeiron*] of the unbounded and indeterminate character of matter, not with reference to size or number, but because of its lack of order and harmony." For, "when numbers and ratios have been generated in it, matter is imprisoned, as it were, and encompassed by lines and by the figures generated by lines, that is solid figures, so that it furnishes the primary kinds and distinct forms of bodies, which are the foundations, so to speak" for the creation of the elements, Air, Earth, Water and Fire. Autobolus explains that the geometrical solids, "octahedra and icosahedra, as well as pyramids and cubes, have equalities among their sides and similarities among their angles, and proportions which could not possibly have arisen from disorderly and erratic matter [outside of] that which defines their shapes and articulates their parts, by geometrical rule." In other words, he describes a creator who is uses geometry (719E).

Autobulus elaborates that "matter is always struggling to break out into unboundedness, and seeking to avoid being subjected to geometry; but reason seizes upon it and encloses it in lines and marshals it in the patterns and distinctions which are the source and origin of all that comes to be." This clear statement about the imposition

[26] Lycurgus is regarded traditionally as the founder of Sparta's *eunomia* ("good order"); Hodkinson 2012 :(719E).

[27] As Minar adds in a note to his translation 1961, 123 note e, in the arithmetical proportion (as in 1:2::2:3), the terms on both sides differ by the same amount; in the geometric (as 1:2::2:4), they differ by the same proportion.

[28] The attribution of the notion of "geometrical" equality to Lycurgan Sparta confirms what some (but not all) ancients thought about the Spartan constitution, namely that it was oligarchic and anti-democratic, because one of the (few) purely intellectual arguments against democracy and for oligarchy was couched in these quasi- or pseudo-mathematical terms. Oligarchs were enabled to claim that they were not against democracy *tout court*, only against democrats' wrong (because absolute, and not relative, proportional) understanding of "equality." The arguments are set out and analysed by Harvey 1965 and 1966. I thank Paul Cartledge for private communication relating to this point. As Cartledge, an historian of ancient Sparta put it: they anticipated Orwell's sarcastic claim that "some animals are more equal than others."

of order on the material world recalls the *Timaeus* (e.g., 53c) and the account of the regular solids, the geometrical bodies which play such a fundamental role in Kepler's cosmology, as we know not only from Swerdlow, but also from the work of his student, Bruce Stephenson (1987, 1994).

As Plutarch-as-interlocutor presents his own contribution to the discussion (from 719F), he praises the others' views "as genuine products of their own conception," each of which had plausibility. However, he offers another explanation of the phrase "God was always geometrizing," which he notes is regarded as truly Platonic (718C). He indulges in a bit of Pythagoreanism, recounting that Pythagoras offered a sacrifice when he solved the following problem: Given two figures, to construct a third equal to one and similar to the other (as in *Elements* 6.25). Plutarch then adds a direct reference to the *Timaeus* (to 48E and 29E, at 720a) when—as the speaker—he states that "you will easily see the point … if you recall the threefold division in the *Timaeus*, of the first principles from which the cosmos came to birth."[29] He explains: one of those we call "God, one matter, and one form. Matter is the least ordered of substances, form the most beautiful of patterns, and God the best of causes" (720B).

Plutarch describes how "God's intention was, so far as possible, to leave nothing unused or unformed, but to reduce nature to a cosmos by the use of proportion [*logos*] and measure [*metron*] and number [*arithmon*], making a unity out of all the materials which would have the quality of the form and the quantity of the matter. Therefore, having set himself this problem, these two being given, he created a third, and still creates and preserves throughout all time that which is equal to matter and similar to form, namely the cosmos." In this way, "by means of reason [*logos*, which could also be translated as proportion], and with reference to the pattern [*paradeigma*], [God] gives limits to that which exists" (720B–C).

What follows next is a statement that has perplexed many readers; it concludes the discussion of what Plato may have meant, when he said that God was always geometrizing. Roger Miller Jones (1912) suggested we read it as "And that is the reason that Plato says that *to peri metron* is better than *to summetron*." This statement can be understood as "that which is according to measure [*peri metron*] is better than symmetry." However, *to summetron* can also be read as "proportion" (as suggested by parallels to other Platonic passages, including some in the *Timaeus*).

Scholars do not agree exactly what Plutarch meant in this passage, and there are a number of issues with the text here as it has come down to us. *Summetron* is etymologically related to *metron*, but the translation of *to summetron* as "symmetry" is not unanimously endorsed; "proportion" is suggested by some translators.[30] The

[29] Trans. Minar 1961, 129.

[30] For example, Fowler 1925, 395 translating Socrates's statement at *Philebus* 66b:

> δεύτερον μὴν περὶ τὸ σύμμετρον καὶ καλὸν καὶ τὸ τέλεον καὶ ἱκανὸν καὶ πάνθ' ὁπόσα τῆς γενεᾶς αὖ ταύτης ἐστίν.

"Second, then, comes proportion, beauty, perfection, sufficiency, and all that belongs to that class."

adjective *summetros* has a number of uses, including connections with pitch and rhythm in music, but the principal meaning is "commensurable," having a common measure."[31] The noun *summetria* in Plato is sometimes translated as "proportion."[32] For Plato, that which possesses measure and symmetry is appealing both intellectually and aesthetically. In the *Timaeus*, measure is the ordering principle that governs the harmonious arrangement of parts, while symmetry is the balance achieved through the distribution of those parts. Plutarch's precise meaning is not entirely clear. There is also the possibility that he was engaging in a bit word play with variations between *metron* and *summetron*.[33]

Measure and symmetry were important in various domains, including mathematics, music and ethics. Measure refers to the quantity or amount, while symmetry is more conceptual, referring to proportion, balance and harmony. Measure may be understood as preceding symmetry. In the *Laws* 4, Plato has the Athenian define the deity as the measure of all things (in contrast to the claim of Protagoras, that man is the measure of all).[34] Theokritos Kouremenos (2019, 103) has argued that "in *Republic* 7 it is the Good that is implicitly defined as measure, when Socrates lays stress on mathematics as important for the training for future philosopher-rulers who must know the Good if they are to know its manifestations (534b3–e1)." In the *Timaeus*, the Creator's use of measure and proportion is described in detail (35–36), involving arithmetical, geometrical and harmonic proportions.

Symmetry had a significant role in Kepler's cosmology. His *Harmonice Mundi* (1619) proudly displayed on the first page of Book 1 a (somewhat truncated) quotation from Proclus's commentary on Euclid's *Elements*, explaining:

> Mathematics also makes contributions of the very greatest value to physical science. It reveals the orderliness of the ratios according to which the universe [*to pan*] is constructed and the proportion that binds things together in the cosmos. … It exhibits the simple and primal causal elements as everywhere clinging fast to one another in symmetry and equal-

---

[31] For σύμμετρ-ος, ον, (μέτρον), Liddell, Scott, and Jones (LSJ) 2011 offer "commensurate with, of like measure or size with" as the primary meaning, with "in measure with, proportionable, exactly suitable" as an alternative gloss. For the latter meaning, their examples include Plato's *Laws* 625d, *Meno* 76d, *Timaeus* 67c; for the meaning "in right measure, in due proportion, symmetrical" they cite *Philebus* 66b.

[32] See LSJ: "for σ. πρός τι also, proportion calculated to produce," in Timaeus 66d. Lloyd 2010 notes that "the principal uses of the noun *summetria* are to convey the senses of 'commensurability' or of 'proportion.'"

[33] As well as *peri metron* and *perimetron*. Jones 1912 suggests that Plutarch conceives that in the *Philebus* (66a–b) Plato meant by *to peri metron* the idea, and by *to summetron* the visible copy. I thank George Boys-Stones, Arthur Harris, and Michiel Meeusen for helping me to further interrogate the meanings here. Betsey Price (personal communication) pointed out that reading *summetron* here as 'proportion' complicates our understanding of Plutarch's position regarding Florus's discussion of Lycurgus and the arithmetical (measured) versus the geometrical (proportional) approach. I am grateful to Arthur and Betsey for comments on an earlier draft.

[34] Plato, *Laws,* 716c4–6; see also *Cratylus,* 386a, *Theaetetus,* 152a. On two kinds of proportion, see Plato, *Laws,* 757b–d.

> ity, the properties through which the whole heaven was perfected when it took upon itself the figures appropriate to its particular region.[35]

But Proclus alludes to the description of the creation of the cosmos in the *Timaeus*, and he refers to the dialogue by name, in a phrase elided by Kepler, explaining that there mathematics turns "divergent and warring factors into friends and sympathetic companions."

Readers since antiquity have sought to understand the passage (36b–d), regarded as both astronomical and cosmological, in which Timaeus describes the fashioning of the celestial sphere by the Demiurge, employing mathematical language that refers to arithmetic (using numbers) and geometry (involving circles). Timaeus offers a vivid description of the way in which the Demiurge constructed circles in the World-Soul, the soul of the entire universe, using mathematical proportions. The motion of the circle of the Same is considered to be a motion of the World-Soul, manifested in the sphere of the fixed stars; the circle of the Different is associated with the motions of the planets, including the Sun. Timaeus explains that the Demiurgos caused the movement of the Same to revolve to the right by way of the side; the movement of the Different to the left by way of the diagonal. This final part of the passage is understood as describing the plane of the Zodiac inclined to the plane of the equator.

As Francis MacDonald Cornford pointed out ([1937] 1975, 73), Timaeus's terms—such as "by way of the diagonal"—would have been unfamiliar to most people. Indeed, Plutarch (*De audiendo* 43a) complained that some young men show off their knowledge of math by mentioning "by the way of the diagonal;" this may have given the impression that Plutarch didn't take such mathematical description seriously.[36] But lest we be in any doubt about Plutarch's view of Plato's use of mathematics, in *On the Generation of the Soul in the* Timaeus (1028A) Plutarch makes it clear that, in his view, "Plato did not as a display of mathematical learning drag arithmetical and harmonic means into a discourse on natural philosophy where they

[35] *HM* 7; *KGW* VI, 13. I thank JoAnn Palmeri, History of Science Librarian and Acting Curator of the John and Mary Nichols Collection at the University of Oklahoma, for bibliographical help.

Proclus 1.8 (22 Friedlein), trans. Morrow p. 19. See also Martens 2000, 35, who uses the translation by Aiton, Duncan, and Field (1997), 7; they do not note Kepler's elision.

> Πρὸς δὲ τὴν φυσικὴν θεωρίαν τὰ μέγιστα συμβάλλεται, τήν τε τῶν λόγων εὐταξίαν ἀναφαίνουσα, καθ' ἣν δεδημιούργηται τὸ πᾶν, καὶ ἀναλογίαν τὴν πάντα τὰ ἐν τῷ κόσμῳ συνδήσασαν, ὥς που φησὶν ὁ Τίμαιος, καὶ φίλα τὰ μαχόμενα καὶ προσήγορα καὶ συμπαθῆ τὰ διεστῶτα ποιήσασαν, καὶ τὰ ἁπλᾶ καὶ πρωτουργὰ στοιχεῖα καὶ πάντη τῇ συμμετρίᾳ καὶ τῇ ἰσότητι συνεχόμενα δείξασα, δι' ὧν καὶ ὁ πᾶς οὐρανὸς ἐτελεώθη, σχήματα τὰ προσήκοντα κατὰ τὰς ἑαυτοῦ μερίδας ὑποδεξάμενος.

Retrieved from: http://stephanus.tlg.uci.edu.ezp.lib.cam.ac.uk/Iris/Cite?4036:011:31159.

[36] Elsewhere in the *Table-Talk* (Book 3.9) an expert on gastronomy jokes about mixing wine in "harmonious" proportions based on the mathematics of ratio that underpins canonics, referring to "the canonicists of the lyre" as his authority. The gastronome is challenged to demonstrate his canonics by mixing a drink, but refuses, saying he is a theorist, not a performer, and he mocks the know-it-all seriousness of experts. See also Creese (2012).

are not wanted but introduced them on the assumption that this calculation (*logos*) is especially appropriate to the composition of the soul."[37] He underscores the fundamental importance of measure and proportion not only to astronomy but to the creation and order of the entire world.

This, I hope, gives us some insight into what Plutarch may have meant by the references to measure and symmetry. Yet, if we try to come to a firm understanding of what *exactly* it was that he meant by *to peri metron* and *to summetron*, we may have missed an important point of the *Table-Talk* itself: that is, we are encouraged to keep talking, discussing, and offering different points of view.

Plutarch's emphasis on the importance of considering different points of view is also highlighted at a certain point in *On the Face on the Moon*, when the interlocuters stop strolling, and sit down to come at the question about the Moon from another perspective. I suggest that we too step back and consider again the context in which his question about God's geometrizing was posed. As Eleni Kechagia has noted: "Plutarch does not … follow the format of Plato's or Xenophon's *Symposium*. Instead of one continuous drinking-party, Plutarch … records an assortment of 'snapshots' from various symposia, each … containing the discussion of a particular question or problem to which alternative answers are offered." She notes that Plutarch himself "refers to his work as a collection of *problemata* (612e, 629e) or *sumposiaka zetemata* (660d, 736c), rather than as a symposium."[38]

In fact, there are a number of questions in *Table-Talk* that might be especially suitable for the convivial setting of drinking and eating; such as "Why are old men very fond of strong wine?" (1.7) and "What is the suitable time for having sex?" (3.6), (6.8). Other questions are less obviously linked to the setting of a symposium, such as "Whether it is possible for new diseases to come into being, and from what causes?" (8.9). Michiel Meeusen has suggested that this latter kind of problem displays "the intellectual 'nourishment of the soul' that these discussions were meant to provide." He also argues that "the fact that they are discussed by a variety of speakers – not only doctors but also philosophers, rhetoricians, grammarians, politicians, priests, etc. – shows how popular" these topics were.[39]

In the *Table-Talk*, it is clear that one need not be a specialist in order to offer an opinion or point of view. We are reminded in Book 9.2 (737D–E) that it was the "custom at the festival of the Muses for lots to be handed round and for those … brought together to propound learned problems to one another.'" Plutarch tells us that a mathematician, Ammonius (also the name of his own teacher), "fearing that some professors of the same subject might be drawn together, directed that … a geometer should put a problem to a teacher of literature and a musician to a teacher of rhetoric, and that afterward they should change round and pay one another back in kind." As an example, following this custom, "the first problem was put to Protogenes the teacher of literature by the geometer Hermeias, who asked him for the reason why alpha is put first of all the letters" (this is the second question in Book 9).

---

[37] That is, the cosmic soul.

[38] Kechagia 2011, 79–80.

[39] Meeusen Forthcoming.

Not only was there no expectation, in the *Table-Talk*, that expert answers would be given to the questions, there was a strong presumption that there is no single "right" answer. As Kechagia has noted "what one certainly does not get in the *Table-Talk* is a systematic presentation … ; nor do we get (at least in the majority of cases) a methodical account … with concrete and firm answers to the questions raised. Instead, it is often the case that the results of the *zētēseis* [investigations] conducted are open-ended."[40] It is the case that "at various places in the *Table-Talk* the speakers acknowledge that solid answers … are hard to find."[41] Kechagia concludes, "The … style of many of the conversations of the *Table-Talk*, with the juxtaposition of alternative or opposing explanations leading to no certainty, may well be considered … as a powerful reminder of the genuine undecidability of at least certain matters."[42] This is, as I have argued in my work on Plutarch's *De facie*, the implicit lesson of that dialogue, and it is repeatedly reiterated in the *Table-Talk*: we may not always have (or agree upon) a single explanation.[43] Furthermore, ending in an *aporia* is reminiscent of many of Plato's dialogues; this may be another form of homage to Plato.

Even we cannot agree on the "right" answer, what can we conclude is the significance of the question about God's geometrizing? Why was Plato "invited" to participate in the discussion? Some writers and commentators regard their predecessors, even though dead, as still having a living voice, encountered through their writings. Some authors write for other readers, some of whom might live long after themselves, just as some authors lived long after their predecessors with whose texts they actively engaged. Some authors and their readers identified themselves to an extent as members of a community—at least of a community whose interests intersected around interactions with particular texts.[44] Recognizing that it is difficult and unwise to generalize across a long time-span and vast geographical spaces, I would like to raise a few questions. What is an intellectual (or even a "scientific") community? Who is a member? Do all of the "members" have to be alive at the same time, in order to participate? Or can even long dead individuals be perceived as "active" members of the group?

A community is distinguished from other groups within society by a sense of common interest. Some of the ancient authors writing on philosophical and mathematical subjects presented themselves as members of a particular community, for example, a certain philosophical school. The philosophical schools, including the Platonic Academy, provided a strong sense of membership within a "community,"

---

[40] See *Table-Talk* 1.8, 19, 2.2, 3.1, 3.4, 3.8, 5.7, 6.5, 7.3, 8.3.

[41] See *Table-Talk* 2.7, 641C; 4.2, 664c; 7.2, 700d.

[42] Kechagia 2011, 99–100.

[43] Taub 2008a, 75–76. As mentioned earlier, Kepler translated and commented upon *De facie* at length.

[44] König and Woolf 2013, 31–33 refer to the "bookworld," that is, a "shared culture of books-once read-and-never-forgotten." They describe a "bookworld" as "an imaginary place, a construct created and shared by the educated, whose possession of it made them into an imaginary community of sorts." Authors (and readers) coexist in timeless proximity in a "bookworld." Such a community may exist over long periods of time, even centuries.

sometimes with people interacting face-to-face. But in certain cases membership in a community was at far remove, and was experienced through an identification with the proponents of particular ideas, and the texts that conveyed them. In many cases, allegiance to a particular philosophical school conveyed a sense of membership not only in a community, but also of operating within a particular tradition, and that tradition was often text-based. Specific texts connect the members of the intellectual community; the participants in the community are the authors, editors, translators, and readers of those texts, who in many cases were divided by vast stretches of time and space. Authors and readers coexist in timeless proximity, and such a community may persist over long periods of time, even centuries.[45] Athenaeus, who gave accounts of discussions at symposia in his *Deipnosophistae* (*The Learned Banqueters*), indicates that it is not necessary to still be alive to be part of the conversation. Referring to the *Deipnosophistae* (but applicable also to *Table-Talk* and other texts), Christian Jacob suggested that there is almost a dialogue of the dead, with individual authors displaying their lively engagement with earlier thinkers (2013, 40; also Taub 2008a, 85).

That a vital sense of membership in a community may transcend historical time and place is shown in various letters to and from Kepler. A debt to their forebears was affirmed in Kepler's letter to Galileo dated 13 October 1597, in which he approvingly commented: "you are following the lead of Plato and Pythagoras, our true masters."[46] And on 1 April 1598, Tycho Brahe wrote to Kepler similarly affirming the significance of their predecessors, stating that "There is no doubt that everything in the universe is related and ordered by divine will (divinities) in a fixed harmony and proportion that may be succinctly described by numbers and figures, as was to some extent seen long ago by the Pythagoreans and Plato."[47]

Plutarch's question regarding Plato's view of God as geometer is linked to the thoughts of later mathematicians and astronomers (including Tycho, Galileo and Kepler), themselves in correspondence about the structure of the world and how it came to be, discussing cosmology, cosmogony and the fundamental importance of mathematics. The focus of Plutarch's question is very long-lived in the minds of many we would regard as mathematicians engaged in astronomy, some of whom styled themselves as Platonists and/or neo-Pythagoreans.

Why did Plutarch choose conversation and discussion, that is, the *Table-Talk*, to confront such a fundamental question as the one we have been considering? I will end by highlighting several points:

The *Table-Talk* is purportedly the recollections of past conversations and shared reading. Plutarch was an avowed Platonist for whom recollection is key to knowledge.[48] (This is true of the presentation in *On the Face on the Moon* too).

[45] König and Woolf 2013, 31–34.

[46] Kepler in Baumgardt trans. 1951, 40–41; cf. Kepler, *GW* 13, *Briefe* 1590–1599, 1945, 145.

[47] Quoted by Swerdlow 2025, chap. 4.

[48] On the importance of recollection, see also Oikonomopoulou 2011, 107.

In the *Table-Talk* an emphasis on shared memories and recollections—sometimes of texts acting as *memoranda*—is often key to the discussions. The community itself and the social nature of grappling with knowledge is emphasized.
Very often no conclusion is reached; rather we see highlighted the value of discussion; agreement is not the point (this is also true in *On the Face on the Moon*).
Finally, there is the question of whether we can in fact have knowledge at all (again, a theme also in *On the Face on the Moon*). The dialogic format allows us to suspend judgment.[49]

The responses to Plutarch's question about what Plato meant by saying that God is always doing geometry leave us with no certain answer. The ending to the discussion of question 2 in Book 8 has left various readers puzzled, and Plutarch does not elaborate on his meaning, perhaps realizing that this would probably require yet another symposium. For Plutarch, getting to the "right" answer was not the point. For him, there were many possible answers, and it is not necessary to settle on just one. Rather, the point of the *Table-Talk*, and of *On the Face on the Moon*, was to gather together as a community, discuss and debate, to share ideas and sociability. And, especially, to remember cherished friends, teachers and colleagues, and all that we have learned from them. The *Table-Talk* suggests that we may not be able to have certain knowledge, but we can raise questions, have stimulating conversations, and recall esteemed colleagues.[50]

Bibliography

Primary Sources

Archytas of Tarentum. 2005. Fragment 2. In *Archytas of Tarentum: Pythagorean, Philosopher and Mathematician King*, ed. Carl A. Huffman, 162–163. Cambridge: Cambridge University Press.
Diels, Hermann and Walther Kranz. 1952. *Die Fragmente der Vorsokratiker: griechisch und deutsch*. 6th edn. Berlin: Weidmann, 1972.

[49] There is an underlying form of scepticism in Plutarch's work; see Karamanolis 2020 and Kechagia 2011, 99–100 and 103, to some extent following the models of Socrates and Plato.

[50] Problem 8.2 was discussed on a special occasion: Plato's birthday.

Noel Swerdlow's lifetime of work represents a singular contribution to our understanding of pre-modern astronomical sciences. When I was invited to contribute to this memorial, I immediately thought I would like to choose a topic from antiquity that would resonate with his work on Renaissance astronomy and his emphasis on the importance of mathematical astronomy. When I wrote to Jed Buchwald and Anthony Grafton, who took on the crucial role of bringing Swerdlow's *Renaissance Astronomy* to publication, asking to have sight of the latest version of the chapter on Kepler, and saw that the first section was titled "God's Plan of the Universe," I felt that I had chosen the right topic. I am grateful to Jed and Tony for sharing that chapter prior to publication. I also note my immense gratitude to Noel for his support, encouragement, and friendship over so many years. I thank Francesca Rochberg and Jed Buchwald for inviting me to participate in the special occasion of paying tribute to Noel Swerdlow, and all that we learned from him.

Kepler, Johannes. 1596. *Prodromus dissertationum cosmographicarum, continens mysterium cosmographicum, de admirabili proportione orbium coelestium, deque causis coelorum numeri* [= *Mysterium Cosmographicum*]. Tübingen: Georg Gruppenbach. [ETH-Bibliothek Zürich, Rar 1367: 1 https://doi.org/10.3931/e-rara-445/ Public Domain Mark].

Kepler, Johannes. 1619. *Harmonices mvndi libri v*. (=HM) Linz: Johann Planck.

Kepler, Johannes [1634] 1870. *Opera Omnia*, ed. Christian Frisch, vol. 8. Frankfurt am Main: Heyder & Zimmer.

Kepler, Johannes. [1619] 1940. Gesammelte Werke (= KGW). In *Harmonices Mundi*, ed. Max Caspar, vol. 6. Munich: Beck.

Kepler, Johannes. [1590–99] 1945. Briefe 1590–1599. In *Gesammelte Werke*, ed. Max Caspar, vol. 13. Munich: Beck.

Kepler, Johannes. 1951. *Life and Letters*. Trans. Carola Baumgardt. New York: V. Gollancz.

Kepler, Johannes [1596] 1981. *The Secret of the Universe* [*Mysterium Cosmographicum*]. Translated by A. M. Duncan with Introduction and Commentary by E. J. Aiton. New York: Abaris Books.

Kepler, Johannes. [1619] 1997. *The Harmony of the World*. Translated with Introduction and Notes by Eric J. Aiton, Alistair Matheson Duncan, and Judith Veronica Field. Philadelphia: American Philosophical Society.

Plato. 1921. *Theaetetus; Sophist*. Trans. Harold North Fowler. Loeb Classical Library. Cambridge, MA/London: Harvard University Press and W. Heinemann.

Plato. 1925. *Philebus*. Trans. Harold North Fowler. Loeb Classical Library. Cambridge, MA/ London: Harvard University Press and W. Heinemann.

Plato. 1926. Laws. In *Plato in Twelve Volumes*, vols. 10 and 11. Trans. R.G. Bury. Loeb Classical Library. Cambridge, MA/London: Harvard University Press and W. Heinemann.

Plato. 1929. *Timaeus, Critias, Cleitophon, Menexenus and Epistles*. Trans. R.G. Bury. LoebClassical Library. Cambridge, MA/London: Harvard University Press and W. Heinemann.

[Plato] *Plato's Cosmology* [Timaeus]. Translated with commentary by Francis MacDonald Cornford. Indianapolis: Bobbs-Merrill; Reprint of 1937 edition. New York/London: Humanities Press/Routledge & Kegan Paul.

Plutarch. 1917. Life of Marcellus. In *Plutarch's Lives*, vol. 5. Trans. Bernadotte Perrin. Loeb Classical Library. Cambridge, MA/London: Harvard University Press and W. Heinemann.

Plutarch. 1961/1969. Table-talk [*Quaestiones Convivales*]. In *Plutarch's Moralia, in Sixteen Volumes*, vol. 8 [Books 1–6]. Trans. Paul. A. Clement and Herbert B., Hoffleit. Loeb Classical Library. Cambridge, MA/London: Harvard University Press/W. Heinemann, 1969; vol. 9 [Books 7–9]. Trans. Edwin L. Minar and F.H. Sandbach, Loeb Classical Library. Cambridge, MA/London: Harvard University Press/W. Heinemann, 1961.

Plutarch. 1976. *On the Genesis of the Soul in the* Timaeus. In *Plutarch's Moralia*, vol. 13, Part 1. Trans. Harold Cherniss. Loeb Classical Library. Cambridge, MA/London: Harvard University Press/W. Heinemann.

Plutarch. 1957/1984. On the Face on the Moon. In *Plutarch's Moralia*, vol. 12. Trans. Harold Cherniss. Loeb Classical Library. Cambridge, MA/London: Harvard University Press/W. Heinemann.

Proclus. 1873. *Procli Diadochi in primum Euclidis elementorum librum commentarii*, ed. G. Friedlein. Leipzig: Teubner.

Proclus. 1970/1992. *Proclus: A Commentary on the First Book of Euclid's Elements*. Trans. G.R. Morrow. Princeton: Princeton University Press.

Secondary Sources

Barker, Peter, and Bernard R. Goldstein. 2001. Theological Foundations of Kepler's Astronomy. *Science in Theistic Contexts: Cognitive Dimensions, Osiris* 16:88–113.

Campbell, John Brian. 2012. Sosius Senecio, Quintus. In *Oxford Classical Dictionary*, ed. Simon Hornblower, Antony Spawforth, and Esther Eidinow, 4th ed. Oxford: Oxford University Press.
Cornford, Francis MacDonald. 1937/1975. *Plato's Cosmology*. Indianapolis: Bobbs-Merrill; Reprint of 1937 edition, New York/London: Humanities Press/Routledge & Kegan Paul.
Creese, David. 2012. Rhetorical Uses of Mathematical Harmonics in Philo and Plutarch. *Studies in History and Philosophy of Science* 43:258–269.
Dick, Steven J. 1980. The Origins of the Extraterrestrial Life Debate and Its Relation to the Scientific Revolution. *Journal of the History of Ideas* 41:3–27.
Dick, Steven J. 1982. *Plurality of Worlds: The Origins of the Extraterrestrial Life Debate from Democritus to Kant*. Cambridge: Cambridge University Press.
Grafton, Anthony. 1997. *Commerce with the Classics: Ancient Books and Renaissance Readers*. Ann Arbor: University of Michigan Press.
Harker, Christina. 2021. The Reception of Plutarch's Universe. In *Plutarch's Religious Landscapes*, ed. Rainer Hirsch-Luipold and Lautaro Roig Lanzillotta, 351–366. Leiden: Brill.
Harvey, F. David. 1965. Two Kinds of Equality. *Classica et Mediaevalia* 26:101–146.
Harvey, F. David. 1966. "*Corrigenda* to *Class. et Med.* XXVI (1965), 101–46." *Classica et Mediaevalia* 27: 99–100
Heath, Thomas L. 1932. *Greek Astronomy*. Cambridge: Cambridge University Press.
Hodkinson, Stephen J. 2012. Lycurgus (2). In *Oxford Classical Dictionary*, ed. Simon Hornblower, Antony Spawforth, and Esther Eidinow, 4th ed. Oxford: Oxford University Press.
Jacob, Christian. 2013. The Web of Athenaeus. Trans. Arietta Papaconstantinou. *Hellenic studies* 61. Washington, DC: Center for Hellenic Studies at Harvard University.
Jardine, Nicholas, and Alain Segonds. 1999. Kepler as Reader and Translator of Aristotle. In *Philosophy in the Sixteenth and Seventeenth Centuries: Conversations with Aristotle*, ed. Constance Blackwell and Sachiko Kusukawa, 206–233. Aldershot: Ashgate.
Jones, Roger Miller. 1912. Note on Plutarch's Moralia 720C. *Classical Philology* 7 (1): 76–77.
Karamanolis, George. 2020. Plutarch. In *The Stanford Encyclopedia of Philosophy*, ed. Edward N. Zalta, Summer 2020 ed. https://plato.stanford.edu/archives/sum2020/entries/plutarch/.
Kechagia, Eleni. 2011. Philosophy in Plutarch's Table Talk: In Jest or in Earnest? In *The Philosopher's Banquet: Plutarch's Table Talk in the Intellectual Culture of the Roman Empire*, ed. Frieda Klotz and Katerina Oikonomopoulou, 76–104. Oxford: Oxford University Press.
König, Jason, and Greg Woolf. 2013. Encyclopaedism in the Roman empire. In *Encyclopaedism from Antiquity to the Renaissance*, ed. Jason König, 23–63. and Greg Woolf Cambridge University Press.
Kouremenos, Theokritos. 2019. *Plato's Forms, Mathematics and Astronomy. Trends in Classics Supplementary Volumes*. Berlin: De Gruyter.
Liddell, Henry George, Robert Scott, Henry Stuart Jones, Roderick McKenzie, and Thesaurus Linguae Graecae Project. 2011. *The Online Liddell-Scott-Jones Greek-English Lexicon*. Irvine: University of California Irvine.
Lloyd, David R. 2010. Symmetry and Beauty in Plato. *Symmetry* 2 (2): 455–465.
Martens, Rhonda. 2000. *Kepler's Philosophy and the New Astronomy*. Princeton: Princeton University Press.
Martens, Rhonda. 2003. A Commentary on *Genesis*: Plato's *Timaeus* and Kepler's Astronomy. In *Plato's Timaeus as Cultural Icon*, ed. Gretchen J. Reydams-Schils, 251–266. Notre Dame, IN: University of Notre Dame Press.
Meeusen, Michiel. Forthcoming. *'What of Our Art?' Medicine and Medical Society in the High Roman Empire*. Cambridge: Cambridge University Press
Neugebauer, O. 1942. On Two Astronomical Passages in Plutarch's De Animae Procreatione in Timaeo. *The American Journal of Philology* 63 (4): 455–459.
Neugebauer, O. 1975. *A History of Ancient Mathematical Astronomy*. Berlin/New York: Springer.
Oikonomopoulou, Katerina. 2011. Peripatetic Knowledge in Plutarch's *Table Talk*. In *The Philosopher's Banquet: Plutarch's Table Talk in the Intellectual Culture of the Roman Empire*, ed. Frieda Klotz and Katerina Oikonomopoulou, 105–130.

Pelling, Christopher. 2011. Putting the -viv- into 'Convivial': The *Table Talk* and the *Lives*. In *The Philosopher's Banquet: Plutarch's Table Talk in the Intellectual Culture of the Roman Empire*, ed. Frieda Klotz and Katerina Oikonomopoulou, 201–231. Oxford: Oxford University Press.

Rothman, Aviva. 2017. *The Pursuit of Harmony: Kepler on Cosmos, Confession, and Community*. Chicago: University of Chicago Press.

Stephenson, Bruce. 1987. *Kepler's Physical Astronomy. Studies in the History of Mathematics and Physical Sciences*. Vol. 13. New York: Springer.

Stephenson, Bruce. 1994. *The Music of the Heavens: Kepler's Harmonic Astronomy*. Princeton: Princeton University Press.

Swerdlow, N. M. 2025. *The Renaissance of Astronomy: Regiomontanus, Copernicus, Tycho, Kepler, Galileo*. New York: Springer.

Taub, Liba. 2008a. *Aetna and the Moon: Explaining Nature in Ancient Greece and Rome*. Corvallis: Oregon State University Press.

Taub, Liba. 2008b. 'Eratosthenes Sends Greetings to King Ptolemy': Reading the Contents of a 'Mathematical' Letter. In *Mathematics Celestial and Terrestrial – Festschrift für Menso Folkerts zum 65. Geburtstag*, ed. Joseph W. Dauben, Stefan Kirschner, Andreas Kühne, Paul Kunitzsch, and Richard P. Lorch. *Acta Historica Leopoldina* 54: 285–302.

Taub, Liba. 2021. Reflection I: The World Soul and Harmony. In *World Soul: A History*, ed. James Wilberding, 67–72. New York: Oxford University Press.

Tolsa, Cristian. 2014. Ptolemy and Plutarch's *On the Generation of the Soul in the Timaeus*: Three Parallels. *Greek, Roman and Byzantine Studies* 54:444–461.

Chapter 5
Ptolemy's *Almagest* and the Philosophical Schools of the Second Century CE: Apuleius of Madaura

Alan C. Bowen

5.1 Introduction

In numerous documents that have survived from antiquity, we find a coincidence of old and new thinking about the heaven. This is especially acute in the astronomical/cosmological theorizing of the first and second centuries CE. Here one finds in striking clarity authors like Cicero, who is plainly aware that the planets make stations and retrogradations, adopting in his *Somnium Scipionis* a cosmology that goes back to the Myth of Er in Plato's *Republic* and his *Timaeus,* a cosmology in which such motions have no place. How is one to understand this? Is it perhaps a mere mark of an obdurate aversion to change, a servile devotion to tradition, or a justifiable or at least excusable indifference? Moreover, in those instances when the new does not prevail and there is but a recasting of the old, what was thought to justify this recasting and why is it typically partial at best?

Such questions loosely frame a project that I have begun recently. Its broad focus is the reception in the later philosophical schools of Ptolemy's *Syntaxis mathematica* (A Scientific Treatise), later known as the *Almagest.* In simple terms, my aim is to understand why Ptolemy, who was apparently active for most of the second century CE, was not addressed explicitly by these schools during his own time, and why he became known later not as the Philosopher, as he might have expected, but only as the astronomer.[1] So far as the second century is concerned, my question is whether there were certain features of philosophical thinking at the time that either rendered it

[1] He is explicitly called this by Elias (sixth century CE) in his commentary on Aristotle's *Categoriae* (Heiberg 1907, 265 fr. 2). On Simplicius's assessment, see Bowen 2013, 84–86.

A. C. Bowen (✉)
Institute for Research in Classical Philosophy and Science, Baysville, ON, Canada
e-mail: bowen@ircps.org

F. Rochberg (ed.), *Studies in Premodern Sciences in Memory of Noel M. Swerdlow*, Archimedes 74, https://doi.org/10.1007/978-3-032-04040-4_5

completely untouched by Ptolemy's project or allowed it to assimilate certain of Ptolemy's lessons without challenging the fundamental integrity of its own views. To borrow a medical metaphor: if the *Almagest* were a virus, my question is What made some philosophers such as the Platonist Apuleius in the second century and their schools seemingly immune, while others such as the Stoic Cleomedes were only susceptible to a mild infection? My expectation is that answering such questions will contribute to an understanding of whether, why, and how members of the later philosophical schools availed themselves of Ptolemy's astronomical works.

On this occasion, I will limit my remarks to the Platonic philosopher Apuleius (b. CE 123/5). Since Apuleius does not mention Ptolemy and the search for any allusions to the Ptolemaic corpus in his works is nugatory, the focus will be on whether his philosophical treatises develop a position that Ptolemy's claims in the *Almagest* could not assail, regardless of whether Apuleius was even aware of them.

But let me be clear: the upshot will not be an explanation of why Apuleius makes no mention of Ptolemy. That he does not is an imponderable fact. Still, this study will, I think, be a useful first step in understanding the larger question of the reception of the *Almagest* in the Greek philosophical schools. After all, even if there is no evidence that Apuleius was aware of Ptolemy's work or why he was not moved to respond if he was, it will be important for our understanding of the later reliance on Ptolemy by Platonizing thinkers to know what a contemporary Platonist thought in writing on topics and in a way that Ptolemy undertakes to discredit and supersede in his *Almagest*, especially in book 1.

Before we get to Apuleius, however, we need to locate the *Almagest* in its intellectual context, noting those points in Ptolemy's presentation that would have been controversial or exceptionable to members of the philosophical schools of the second century CE. Once this is done, it will prove easier to see how Apuleius's work stood vis-à-vis the project of the *Almagest*.

5.2 Ptolemy's *Almagest* in Context

For present purposes, we will first consider the *Almagest* as a presentation of a project that effectively redefines what is to count as science in the Greco-Roman world. Then, we will appraise the relation of this project to that in his *Apotelesmatica* (Celestial Influences), also known as his *Tetrabiblos* because it comes in four books.

5.2.1 The Project of the Almagest

5.2.1.1 The Philosophical Background

From the late second century BCE onward, Greek and Roman philosophers maintained with minor adjustments a view of science initially expressed in Aristotle's *Phys.* 2.2 (cf. *Meta.* 6.1). There is clear evidence of this view in a report about

Posidonius of Apamea (*ca.* 135–*ca.* 51 BCE), a Stoic, as recorded apparently by Geminus, a Peripatetic perhaps, in the first century BCE in the course of commenting on one of Aristotle's works. The contention of interest here is that those sciences of phenomena in the physical world which use mathematics to secure knowledge require physical theory (φυσική), a higher science, for their validation. That is, given that each of these sciences must proceed from undemonstrated first principles—the argument is that any attempt within such a science to demonstrate its first principles will be circular—it follows that each of these sciences is by itself only a system of consistent propositions, not a body of knowledge disclosing truth. To become sound, that is, to constitute knowledge, the demonstrations within a science such as astronomy must not only follow properly from the first principles of that science, these first principles must themselves be demonstrated by argumentation within physical theory. Thus, for example, the astronomer's initial assumptions that the Earth is at the center of the cosmos and that the Earth is spherical were thought to be validated only by argumentation in physical theory drawing on the premise that the inherent motion of the elemental body earth is downward to the center of the cosmos.

Now, the astronomer's first principles were called hypotheses (ὑποθέσεις) in part for the very reason that they were seen as requiring validation *ab extra*. Among these hypotheses from the first century BCE onward were included the various geometrical hypotheses used to account for the planetary phases and positions. The philosophical problem posed by these hypotheses was that, for the most part, traditional Aristotelian physical theory was not able to account for them. For example, given Aristotle' contention in his *De caelo* that celestial bodies are each made of ether, a fifth elemental body the nature of which is to move in a circle about the center of the cosmos, what was one to make of an eccentric hypothesis of solar motion that seemed so promising as a way of accounting for the differences in the lengths of the four seasons? Indeed, more troubling, was the realization that the planetary motions include stations and retrogradations. How was one to account for the various hypotheses that make sense of this? And worse: How was one to determine which of the competing hypotheses, if any, was actually the case? This is the question posed by Posidonius, reported by Geminus, and quoted faithfully by Alexander of Aphrodisias, a Peripatetic philosopher, who was active at the close of the second century CE.

5.2.1.2 Ptolemy the Philosopher

Book 1 of the *Almagest* is an introduction to what comes subsequently in that it supplies a way of arguing for the most fundamental hypotheses of Ptolemy's astronomical science, e.g., that the cosmos and Earth are spherical, and presents as well as some technical requirements for quantitative demonstration. What is odd is its first chapter, entitled "Preface" in the manuscript tradition. The burden of *Alm.* 1.1, often ignored or dismissed for its being "philosophical," is, as I have argued elsewhere (Bowen 2026a, b), to establish a project that entails a complete rejection of past and contemporary understandings of what is to count as a science of the heaven, to say nothing of science more generally.

The project that Ptolemy advocates in *Alm.* 1.1 rests on the assertions that whatever exists must have matter, form, and motion—which means that it must be perceptible by the senses—and that matter, form, and motion can only be separated in thought from what exists. This and the tenet that the quality of a body of knowledge is like that of its objects are what underlie his rejecting as mere conjecture (εἰκασία) all theological theoretical philosophy (*scil.* theology and metaphysics) and all physical theoretical philosophy (*scil.* physical theory, φυσική). This denial that theology or metaphysics and physical theory constitute knowledge amounts to a repudiation of the philosophical schools, both antecedent and contemporary, as properly philosophical. Yet, Ptolemy says only that what they have to offer is mere conjecture. He does not announce, for example, that he is rejecting Aristotle's metaphysical and physical theorizing, though such rejection is entailed by what he writes. Rather, this rejection is but a step in his establishing that

> the kind (of theoretical philosophy) that can account for quality in respect of its kinds and its locomotions, one should define as scientific (μαθηματικόν), because it can inquire into shape, quantity, size, and, further, into place, time, and the like, where being of this sort falls, as it were, between those other two[2] not only by virtue of the fact that it can be thought of both through sense-perception[3] and apart from sense-perception, but also by virtue of its being a property of absolutely all things both mortal and immortal, since it changes along with those things that are ever changing in respect of their inseparable form,[4] whereas for things that are eternal (that is, of an etherlike nature), it preserves unchanged the (aspect) of their form that cannot be changed (Heiberg 1898–1903, 5.25–6.11; Bowen 2026b).

In other words, rather than engage his contemporaries directly, he asserts a broad criticism and then contends without supporting argument that:

- the only way to get real knowledge of phenomena is through the application of arithmetic and geometry to what is directly observed, and
- such cognition is the only one that counts as knowledge because it alone is demonstrably true of what is real, i.e., what exists.

Further, for him, the paradigmatic science is the one that focuses on the motions of the eternal, etherlike objects in the heaven, where, as will become clear, the observations are dated and involve measurement, and where the application of mathematics

[2] *scil.* Theology/metaphysics and physical theory.

[3] 1.6.5–6 οὐ μόνον ... νοεῖσθαι. The means that mathematics, understood as the science of magnitude, which thus includes arithmetic and geometry in particular, is *not* an instance of the scientific kind of theoretical philosophy, since mathematics does not proceed "through (or by means of) sense perception" but only "apart from sense perception." As Ptolemy will shortly make clear, arithmetic and geometry are ancillary to scientific (or mathematical) theoretical philosophy in that the sciences comprising this kind of theoretical philosophy (e.g., astronomy and optics), rely on arithmetic and geometry for procedures in constructing demonstrations, demonstrations that also involve observation.

[4] 1.6.8–9 κατὰ τὸ εἶδος τὸ ἀχώριστον. Thus, for instance, Socrates, whose inseparable form is his being a man, is always changing in shape, configuration, and position.

entails developing geometrical hypotheses to account for what is observed and deriving arithmetical tables from these hypotheses. That Ptolemy does not name this science in the *Almagest* has the obvious advantage of not calling to mind the ἀστρολογία taught in the philosophical schools and thereby putting his project on par with theirs.

Yet, even Ptolemy's science of the heaven must have starting points or hypotheses, as his continued use of the term shows. So, if this new science is to be demonstrably true on its own, then these hypotheses will have to be validated by iteration, a requirement built into the project as presented in *Alm.* 1.1. That is to say, if cognition is to count as knowledge, it must be directed to what exists or, equivalently for Ptolemy, to what is perceived by sense. But this means that the study of the heaven can only qualify as knowledge if the hypotheses used to account for a set of dated observations are themselves validated by further observations, given that they are otherwise merely consistent with the original observations. To validate geometrical hypotheses on Ptolemy's terms thus requires subsequent observations typically taken over long intervals, given that the periods of some celestial motions are sufficiently great that often one can detect them, or changes in them, only by comparing observations taken many, many years apart. So understood, the *Almagest* is the first iteration of this project and, as such, its main burden is to show how to carry the project out.

5.2.2 *The* Almagest's *Rhetorical Liability*

But the claim in the *Alm.* 1.1 that the only form of theoretical philosophy that produces real knowledge involves the application of arithmetic and geometry to what is observed presents in turn a rhetorical challenge and a liability in addition the one incurred by not naming this new science of the heaven ἀστρονομία or ἀστρολογία.

The challenge facing Ptolemy was that of persuading readers to take this project up when these readers necessarily included not just his contemporaries but the host of those yet unborn. So, in times when philosophers maintained that any changes in astronomical theorizing must be sanctioned by arguments in physical theory if they are to constitute what is called knowledge, how was Ptolemy to present the view that real knowledge of the heaven is secured by means of dated observations, the construction of geometrical hypotheses to account for them, the derivation from these hypotheses of arithmetical tables, and the iteration of these steps generations later? The answer is, I think, obvious: there was no way; and this, I submit, is why Ptolemy does not engage the philosophical schools directly at all.

Ptolemy's rhetorical strategy in the *Almagest*, however, has a crucial liability in its search for readers to take up his project. So far as one can tell, there were two main groups with some competence in mathematics in the Greco-Roman world. In the first were philosophers and their students. Indeed, during the first and second centuries CE, learning some mathematics was part of philosophical training. Granted, such learning was apparently rudimentary at this time (see Rawson 1985,

156–69). Still, the point is that the number of those with any chance of taking up Ptolemy's project were already imbued with philosophical presumptions at serious odds with the claims underlying his project. Thus, in declining to address this fact and downplaying his challenge to philosophy as practiced in this time and before, Ptolemy's science of the heaven risked dismissal as having no philosophical interest, despite his contention that what he proposed was philosophy proper. In effect, Ptolemy's presentation effectively offered adherents of the philosophical schools, one group with an ostensive interest in mathematics, grounds or cover for silence and neglect of what he had written. Declining to use terms such as "ἀστρολογία" or "ἀστρονομία" for his science and thus to confront what contemporary philosophers took astronomy to be did not help.

The others with some competence in mathematics were the practicing astrologers (ἀστρολόγοι, *astrologi*). These are usually thought to be the intended beneficiaries of Ptolemy's *Canones manuales* (Handy Tables). Yet, Ptolemy's presentation of the *Almagest* had the same liability with them as it did with the philosophers. After all, given the evidence from the second century, it would seem the astrologers' numerical tables were typically composed using Babylonian arithmetical schemes or some adaptation of them (see, e.g., Jones 1999, 113–19). Thus, Ptolemy's not declaring at the outset that the *Almagest*'s science of the heaven culminates in numerical tables enabling the determination of where the celestial bodies are at any time, coupled as it was, with the lack of any argument about the superiority of his trigonometric tables, might well have supplied those committed to other procedures in producing tables cover for silence and neglect of his project as well. In any case, as Jones notes, "Whatever was the readership that Ptolemy imagined for this work (*scil.* the *Almagest*) it was not to be found among the practicing astrologers" (Jones 1999, 118).

5.2.2.1 A Caveat

In focusing on what falls under the rubric of Ptolemy's philosophy, I have limited my remarks to the *Almagest* and will continue to do so. There are several reasons for this. The first is that, as I understand it, the *Almagest* redefines philosophy in a way that is inconsistent with the analyses offered in earlier treatises such as the *Harmonica*.[5] After all, if philosophy is the love of wisdom, and wisdom entails a grasping of truth, then philosophy fails if it is merely conjectural. But, given the criticism of theological or metaphysical theorizing in *Alm.* 1.1 (esp. Heiberg 1898–1903) and, in particular of the positing of an unmoved mover, one can only conclude that the talk of god and gods in *Harm.* 3.3 and 3.7, for example, must be equally conjectural and not properly scientific—unless there is the sort of argument for harmonic science that Ptolemy makes for ἀστρονομία (see Bowen 2026c).

---

[5] On the sequence of Ptolemy's works, see Jones 2020, 25–27.

The issue here is important and merits consideration at greater length than I will give it on this occasion. To get a fuller sense of why and how the *Almagest* was read and understood in later times and in this way to grasp the mechanics, as it were, of its later reception, one must first determine:

- whether the philosophy as it appears in Ptolemy's other works is actually coherent with the position regarding knowledge and its objects outlined in the *Almagest* (see §2.3 below, esp. n11), and then
- whether his later readers were aware of this.

In short, tracing the reception of Ptolemy's *Almagest* involves more than a consideration of parameters and geometrical hypotheses in later texts. It also involves addressing Ptolemy's philosophy more generally and the question of its coherence through the Ptolemaic corpus. Only then will we achieve full understanding of Ptolemy's project in the *Almagest* and a finer grained account of how the *Almagest* and its project were subsequently understood.

Nevertheless, for present purposes, it makes good sense to focus solely on philosophy as developed in the *Almagest*. After all, the fact that the *De mundo* makes neither reference nor allusion to the *Almagest* renders moot the questions: Is Ptolemy's philosophy coherent throughout the corpus of his writings? and Was Apuleius aware of the consistency or inconsistency of Ptolemy's philosophy?

To understand the *Almagest* fully, however, one should also consider its relation to a subsequent treatise, Ptolemy's *Apotelesmatica*.

5.2.3 The Project of the Apotelesmatica

For some time now, scholars have distinguished Ptolemy's earlier treatise, the *Almagest*, from his *Apotelesmatica* by designating the former as an exercise in mathematical astronomy and the latter as astrology. Though this is consistent with Ptolemy's own view that the project of the *Almagest* can be pursued for its own sake, one should not lose sight of the fact that Ptolemy is also careful to present these two treatises as comprising a connected project, astronomical prediction (δι' ἀστρονομίας πρόγνωσις).

This connection is evident in how Ptolemy writes of the two treatises. His title for the work later known as the *Almagest* is "Μαθηματικὴ σύνταξις" (Scientific Treatise) and it was referenced as such with minor variations in antiquity. The title given by Ptolemy to what was later titled "Apotelesmatica" and "Tetrabiblos" is not known. For his part, Ptolemy simply refers to it as "(our) treatise" (σύνταξις) in Hübner 1998, 3.407 and 3.926 (ἐν τῇ ἀρχῇ τῆς συντάξεως: in the beginning of our treatise). He also uses the same term in citing the *Almagest* at Hübner 1998, 2.608: κατὰ τοὺς διωρισμένους ἡμῖν ἐν τῇ πρώτῃ συντάξει (according to the (configurations) that we have defined in our first treatise) when he refers to a passage in *Alm.* 8.4 (Heiberg 1898–1903, 2.189.6–193.13).

Yet, perhaps in recognition that there are still the profound differences between the treatises, Ptolemy uses different terms to designate the practitioners of these two disciplines of prediction. In the *Apotelesmatica*, they are ἀστρολόγοι; whereas in the *Almagest*, those whom Ptolemy seems to count as his predecessors are μαθηματικοί (scientists; Heiberg 1898–1903, 1.269.18–21, 2.210.8–19).[6] Nevertheless, Ptolemy presents the two disciplines as constituting a single, predictive undertaking, which he typically labels as "ἀστρονομία" in the *Apotelesmatica* (cf. Hübner 1998, 1.32, 222–23, 356–59, 363–71; 2.19–25). But perhaps more to the point, this was how Ptolemy was subsequently understood. Porphyry, for example, uses "ἀστρολογική" to designate the discipline set out in the *Almagest* and "ἀστρολόγοι" for its practitioners, though he still gives the title of the *Almagest* as "Μαθηματικὴ πραγματεία" (Düring 1932, 7–21).

This connection between the two treatises and their fundamental distinction is, however, more than a matter of language. As he writes, the first, that is, the *Almagest*, is the one in which we understand (καταλαμβανόμεθα) the configurations (σχηματισμοί) at any time of the Sun, Moon, and stars in relation to one another and to the Earth (1.34–37). Then second is the one in which we investigate (ἐπισκεπτόμεθα) the changes in our surroundings rendered through the physical character (φυσικὴ ἰδιοτροπία) of these configurations (1.37–39). The first alone is a distinct theory and choiceworthy for its own sake (1.39 δι' ἑαυτὴν αἱρετὴν θεωρίαν). The second is not as self-sufficient (αὐτοτελές) as the first. Indeed, Ptolemy's aim is to set it out in a manner that befits philosophy[7] so that a lover of truth (φιλαλήθης) who uses it does not compare its cognition (κατάληψις) to the certainty (βεβαιότης) of the first and what is always the same[8] in citing as a pretext (προσποιούμενος) the weakness and inscrutability (τό δυσείκαστον) of material quality[9] nor shrink from the investigation insofar as (it is) possible (κατὰ τὸ ἐνδεχόμενον ἐπίσκεψις), since most events in general show so clearly their causation from the heaven (ἀπὸ τοῦ περιέχοντος αἰτίαν; Hübner 1998, 1.44–53).

In short, in the terminology of the *Almagest*, the first marks and provides cognition of what is always the same and is thus always the same itself. It is the only theoretical undertaking that constitutes knowledge.[10] In the *Apotelesmatica*, such scientific (or mathematical) theoretical philosophy is simply termed philosophy. In contrast, what the second treatise provides can at best be conjecture (εἰκασία; cf. Hübner 1998, 1.48–50, 161–69).[11]

---

[6] Not mathematicians. See n3 above. Cf. Bowen 2020, 71–78.

[7] Hübner 1998, 1.44–45 κατὰ τὸν ἁρμόζοντα φιλοσοφίᾳ τρόπον, that is, in a way that befits the definition and presentation of philosophy in the *Almagest*.

[8] Cf. Heiberg 1898–1903, 1.6.17–7.3.

[9] Cf. Heiberg 1898–1903, 1.5.19–24, 6.14–17.

[10] In *Alm.* 1.1, this is κατάληψις ἐπιστημονική (scientific cognition; Heiberg 1898–1903, 1.6.13).

[11] Thus, the *Almagest* and the *Apotelesmatica* are presented as philosophically consistent at the cost of relegating the latter as whole with its talk of daemons, gods, and so forth to the rank of conjecture alongside physical theory and theology/metaphysics. The burden of the next chapter in book 1 of the *Apotelesmatica* is to show that conjecture in case of apprehension through astronomy can

For Ptolemy, then, what he calls ἀστρονομία and others at the time, ἀστρολογία, is a predictive discipline that was first set out in the *Almagest* and then completed in the *Apotelesmatica*. Thus, two questions for the historian interested in the reception of Ptolemy's work are:

- Did his readers understand the epistemic distinction that Ptolemy makes between the *Almagest* and the *Apotelesmatica*?
- Did they accept the assertion that, though the *Almagest* is worthy for its own sake, within the discipline of astronomy, it is foundational to the *Apotelesmatica* and stands in relation to that treatise only?

5.2.4 Some Issues in Ptolemy's New Science

Ptolemy's conception and presentation of his new science of the heaven as laid out in the *Almagest* is strikingly modern in several respects, though there are key differences which point to an alien intellectual climate. Certainly, to philosophers of the second century CE, we may reasonably conjecture that Ptolemy's framing of his project in *Alm.* 1.1 exposed several issues that they might have found controversial or even wrongheaded. I say "conjecture" because we have no text dating from that period which tackles Ptolemy head on.

Be this as it may, to get some sense of what reception in the philosophical schools might have involved, we should consider on our own how members of any one of these schools might have viewed the substance of what Ptolemy proposes. To begin, one question, we may presume, would be whether to decouple the *Almagest* and the *Apotelesmatica*, thus opening the possibility of allowing that the astronomer details the structure of the cosmos while denying the validity of astrological divination. After all, there was a presumption among philosophers of the time that astronomy dealt with the heaven as it really is—why else insist on deriving the science's starting points from φυσική, the science of nature as it were? Moreover, several schools of the second century rejected astrology out of hand for the reason that evaluating and choosing an action or a way of life was not a matter of divination but one of understanding what is right and what constitutes a good life, an understanding that could arise only in a process of education which inculcates awareness of the cosmos, its divine order, and one's place in it.

Then, there are Ptolemy's claims that whatever exists must have matter, form, and motion; and that each of the three can be separated from what exists only in thought and never in reality. These claims plainly introduce a complex of issues. For

be cognitive (1.63 καταληπτικὴ ἡ δι' ἀστρονομίας γνῶσις) and to what extent this is so. In the third chapter, Ptolemy argues that it is also beneficial. So, it would seem that, for Ptolemy, not all conjecture is bad or misleading, perhaps only the sort found in the branches of what was traditionally called philosophy. Note Ptolemy's implicit but modest "rehabilitation" of such conjecture at Heiberg 1898–1903, 1.7.5–10 (theology/metaphysics), 1.7.10–17 (physical theory; Heiberg 1898–1903, 1.6.13).

example, the claim that all being is perceptible by sense introduces the closely related issue of what knowledge is. For ancient philosophers in general, explanations constituted knowledge when they followed demonstratively from causes (αἴτια). They also posited that there are different sorts of explanation. Thus, ancient philosophers did agree with Ptolemy that demonstrations of the fact *that* something is the case (τὸ ὅτι) constitute good argument of one sort. An example would be an account of the Earth's position at the center of the cosmos along these lines:

> The Earth is either (a) at the center of the cosmos or (b) on the celestial axis but not equidistant to its poles or (c) neither on this axis nor equidistant to its poles. Observations concerning the length of daytime during the year and the occurrence of lunar eclipses show that alternatives (b) and (c) are not the case.[12]
>
> ∴ The Earth is at the center of the cosmos. (cf. *Alm.* 1.6; Cleomedes, *Cael.* 1.6)

But what ancient philosophers prized above all were demonstrations showing *why* something is the case (τὸ διότι), as in:

> The natural or inherent motion of the elemental body, earth, is in a straight line to the center of the cosmos.
>
> Since the order of the cosmos is eternal, earth as a whole, that is, Earth, cannot have any enforced motions.
>
> But the Earth does not have any natural or inherent motions.
>
> ∴ The Earth is immobile and lies at the center of the cosmos. (cf. Aristotle, *De caelo* 296a24–297a8; Cleomedes, *Cael.* 1.6.40–43)

Indeed, only when such causal demonstrations of the why (τὸ διότι) were in hand would these philosophers allow that one actually had knowledge.

Now, since Aristotle, it was customary to hold that astronomers in the main offer explanations of the fact, so that particular feature of Ptolemy's project in the *Almagest* was not problematic. What would be anathema, however, is Ptolemy's implication that *only* such explanations were possible and, in fact, required if a science was to count as knowledge. For them, the causal accounts in physical theory and theology, which typically introduced objects not perceptible by sense, were absolutely necessary if one was to have knowledge of the heaven. This is the real force of the requirement that the starting points of astronomical theorizing require demonstration in a higher science.

Finally, there is Ptolemy's division of theoretical philosophy into its physical, scientific, and theological branches. Though he attributes this tripartition to Aristotle, what he makes of it is anything but Aristotelian. As Aristotle maintains in *Meta.* 6.1, optics, astronomy, and the like are sciences (μαθήματα) in that branch of intellectual (διανοητική) or theoretical (θεωρετική) knowledge (ἐπιστμήμη, φιλοσοφία) that he terms scientific or mathematical (μαθηματική). In *Phys.* 2.2, he asserts that these same sciences are "more physical" (φυσικώτερα). His point, I take it, is not that these sciences are mathematics (as opposed to being mathematical) or that they belong to physical theory, which is in the first instance an investigation of the

[12]According to Ptolemy, lunar eclipses would not be restricted to occurring when the Sun, Moon, and Earth are not in, or near being in, a straight line.

principles and causes of beings that contain within themselves the principle of their motion and rest (*Meta.* 6.11025b19–21). The point, as he indicates, is that these sciences differ from arithmetic and geometry. The reason is that, while the arithmetician and geometer study the quantity of physical objects and the form of their limits, they do not treat these quantities and limits as physical because in defining their objects of study, they separate in thought these quantities and limits from the matter in which they occur. In contrast, the "more physical" scientist studies the mathematical form but as enmattered or as subject to change and not as mathematical. Thus, whereas the geometer sees a triangular object and separates in thought for study its triangular shape, the astronomer, we may infer, separates in thought the circularity, say, of a planetary motion and then reintroduces change by construing that circle as the trace of a moving point.[13]

This characterization of the more physical sciences thus introduces yet another consideration underlying the contention that the starting points of astronomical theorizing require validation in physical theory. After all, in the example just given, the knowledge that Aristotle's astronomer has of the planet's circular motion will only be secured once it is demonstrated why the planet moves in a circle.

Ptolemy, for his part, in introducing Aristotle's tripartition of theoretical philosophy, makes no allowance for the claim that some sciences in the scientific or mathematical division are more physical than others by virtue of the way in which they define their objects and, thereby, make assumptions subject to demonstration in physical theory. Instead, he defines physical philosophy solely in terms of the ever changing quality of physical objects. Moreover, he suppresses the assertion that physical theory can deal with eternal objects subject only to locomotion, and thus buries the critical contention that the physical theorist and the astronomer often talk of the same objects. In defining physical theory as he does, Ptolemy effectively repudiates Aristotle's assertion that it would be bizarre if it belonged to the physical theorist to know what the Sun is or the Moon but to know none of their attributes[14] especially because those concerned with nature (φύσις) manifestly speak about the shape of the Moon and of the Sun too, and, what is more, about whether the Earth and the cosmos are sphere-shaped (*Phys.* 2.2193b25–30).

Let us now turn to Apuleius and his treatises, the *De mundo* and *De Platone et eius dogmate*. Neither treatise mentions Ptolemy or cites the *Almagest*. Indeed, neither has passages indicating even indirectly any borrowing from Ptolemy. Obviously, one might conclude that Apuleius was, therefore, unaware of Ptolemy's work or that he willfully ignored it. But, such conclusions are sterile in themselves and lack any

[13] In *Phys.* 2.2194a7–12, Aristotle states that optics, harmonic science, and astronomy count as more physical sciences and clarifies this with an example from optics. That is, he contrasts the geometer's studying physical lines but not as physical and the optical theorist's studying the geometer's lines as physical rather than as geometrical. In other words, the optical theorist views the geometer's lines as visual rays that advance from an origin and can only go so far before petering out.

[14] *scil.* the circles, planes, and vertices defined by the rotation of the celestial sphere and their proper motions.

means of arbitration. Moreover, they tell us nothing about what the reception of Ptolemy's work involved. To get to that, we will need to determine whether Apuleius's treatises evince a sufficiently articulated account that could stand in opposition to Ptolemy's presentation in *Alm.* 1.1. Taking up this question will not show us why Apuleius makes no mention of Ptolemy; that is indeterminable. Still, it is manifestly answerable and it will help us to identify modes of argumentation and theorizing that figure critically in the response to Ptolemy in the later philosophical schools.

5.3 Apuleius on the Heaven

5.3.1 Apuleius's Platonism

Apuleius refers to himself on several occasions as a Platonist (see Vallette 1960, 77, 149, 164). Indeed, in his *De Platone et eius dogmate*, he proposes to present to Faustinus[15] what Plato perceived (*sentire*) about moral philosophy, specifically, the principles by which one may attain a blessed life (Magnaldi 2020, 57–58). But this does not mean that he was Platonist in the sense that his contemporary, Alexander of Aphrodisias, was an Aristotelian. For his part, Alexander, who presents his own thought as a development of Aristotle's, maintained that to understand Aristotle one had to explain what Aristotle wrote on Aristotle's terms. This is not how Apuleius reckons himself to be a Platonist. As Richard Fletcher (2014, chap. 3) explains, Apuleius's Platonism is evident in writings that take a Platonic theme and develop it by introducing the thoughts of others and his own, as though the outcome were a seamless whole detailing the reaches of Plato's own thought.[16] This is evident in the second part of the *De mundo*, where, after declaring in the first that he will follow Aristotle and Theophrastus, Apuleius expounds at length his own ideas of the divine Cosmic Ruler and then concludes with a summation that explicitly mentions Plato, alludes to *Phaedr.* 246e4–5, and conflates *Legg.* 715e7–716a3 with 730c2–3 (38.4–5):

> He will not be deceived who hears these words of Plato. Indeed, he says, God, as the ancient account has it, permeates the beginnings, the ends, and the middles of all things and is borne above them in his luminous swift chariot. Avenging Necessity accompanies this same God always and everywhere, the future punisher of those who depart from sacred law. He offers clemency to anyone who discerns (this law) from a tender age and even from the cradle itself, who stands in awe and gives himself and surrenders to it entirely. (Magnaldi 2020, 38.4)

So viewed, Apuleius's Platonism is more like Aristotelianism in the first century BCE, when qualifying as an Aristotelian required little more than taking Aristotle as

[15] On who Faustinus might have been, see Bowen 2022, 21 *ad* pr. 1.

[16] Calling this syncretism, as many are wont to do, actually misses the point.

"a starting point" in the development of one's own thought, agreement being optional (cf. Falcon 2016).

5.3.2 Reading Apuleius's De mundo

Apuleius's *De mundo*, especially chaps. 1–3, is, for present purposes, our primary text. Though typically described as a translation of the earlier Greek treatise, Περὶ κόσμου,[17] the *De mundo* is in fact a paraphrase, and a loose one at that given its numerous elisions and additions. As Jean Beaujeu puts it in his edition, "In the entire work, one barely counts a dozen phrases translated word for word."[18] Accordingly, to understand Apuleius's adaptation of the Περὶ κόσμου, which he likely thought was by Aristotle, one should bear in mind that it was prepared for readers of his own time, the second century CE, and thus at least one century after the composition of the Greek text. This means that to interpret Apuleius's departures from the original Greek, one should not suppose that the intellectual significance of the Latin version is exhausted either by the Περὶ κόσμου itself or by a study of Apuleius's sources. One really must also take Apuleius's departures from the Greek original as tokens of his engagement with his own intellectual context. But this will entail examining what Apuleius says to his contemporaries in the light of what is currently known of that context. In our present inquiry, this specifically means detailing how Apuleius's *De mundo* fares in the face of the project that Ptolemy announces in *Alm*. 1.1.

5.3.3 Apuleius on Philosophy

In the preface to the *De mundo*, Apuleius presents Philosophy as a supreme, personified discipline by which alone inquiry about the divine cosmos as a whole and about the heaven (*caelum*) in particular can secure knowledge. It is the only guide that enables us to know what lies beyond the human power to visit and inspect. So much is found also in the Περὶ κόσμου. Where the *De mundo* differs is in emphasizing that Philosophy is able to discover virtue and banish vice, thus making it *the* moral authority (pr. 1). Both treatises also agree that the philosophical soul recognizes and identifies the divine things in the cosmos, which it then shares with the rest of mankind. The *De mundo* differs, however, in comparing philosophers to prophets (*prophetae*) who tell or announce (*effari*) the truths that they alone have seen by divine grace or favor (*beneficium*; pr. 6), rather than saying only that they

[17] On the date of this treatise, see Bowen 2022, 3 n3.

[18] *Dans tout l'ouvrage, one compte à peine une dizaine de phrases extradites mot à mot* (Beaujeu 1973, 113).

interpret (προφητεύειν) and share what they have grasped (καταλαμβάνειν; 391a15–16). The contrast here is between the inspired and solemn relaying of a vision of divine truths and the relaying of divine truths that have been grasped in thought and not just seen. In short, the slight differences in language point to a difference in the idea of Philosophy itself: for Apuleius, Philosophy describes reality and does not deal in reasons for it or arguments about it. Moreover, communicating philosophical knowledge is largely a matter of revelation.

The preface to the *De mundo* also differs from the Περὶ κόσμου in the statement of its aim. Where the latter proposes to theologize (θεολογίζειν) about all things together and how they are disposed, that is, about their nature (φύσις), position, and motion (391b3–5), the former will address the entire, divine system and, in comprehending its elemental bodies (*naturae*) and their functions (*officia*), explain both why and in what way they move, all the while following Aristotle and Theophrastus (pr. 10). This emphasis on the motion of the natural bodies and, implicitly, on ether, the material of the heaven is, for us, a crucial difference in that it leads into how Apuleius's understands the cosmos and, thus, the heaven itself.

5.3.4 Apuleius on the Cosmos

In chapter 1, Apuleius presents the cosmos in two ways (1.1). In the first, it is a cohesive structure of the heaven, the Earth and the nature of those things that belong to each, by which he means the elemental bodies: ether, fire, air, water, and earth. In the second, it is an orderly arrangement adorned by the grace of God and in the safekeeping of the gods. This is pretty much what one finds in the Περὶ κόσμου (391b9–12). That Apuleius speaks directly of the adornment of the *mundus* signifies little more than the fact that the Latin *mundus* does not have the same aesthetic connotation as the Greek "κόσμος". It is important to note, however, that, where the Περὶ κόσμου mentions a single God who preserves the cosmos and for whose sake it is, the *De mundo* mentions a God as well as gods and assigns to the latter the duty of maintaining and preserving the cosmos. These turn out to be the seven planetary gods (2.4). So, why does Apuleius need planetary gods in his account of the cosmos, why does he make it explicit that the planets are divine beings when the author of the Περὶ κόσμου, who also affirms that the heaven is the dwelling place of the gods (391b15–17), does not?

The answer must begin with our recognizing that Apuleius's two accounts of the cosmos mark a division of the *De mundo* into two parts, the first dealing with the cosmos as a material structure and the second, with its structure as a system maintained by God or gods.[19] In this treatise, then, the division affords two ways of looking at the planets. One may view them either as material bodies or as divine, living beings. Consequently, we may expect two accounts in the *De mundo* of why the

[19] There is a similar division in the Περὶ κόσμου.

bodies in the heaven move as they do, one that focuses on their material nature and the other that sees them as gods subject somehow to God.

5.3.5 Apuleius on Ethereal Bodies

So, let us consider first what Apuleius says about the planets as material bodies. In 1.7, he affirms that the matter constituting the heaven and all bodies in it is ether; and about ether he says only that it is a fifth elemental body, that it is primary among the elemental bodies, and that it is both divine and indestructible. (Note that whatever is indestructible is eternal and, hence, divine but not necessarily a god.) He does not say, for example, that it is the nature of ether to move in a circle about the center of the cosmos, though in point of fact, he does assumes this as we shall see.

Regarding the motions of the planets viewed as ethereal bodies, Apuleius asserts:

> (2.1) Now, in part, the innumerable host of celestial bodies (*astra*) glides forward effortlessly in concert with a region (*regio*) of the fixed sphere about which there goes a sign-bearing band (*circulus*) that is placed round in an oblique encirclement and illuminated by 12 signs. In part, (it also moves) in concert with the wandering stars (*stellae*), which neither have the motions of the former[20] nor (motions that are) very alike or smooth (*aequales*) among themselves. But, while they are fastened to different spheres (*globi*) in disorder, they do preserve order (*ordo*), as I would say. Some are farther away; some, closer ... (2.3).[21] But seven (stars) known widely by the names of the gods, are fastened on the same number of spheres; so that an upper one is larger than a lower one, and linked by bonds back and forth, they are contained in the embrace of that sphere (*orbis*) which is called fixed (*inerrabilis*).[22]

Plainly, Apuleius holds that the planets do not have the daily rotation as their proper motion, in effect relegating it to being a motion forced on them by the rotation of the celestial sphere. Moreover, he correctly affirms that the planets do not have the same proper motions as one another. But then he adds that among the planets themselves, their own motions are not smooth (*aequales*). I take this addition, which is not found in the Περὶ κόσμου, to mean that the proper motion of each planet includes regularly occurring irregularities. Thus, for example, he has in mind, I think, the fact that the Sun's daily progress is not smooth but is sometimes greater and at other times less. As for the five planets, such variations in speed or daily progress occur as each makes its characteristic stations and retrogradations.

Let us be clear that Apuleius does not retail the account of planetary homocentric spheres found in Aristotle's *Meta.* 12.8 as some have supposed.[23] That is, his aim is different. When he writes, "so that an upper one is larger than a lower one," he is only pointing in a compressed way to why these circles or spheres are of different radii. After all, if the planetary gods are all ethereal, it follows that they all move at

---

[20] *scil.* The stars fixed on the celestial sphere.

[21] See Bowen 2022, 25 *ad* 2.1.

[22] See Bowen 2022, 26 *ad* 2.3.

[23] So, e.g., Beaujeu 1973, 311; and Thein 2021 on the corresponding passage in the Περὶ κόσμου.

the same speed by virtue of the sameness of their bodily composition. But in talking of planetary motion, we must distinguish linear and angular motion. If we suppose that the planetary gods have the same linear speed by virtue of their being ethereal, then, if each has its own circle or sphere, as he says, this linear distance that each travels in a given interval of time must be an arc on this circle. But the length of this arc is the product of the radius of the circle and the angle traversed as measured from its center. This means that the planets must be arranged so that upper planetary circles or spheres are larger than the lower planetary circles or spheres, the very point that Apuleius makes.

So, has Apuleius's recognition of unsmoothness in the planetary motions brought us to an impasse? After all, there is nothing in any ancient account of ether that envisages an ethereal body's really stopping and then going backward or forward by virtue of its material composition. Indeed, if we attend to the bodily composition of the planets and follow the argument in Aristotle's *De caelo*, one can account only for the fact that each moves on a single circle or sphere about the cosmic center at different distances from this center. Yet, Apuleius apparently holds as well that the seven planetary motions each involve unsmoothness.

5.3.6 Apuleius on the Planetary Gods

To resolve this, we must recall Apuleius's second definition of the cosmos as an orderly arrangement adorned by the grace of God and in the safekeeping of the gods. The gods are evidently the planetary gods. But the God by whose grace the *mundanes* is adorned and who, it is hinted, may be its creator,[24] is the Cosmic Ruler (*Rector mundi*), the subject of the second part of the *De mundo* (ch. 19 *ad fin.*). This second part conveys by means of several metaphors how the Cosmic Ruler leads the planetary gods, or commands and legislates what they do (chaps. 24–36). One assumption common to several of these metaphors is that those subject to the Cosmic Ruler are intelligent and readily assent to the action required. For Apuleius, then, as gods or divine beings, the planets recognize themselves as subject to the Cosmic Ruler and assent to its dictates, dictates that enjoin the cosmic order.

The upshot is to account for the planetary motions one must consider both their bodily composition and their being divine, intelligent living creatures with volition.[25] What Apuleius apparently imagines is that while, each of the planets moves on a single circle about the center of the cosmos by virtue of its ethereal nature, by

[24] Cf. 1.4 in which the heaven is likened to the work of a craftsman on a lathe. This is confirmed in 24.4.

[25] See van Buren 2023, for argument to a similar conclusion that Cicero's critical citations of Aristotle's lost *De philosophia* in his *De nat. Deor.* 1.13.33 and 2.15.42, where it is said that Aristotle held that the stars move as they do because they are rational and have volition, and Aristotle's *De caelo*, which ascribes their motions to their bodily makeup, are not in opposition but complement one another and are in harmony.

virtue of its being a divine living creature subject to the Cosmic Ruler, each planet in maintaining the cosmic order also introduces characteristic irregularities in its motion that involve speeding up and slowing down in the case of the Sun and Moon, or actually stopping and going backward, stopping, and then going forward in the case of the five planets.[26] That is, Apuleius supposes that what we see is what the planetary gods really do, and that only through Philosophy do we realize this. In contrast, we have Ptolemy and others who, though they may allow that the planets are divine, do not concede that the irregularities in their motions are due to their being gods (cf. Geminus, *Intro. ast.* 1.20–21) but regard these motions as apparent only and, thus, as something to be explained away by means of astronomical hypotheses that indicate the underlying reality.[27]

5.3.7 Apuleius and Ptolemy

Let us return to the issues in Ptolemy's astronomy that would have concerned contemporary philosophers. To begin, Apuleius would agree that there is no basis for the claim that the science of astronomy is dependent on Philosophy for its validity. For him, the relation between Philosophy and astronomy is that of whole and part, and parts are not usefully viewed as subordinate to the whole of which they are parts. Granted, for Apuleius, Philosophy and astronomy address the same things. Still, what they actually say would be different much as geography with the map described in chapters 6 and 7 differs from the popular accounts of the locations described in 1.7–9. In other words, while the vision of a particular natural wonder is admittedly part of the vision of the Earth as a whole, what one says about the former has little or no bearing on what one says of the latter. In the same vein, then, astronomy is not a guide to what is real in the sense that it supplies truths that Philosophy may or must address and accommodate. Further, in that Apuleius treats the Philosopher as a prophet who conveys a vision of reality but not any argument, the implication is that Philosophy cannot validate anything beyond stating baldly the way things are.

But what might astronomy then be for Apuleius? As a science, it is part of our knowledge of the cosmos. If, as ἀστρολογία, astronomy is knowledge of the heaven that has astrology as its main outcome, the question is, Would astronomy so conceived have any legitimate use to Apuleius? His remark in 1.2 that Philosophy alone has been granted arbitration of matters divine and human does suggest that Philosophy is the sole guide to what we should and should not do. If so, then it would seem that for Apuleius astrology and, thus, Ptolemy's grand predictive

[26] Thus, Apuleius embraces what Geminus (*Intro. ast.* 1.20–21) rejects both as shameful (one should not impute a varying speed to an imperishable nature in its travels that one would scarcely attribute to a decent, responsible man) and as physically impossible (imperishable nature is unlike human nature in its not being subject to causes of speeding up and slowing down).

[27] On realism in Greco-Roman astronomy, see Bowen 2020, 91–94.

project, his ἀστρονομία, would be both pointless and unacceptable, which may explain in part why there is no mention of astrology in the *De mundo*.[28]

5.4 Conclusion

If we suppose that disagreement between two thinkers can have purchase in their thinking only if it is not total, that is, only if there are significant, relevant commonalities in their thinking, it would appear that, in Apuleius, we have found a Platonism that is immune to the attempts by other philosophers, and Ptolemy in particular, to establish astronomy as a valid science by way of deriving its first principles or hypotheses in one way or another, given that, for Apuleius, there is simply no occasion for any such derivation. That Apuleius presents a view of the heaven which in some respects goes back to Plato's Myth of Er, however, is not reason to view his own account as an obdurate adherence to older wisdom, especially given that it recognizes motions unknown in the fifth century BCE. As we have seen, he offers a coherent account of the very celestial phenomena that other philosophers and astronomers propose to explain away. Indeed, from Apuleius's standpoint, there is no basis for worry about what the science of astronomy is or should be. His Platonism is untouched by any of the concerns that Ptolemy and Stoics such as Posidonius raise. In short, Ptolemy's implicit criticism and overall project fail to have any purchase in Apuleius's Platonizing world view. So, even if Apuleius had been aware of Ptolemy's project in the *Almagest*, given his account in the *De mundo* and *De Platone et eius dogmate*, there was no compelling reason for him to address it, since, in that this project was presented by way of a weakly argued, indirect attack on that account, it was a cognitive misfire.

References

Beaujeu, Jean. 1973. *Apuleius. Opuscules philosophizes et fragments. Text établi, traduit et commenté*. Paris: Les Belles Lettres.

Bowen, A. C. 2013. *Simplicius on the Planets and Their Motions: In Defense of a Heresy*. Leiden/Boston: Brill.

Bowen, A. C. 2020. Hypothesis in Greco-Roman Astronomy. In *Hellenistic Astronomy: The Science in Its Contexts*, ed. A. C. Bowen and F. Rochberg, 71–94. Leiden/Boston: Brill.

[28] In his novel, the *Metamorphoses*, Apuleius presents astrology as the Chaldean art. At *Metamorphoses* 2.12–14 (Helm 1961), Apuleius relates the tale of Diophanes, a Chaldean who, given the description of his activities, is plainly an astrologer. The point of the story is that astrology is a fraud: Diophanes profits enormously when selling his advice to others but cannot save himself from grievous mishap and loss, which he compounds by bemoaning it in public and making himself ridiculous. In 2.14–15, Lucius, the protagonist of the novel, states annoyance at having to listen to tales about instances in which Diophanes's predictions came to pass, thus wasting an evening.

Bowen, A. C. 2022. Apuleius on the Heaven: A Question of Authority. *Aestimatio ns* 3 (1): 1–37. https://jps.library.utoronto.ca/index.php/aestimatio/article/view41806/3198.
Bowen, A. C. 2026a. The Preface to Ptolemy's *Almagest*. In Bowen and Gannagé 2026.
Bowen, A. C. 2026b. Ptolemy, *Almagest* 1.1: An Annotated Translation of Heiberg 1898–1903, 1.4.6–8.16. In Bowen and Gannagé 2026.
Bowen, A. C. 2026c. The Epistemic Worth of Ἀστρονομία according to Ptolemy. In Bowen and Gannagé 2026.
Bowen, A. C. and E. Gannagé. 2026. *The Philosophy of Ptolemy and Its Byzantine, Arabic, and Hebrew Reception*. Leiden/Boston: Brill.
Düring, I. 1932. *Porphyrios Kommentar zur Harmonielehre des Ptolemaios*. Gothenburg: Elanders Boktryckeri Aktiebolag.
Falcon, A. 2016. Aristotelianism in the First Century BC. In *Brill's Companion to the Reception of Aristotle in Antiquity*, ed. Falcon, 101–119. Leiden/Boston: Brill.
Fletcher, R. 2014. *Apuleius' Platonism: The Impersonation of Philosophy*. Cambridge, UK: Cambridge University Press.
Heiberg, J. L. 1898–1903. *Claudii Ptolemaei syntaxis mathematica*. 2 parts. Leipzig: Teubner.
Heiberg, J. L. 1907. *Claudii Ptolemaei opera minora astronomica*. Leipzig: Teubner.
Helm, R. 1961. *Apuleius, Metamorphosen oder Der goldene Esel*. Berlin: Akademie Verlag.
Hübner, W. 1998. *Claudius Ptolemaeus. Apotelesmatika*. Stutgardt/Leipzig: Teubner.
Jones, A. 1999. *Astronomical Papyri from Oxyrhynchus (P. Oxy, 4133–4300a). Edited with Translations and Commentaries*. Philadelphia: American Philosophical Society.
Jones, A. 2020. The Ancient Ptolemy. In *Ptolemy's Science of the Stars in the Middle Ages*, ed. D. Juste, B. van Dalen, D. N. Hasse, and C. Burnett, 13–34. Turnhout: Brepols.
Magnaldi, J. 2020. *Apulei opera philosophica*. Oxford: Clarendon Press.
Rawson, E. 1985. *Roman Intellectual Life in the Late Roman Republic*. London: Duckworth.
Thein, K. 2021. On the Heavenly Sphere (Chapter 2, 391b9–392b8). In *Pseudo-Aristotle: De mundo (On the Cosmos)*, ed. P. Gregorić and G. Karamanolis, 33–61. Leiden/Boston: Brill.
Vallette, P. 1960. *Apulée. Apologie, Florides. Texte établi et traduit*. Paris: Les Belles Lettres.
van Buren, F. 2023. Circular Motion and Circular Thought: A Synthetic Approach to the Fifth Element in Aristotle's de Philosophia and de Caelo. *Apeiron* 56:15–42.

Part III
India

Chapter 6
Visual Representations and Intertextualities in Second Millennium Sanskrit Eclipse Reckoning

Clemency Montelle

6.1 Introduction

The *Parvadvayasādhana* ("Computation of the two syzygies") of Mallāri (fl. late sixteenth century) is a short Sanskrit treatise dedicated to the computation of lunar and solar eclipses. With an epoch date of 1588, this composition is preserved in at least 25 manuscripts (Pingree 1970–1994, 4:365 and 5:284–85), indicating its widespread popularity. As was conventional in the Sanskrit astral sciences, Mallāri divides his exposition of eclipse reckoning into three distinct parts. The initial section, comprising 14 verses, covers lunar eclipses (*vidhu-grahaḥ*), and the following section, solar eclipses (*ravi-grahaḥ*) contains 11 verses. The final section, the shortest of the three, sets out procedures pertaining to the graphical representation of eclipses (*parilekha*) in only five verses.

Virtually all aspects pertinent to eclipse calculations are addressed within this concise discourse. Mallāri covers the conventional topics in eclipse reckoning including determining apparent diameters, latitudes, duration, magnitude, key phases, parallax, and the so-called direction (*valana*) of impact along with other key phases in an eclipse. Certain manuscripts include commentaries written by later authors that feature exemplar calculations (*udāharaṇa*). These example calculations lay out the implementation of algorithms and data provided in the text to produce the details for specific datable eclipse occurrences. In this way the textual tradition, with its legacy of scribal copying and associated prose commentaries including calculations provide rich documentation of intertextuality in the astral sciences.

There are many interesting features of this concise work which underscore some distinctive aspects of the tradition of the Sanskrit exact sciences in the second

C. Montelle (✉)
School of Mathematics and Statistics, University of Canterbury, Christchurch, New Zealand
e-mail: clemency.montelle@canterbury.ac.nz

F. Rochberg (ed.), *Studies in Premodern Sciences in Memory of Noel M. Swerdlow*, Archimedes 74, https://doi.org/10.1007/978-3-032-04040-4_6

millennium. Among them are: the versified format in which technical material is encoded, assumptions around astronomical modeling of the paths of the luminaries, and tradition aspects of timekeeping, trigonometry, and foundational numerical parameters. However, this composition deviates from many standard works of this kind. For one, it is the first known treatise dedicated exclusively to eclipse phenomena. Prior to this text, eclipse reckoning as a topic in the astral sciences was always included in larger astronomical works. A second notable feature is its emphasis on numerical data. Rather than include abstract algorithms and techniques for computing the circumstances and details of an eclipse, the majority of this work is versified numerical tables, namely precomputed data (ultimately produced from algorithms) woven into a memorable metrically defined format. From this encoded data, the requirement for actual computation vanishes. Once the user has established the argument with which to retrieve a tabular entry, all computation is restricted to simple linear interpolation to produce values for nontabulated arguments. In this way, the *Parvadvayasādhana* emblematizes the new directions astronomy was taking, from fundamental algorithms which required direct engagement with calculation, to optimized tabular data for handy and efficient retrieval of appropriate numerical results for a phenomenon in question, thus relieving an astronomer of long often laborious computations.

While this numerical data is cleverly encoded in verse, as Figs. 6.1 and 6.2 show, scribal practice was often to produce an aligned numerical table with rows and columns alongside the versified version. Interestingly enough, this feature appears in all known extent manuscripts of the *Parvadvayasādhana*, suggesting that this practice was widespread for this particular work. It may also hint at the limitations of working with numerical data encoded in a versified format. While memorable, versified data has many drawbacks. A significant constraint is the absence of an argument paired with its corresponding entry. Another limitation is that retrieval of any entry requires one to "recite" the verse from the beginning until the desired entry is

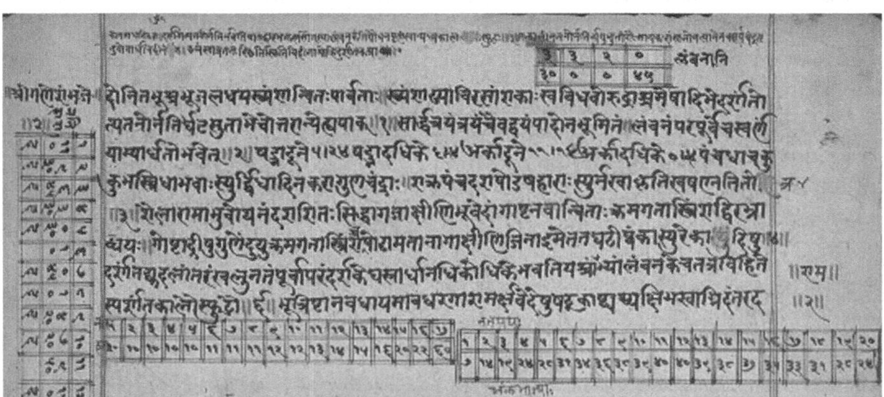

Fig. 6.1 An excerpt from a manuscript of Mallāri's *Parvadvayasādhana* depicting the layout on the page of the Sanskrit text and graphical tables (University of Pennsylvania Rare Book and Manuscript Library, Philadelphia; UPenn 390 1888, f. 2v.)

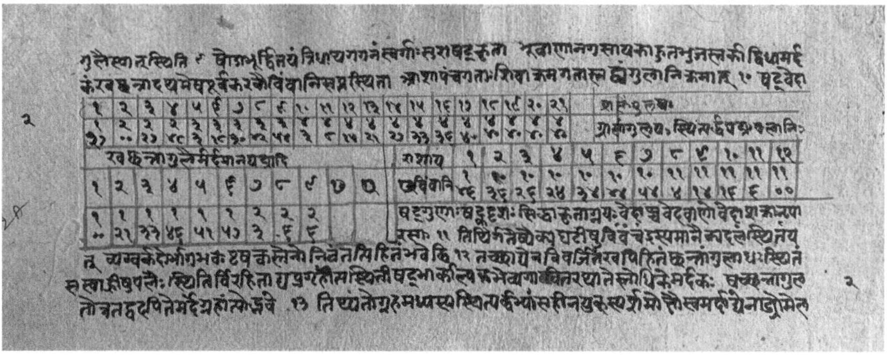

Fig. 6.2 An excerpt from a manuscript of Mallāri's *Parvadvayasādhana* depicting the layout on the page of the Sanskrit text and graphical tables (Oriental Research Institute, Mysore, C4464, f. 2v)

reached. A ruled table in contrast, allows one to simply scan to the appropriate argument and corresponding entry in question. In this way, the *Parvadvayasādhana* is a transitional text between fully algorithmic and a graphically aligned numerical table.

Another way in which this work is particularly valuable for historians is its focus on the graphical representation of eclipses. While the tradition of eclipse diagrams stretches back to the first astronomical texts in the middle of the first millennium, the *Parvadvayasādhana* and its associated commentaries and worked examples provide some fascinating evidence for the study of graphical construction and diagrammatic procedures more broadly in the Sanskrit astral sciences. In particular, in some manuscripts, the worked examples which compute the circumstances and features of a dated eclipse included in commentaries culminate in a unique graphical rendering of the eclipse itself. Interestingly enough, despite the fact that a whole distinct section on graphical representation is devoted to instructions on how to construct an eclipse graphically, the inclusion of an actual drawn diagram in a manuscript is a highly uncommon feature in the Sanskrit corpus. Because of this, these graphical representations included in the commentary are of considerable interest to modern historians. This aspect will be the subject of this contribution here.

6.2 The Stated Motivation of This Work

A notable feature of Sanskrit works, including those in the genre of astral sciences, is an introductory benediction verse which pays homage to the authors deities and patrons.[1] Very occasionally this is followed by any number of "nontechnical" verses which may give a bit more context to the work. These may take the form of an encomium, a historical discourse, or some sort of motivation for composing the text. These introductory passages, where they occur, are of great value to the historian as

[1] This is known in Sanskrit as the *maṅgalācaraṇa*. For further details, see Minkowski 2006.

they provide a glimpse into the broader circumstances of the work, their authors, and their audience, details which are usually very hard to come by otherwise.

Here, in the *Parvadvayasādhana*, after a one-verse benediction, Mallāri includes three verses which provide a poetic outline of the motivation of the work. The passage is almost certainly fictional; nonetheless it is no doubt indicative of practices and people involved in Sanskrit science more broadly. Mallāri's passage is notably theatrical in its nature. It recounts a conversation between an astronomer and the king, involving dialogue on the circumstances and features of an eclipse when it is unable to be observed directly. It opens by invoking a scene where an astronomer (the "twice born") is seated on his horse, being part of the royal retinue of the king. The king inquires about an upcoming solar eclipse, noting that because of the enormous amount of dust kicked up by his correspondingly regal and extensive retinue of humans and animals, he is unable to see the sky:

rājñā sukhāsanagatena caladdhayasthaḥ
pṛṣṭo dvijo na hi ravir na vidhur grahaḥ kim |
hastyaśvayānarathapanikhuronhadh ūliv
yāptāntaraṃ ca gaganaṃ grahaṇena hīnam ||2||

The twice born, who is seated on his moving horse, on being asked by the king who is positioned comfortably [on his palanquin] "I [see] neither the sun nor the moon. Is it [an instance of a solar] eclipse?" [The astronomer replies:] "The entire space is filled with the dust arising from the hooves of moving elephants and horses, [from the feet] of the foot shoulders, [and from the rims] of the chariots, without there being an eclipse."

The king presses the astronomer further, asking for specific details of the eclipse. The astronomer feels apprehensive about his ability to compute the circumstances, given his position on a horse, outside of his usual spaces for performing computations:

ityukte gaṇakena bhūmipatinā pṛṣṭaṃ punaḥ syāt kadā
parvācakṣva savistaraṃ grahamukhaṃ kasyāṃ diśi syād iti |
śrutvā rājagiraṃ hayopari kathaṃ vācyo graho mānase
kṛtvenhaṃ nṛpabhītivihvalatanur jātas tadā daivavit ||3||

When the astronomer replied so, it was asked by the king "Tell me in detail when will an eclipse occur again and in which direction is the first contact?" The astronomer, having heard these words of the king, [wondered] how, [being seated] on a horse, an eclipse can be computed mentally and stated. Then the body of the astronomer started trembling out of fear of the king [for not answering such a question].

Having set up this scenario, Mallāri then situates the key feature of his work on eclipses—a set of numerical tables (*pañcāṅga*) encoded in verse, that are notably short and succinct, which thus provide a quick and easy way to compute the circumstances of an eclipse (and, notably, while still being able to use one's hands to control one's horse!):

tasmād vacmi ravīnduparva
gaṇakānandārthamatyadbhutaṃ pañcāṅgaṃ
gaṇakena dhāryam akhilaṃ haste dvitīye guṇaḥ |
aśvākarṣaṇajas tato grahayugaṃ vācyaṃ vinodāya bhū-
pālānāṃ sakalaṃ sphuṭārthasahitair vṛnaiḥ kharāmapramaiḥ ||4||

> Therefore [to avoid such distressing moments for astronomers], for the purpose of the pleasure of computing the instances of solar and lunar [eclipses], I am composing an awesome pañcāṅga, [which] can be held in its entirety by the mathematician in one hand, the reins for pulling the horses in the second [hand]. [Then] using that [pañcāṅga], both the eclipses may be fully articulated for the amusement of the king [with the help of] thirty verses containing clear expositions.

Mallāri's motivation is thus to provide a "handy" and simple guide that can be literally wielded in one hand to provide the astronomer with all the necessary details so he may fulfill the wishes of the king.

This passage is obviously intended to delight the reader with a dramatic and amusing episode. But it is more than a theatrical scene setter. Through the artistic layers, it hints at how the relationship between a patron and practitioner might function and the sorts of tasks an astronomer might be asked to deliver from a patron or client. In this particular episode, details such as an astronomer being part of a royal retinue, his social status as being "twice born," being directly addressed by the king, the apparent interest of a monarch in technical details of an eclipse, and the drivers for writing concise astronomical works or compiled tables of precomputed data are all substantiated in this passage, in ways that are very difficult to glean in other sources. As such, this short passage provides valuable ethnographic evidence of the circumstances and practices of a royal astronomer and the sorts of tasks they were expected to perform. Of course, historians must be wary of generalizing from these instances about the social settings of scientific activity; however, treated with historical sensitivity, they offer precious glimpses into aspects of scientific practice.

6.3 Visual Renderings among Scribes

As well as the opening of the work, the ending too provides insight into forms of practice in the astral sciences in ways which speak to the expectations from clients or patrons on the practitioner and the more visible public facing outputs of scientific activity. A key component of eclipse reckoning in the Indian astral sciences tradition is the so-called *parilekha* or "diagrammatic representation/construction" section. This part of the work is usually a separate section placed toward the end. The eclipse diagram is an explicit aspect of visual practice in astronomy that is not at all frequent in Sanskrit science—hence its significance. It is one of the few instances in which astronomers are instructed to produce a visual artifact of a physical phenomenon based on their computations for a datable astronomical event.[2]

Instructions for producing eclipse *parilekha* can be found in Sanskrit texts from the mid-first millennium onward, in works such as Varāhamihira's *Pañcasiddhāntikā* (sixth century), Brahmagupta's *Brāhmasphuṭasiddhānta* (628 CE), and Lalla's *Śiṣyadhīvṛddhidatantra* (eighth century), to name a few. The comprehensive and

[2] Another standard instance is the lunar crescent diagram, typically a separate subject as well. See, for instance, Brahmagupta's *Brāhmasphuṭasiddhānta*, Chap. 7.

detailed nature of these procedures suggest that this practice likely was much older, based on a long tradition developed and refined over centuries. Some authors provide insight into the underlying motivation for such a practice. For instance, Brahmagupta esteems that proficiency in these procedures indicates full mastery of the topic of eclipses and other astronomical subjects. He also reveals with whom it is appropriate to share this knowledge:

> Since one who knows diagram construction is the master of those who know everything in treatises on eclipses, planetary conjunctions, and occultations, I will explain diagram-construction. ... Diagram construction is to be given to a devoted student of long-standing and exceptional merit, or to a protector or a friend or a son. (from Brahmagupta's *Brāhmasphuṭasiddhānta* qtd. in Plofker 2021, 309–10)

Textual references like this suggest that preparing an eclipse diagram was a special skill, not just to demonstrate mastery of a subject, but likely also as an output to share with one's patron or the person commissioning the specific eclipse computation.

Indeed, surveying the various expositions on this topic in Sanskrit works over the ensuing centuries reveals a great variety of practices, from simple allusions to diagram construction in texts (which apparently assume the practitioner already knows what to do), to detailed instructions to produce highly elaborate diagrams with specific features. A selection of excerpts from *parilekha* chapters can be read in translation in Subbarayappa and Sarma (1985, 217–24, 238). However, a major challenge for modern historians is the overall paucity of actual drawn diagrams surviving in manuscript sources to compare with the textual instructions of construction.

General aspects included in a set of eclipse diagram instructions typically involve the production of a perpendicular grid with cardinal directions marked off, circular disks whose radii are commensurate with the eclipsed and eclipsing body, the so-called angle of inclination or deflection (Sanskrit: *valana*) for various key phases of the eclipse (first impact, mideclipse, release, and beginning and end of totality where appropriate), the path of the eclipsed body, the lunar latitude, and the magnitude. These sections may include very specific instructions pertaining to graphical rendering: the tools of construction could be itemized, geometric graphical objects identified, such as circles, points, radius, intersections, fish figures, and their connection to specific astronomical phenomena (latitude, first contact, mideclipse, release). Even the place to inscribe the diagram may be indicated (such as a flat piece of earth or a well-prepared surface).

As an example, the following passage is from the *parilekha* section of a seventeenth century eclipse text called *Karaṇakesarī*; it is a set of versified instructions to produce an eclipse diagram which has been translated into prose (Montelle and Plofker 2013, 32–36):

> When one has produced a circle by means of a [radius] string half [the size] of the eclipsed body, and [another circle by a radius string equal in length to] the[half] part of the sum of their measures [i.e., diameters of eclipsed and eclipser: this is the circle of first contact and release] with cardinal directions established. ... By means of circles constructed from a centre in the [appropriate] direction at the tips of the latitude [line segments] now [at that time], with [radius equal to] the [half] part [of the diameter] of the eclipser, the forms [of

> the eclipser] at contact, mideclipse and release should be [shown]. Now, producing a circle with [radius equal to] half the difference of the amounts of the diameters of the two bodies], one should draw [it] from the point of the tip of the mideclipse latitude. Now [there are] two centres at the two intersections of that [circle] and the path [of the eclipser]. [Upon] drawing a circle [at each center] with radius [equal to that of] the eclipser, the beginning of totality and end of totality too are to be known.

The means of construction (a string) is identified to produce circles the size of which are to be determined for the specific instance in question (the diameter of the eclipsed and the eclipser). Aspects such as latitude and specific orientation of the disks in the night's sky with reference to cardinal directions are to be rendered. Distinct phases of the eclipse (first contact and release, mideclipse, and beginning and end of totality if appropriate) are to be documented as well.

Unfortunately, in all the extant manuscripts of this particular text, no actual diagrams have ever been rendered by a scribe. Plausible modern reconstructions have been produced by carefully following the instructions, and the following have been proposed as what the diagrams may have looked like (Fig. 6.3):

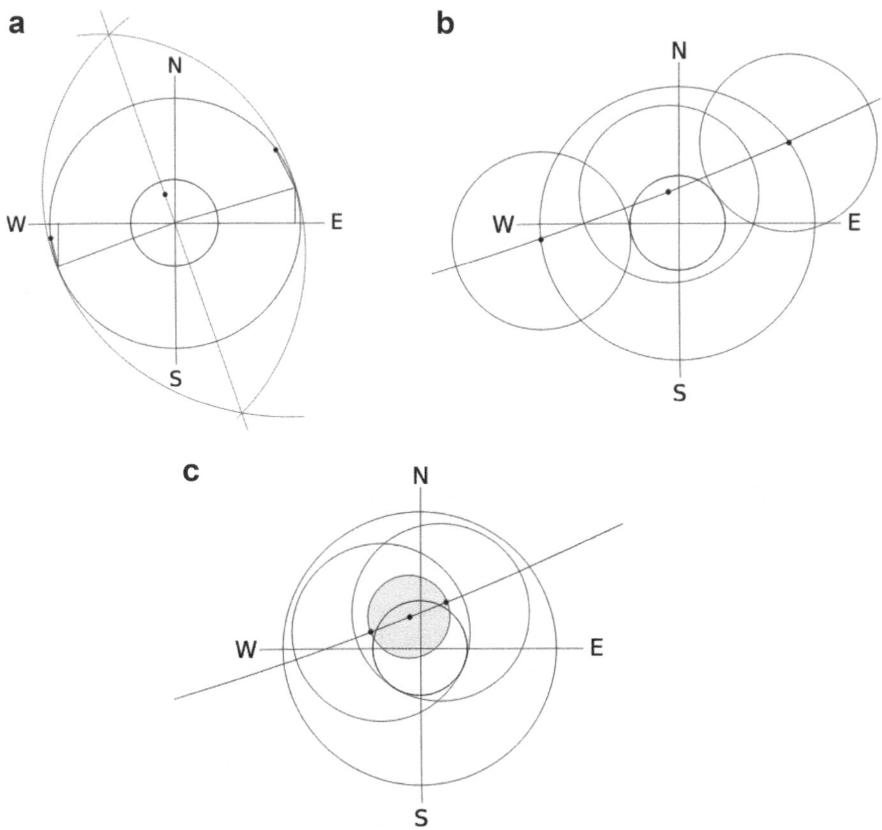

Fig. 6.3 Modern reconstructions following the instructions set out in the *parilekha* section of an eclipse text (Montelle and Plofker 2013, 32–36)

In the absence of scribal diagrams included in these manuscripts, modern scholars can only guess at the result. Indeed, a major challenge for modern historians is the overall paucity of drawn diagrams to compare with the textual instructions of construction. In this light, the significance of eclipse diagrams included in commentaries of the *Parvadvayasādhana* can be better appreciated. These are rare instances of graphical representations arising from computations included alongside textual passages. Such examples are thus of considerable interest to historians as they allow us to further our understanding of *parilekha* techniques in the Sanskrit tradition. In what follows we will study in detail these visual artifacts, account for the procedures and where relevant compare them to their textual descriptions.

6.3.1 Analyzing Eclipse Diagrams Included in Commentaries of the *Parvadvayasādhana*

In the Parvadvayasādhana, the *parilekha* material is given its own five-verse section. As previously noted, the importance of this feature of eclipse reckoning as an "output" is directly emphasized at the beginning of the work in the introductory verses where the king poses the question:

> Tell me in detail when will an eclipse occur and in which direction is the first contact? (Verse three)

It seems likely that part of this "reporting" on the details of the purported eclipse would have involved the presentation of a diagram. Namely, this visual representation is an output of the more technical algorithmic practices outlined earlier in the work and provides a medium through which eclipse science can be appreciated by nonspecialists.

The *Parvadvayasādhana*'s *parilekha* section is notably quite different to the *parilekha* section in the *Karaṇakesarī* cited above. Rather than give explicit instructions for construction, instead it provides algorithms and data to determine the key features of an eclipse, such as direction of impact, magnitudes at specific times and location of mideclipse, as follows:

> **Verse one** Determining the magnitude of the "inclinations" or "angle of impact" (*valana*) (which is divided up into two components, the so-called *ayana-valana* and *akṣa-valana*) by means of precomputed tabular values
>
> **Verse two** Rendering the total *valana*, the latitude, the magnitude at mideclipse, and the two instants of beginning and ending of the eclipse, the middle of totality, the beginning and ending of totality for both a solar and lunar eclipse
>
> **Verse three** Versified tabular data relating the magnitudes of the eclipse for the graphical rendering
>
> **Verse four** Depicting the location of mideclipse in the graphical representation
>
> **Verse five** An algorithm for obtaining the obscuration at a desired instant

While these five verses cover the content and numerical parameters required to produce the graphical rendering of the eclipse (albeit in a very terse way!), what is immediately obvious is that there is not an actual drawn example included as part of the text to guide the astronomer or give them a general idea as to what they are aiming for. In addition, this particular text, unlike some *parilekha* sections in other works, doesn't contain a single reference to the actual process of execution or drawing, or the manual procedures one should undertake to produce a graphical representation. The verses in the *Parvadvayasādhana* present themselves as particularly opaque as to what the resulting graphical representation and its features should look like and how exactly the magnitudes and eclipsed and eclipsing bodies are to be rendered. How should one mark off units of measurement? How exactly is the eclipse path to be drawn? How should the magnitude at mideclipse be rendered? The text assumes practitioners come equipped with such background knowledge. Furthermore, it appears this practice was based on an activity which was largely ephemeral—diagrams just weren't preserved once they had served their purpose. This makes it challenging for the modern historian to fully account for the practices surrounding graphical representation. Answers to these questions are generally based on extended speculations and hypothetical guesswork.

However, the commentaries come to the rescue in a special and significant way. Out of the number of commentaries available for consultation in this study, at least three of them include worked examples. Alongside detailed step by step numerical computations, some scribes have included actual graphical representations of eclipses corresponding to the worked examples. In one particular case we even have an instance of the same eclipse graphical representation being rendered by two different scribes.

In what follows we will consider four eclipse diagrams produced by scribes in their commentaries to the *Parvadvayasādhana* to bring to light some of the technical processes and practices related to diagram execution and features. In particular, we will consider two pairs of diagrams to highlight some of the salient features of these graphical renderings. The first example we will consider will be a pair of distinct eclipse *parilekha* drawn by the *same* scribe. The second example will be a pair of eclipse *parilekha* of the *same* eclipse rendered by two different scribes. We will highlight some of the commonalities and differences to consider the rich intertextualities of this genre. Such comparisons will form fascinating evidence from which to derive some more general observations about scribal practice and conventions related to eclipse diagram rendering. It will allow us to reflect on generic versus individual features of scribal practice and the manual steps and artifacts that were undertaken by practitioners to display the output of their computations.

The four eclipse diagrams and the manuscript pages they appear on can be seen in Figs. 6.4, 6.5, 6.6, and 6.7.

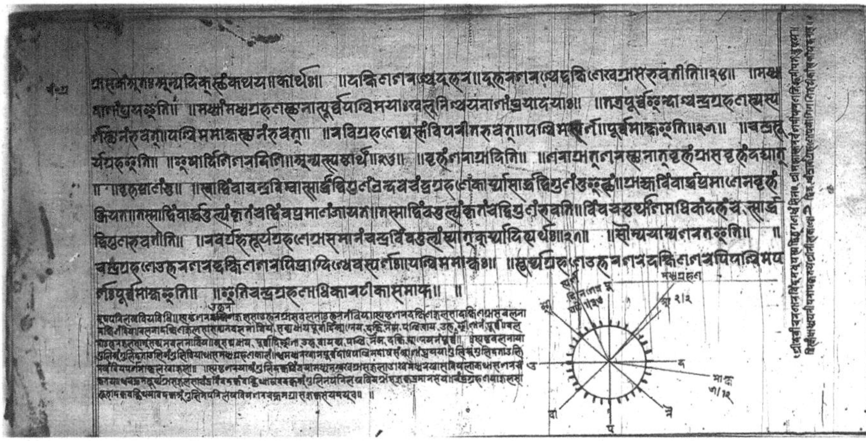

Fig. 6.4 An excerpt from a manuscript of Mallāri's *Parvadvayasādhana* depicting a lunar eclipse diagram (Kathmandu 433–53, f.8r)

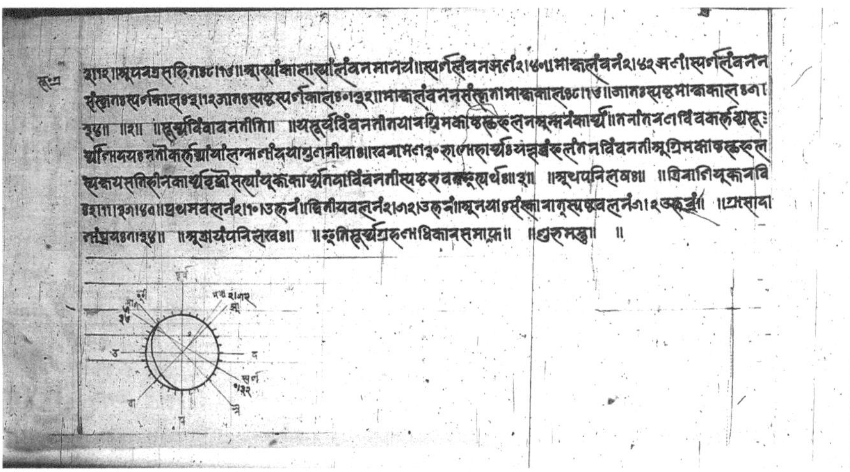

Fig. 6.5 An excerpt from a manuscript of Mallāri's *Parvadvayasādhana* depicting a solar eclipse diagram (Kathmandu 433–53, f.9r)

6.3.1.1 A Pair of Eclipse Diagrams: One Lunar and One Solar—Executed by the Same Scribe

The first pair of eclipse diagrams are found in a manuscript of the *Parvadvayasādhana* along with a prose commentary: MS Kathmandu (1960) 27 (III 577) 1–1208, which was copied in 1689 CE.[3]

[3] For details on this manuscript, see Pingree 1970–1994, CESS IV: 365.

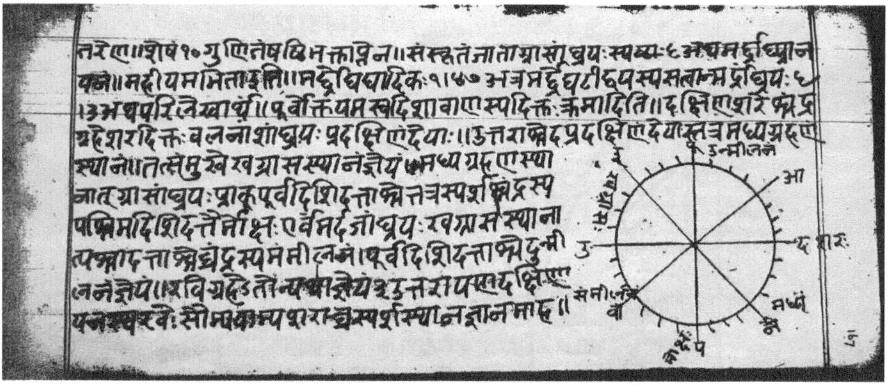

Fig. 6.6 An excerpt from a manuscript of Mallāri's *Parvadvayasādhana* depicting a lunar eclipse diagram (Oxford CSS I 90 d.800(7), f. 9r)

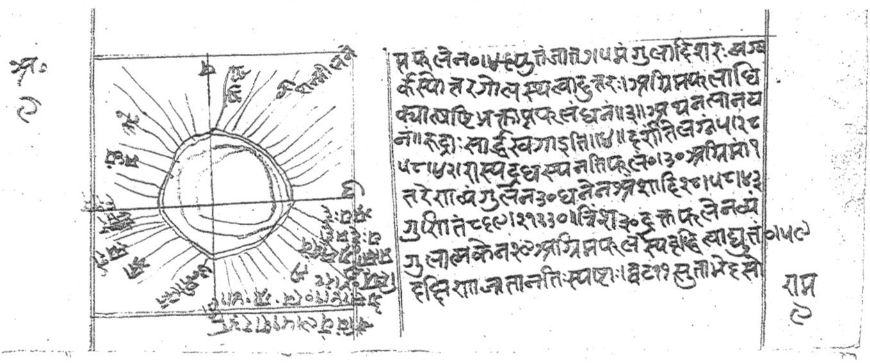

Fig. 6.7 An excerpt from a manuscript of Mallāri's *Parvadvayasādhana* depicting a lunar eclipse diagram (Benares 4804 f. 9v)

This commentary includes two worked numerical examples along with a diagram for each. The first example concerns a lunar eclipse (f. 5r f). The example opens with the precise date and time of the eclipse. Invoking traditional Indian chronology, this is given as "Śaka 1666, Aśvinī, the bright-(fortnight), full moon … 'Mercury'-day." This date corresponds to 1744 CE, 21 October, which was a Wednesday. According to modern records, a lunar eclipse indeed did occur on this day.[4]

After a lengthy passage of numerical computations, the following diagram (see also Fig. 6.4) is displayed:

[4] See https://eclipse.gsfc.nasa.gov/5MCLEmap/1701-1800/LE1744-10-21P.gif

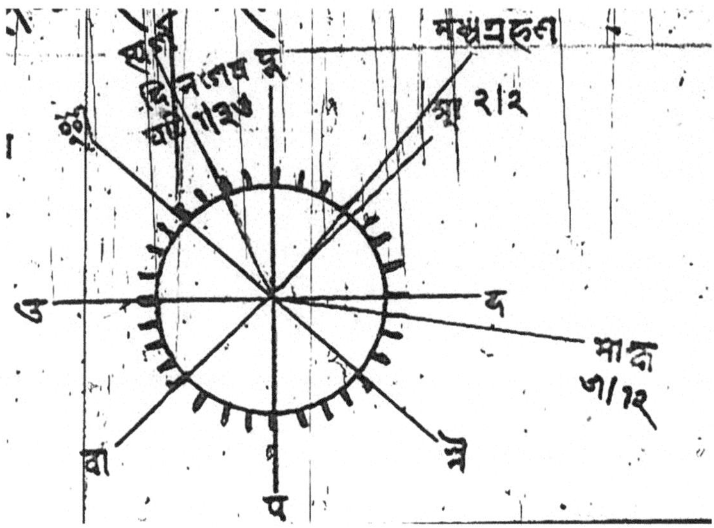

This is a carefully drawn diagram. A circle has evidently been drawn with a compass. The cardinal directions are identified on a ruled set of perpendicular lines and labeled with their first-letter abbreviations, as have been the intercardinal directions. As is conventional in Sanskrit tradition, the North-South direction is a horizontal line, and North appears on the left.

Three main phases have been indicated on the diagram along with a numerical value, precise to two significant places: *sparśa* "first contact" 1|37, *madhyagrahaṇa* "mideclipse" 2|2, and *mokṣa* "release" 5|12. Each of these phases has been marked off with its appropriate angle of inclination: first contact, six notches east of the north point; mideclipse, three notches south of the east point, and release, one notch west of the south point. No further information has been graphically rendered.

The second example concerns a solar eclipse (f. 7v and following). This example also opens by giving the precise date and time of the eclipse. This is stated as "Śaka 1667, Caitra, dark-fortnight, new-moon, … 'Moon'-day." This apparently corresponds to 1745 CE, April 17th which was a Saturday. The closest solar eclipse is 14 days before on April 2nd which was a Friday."[5]

Again, after detailed numerical computations, the following diagram appears:

[5] See https://eclipse.gsfc.nasa.gov/SEsearch/SEsearchmap.php?Ecl=17450402

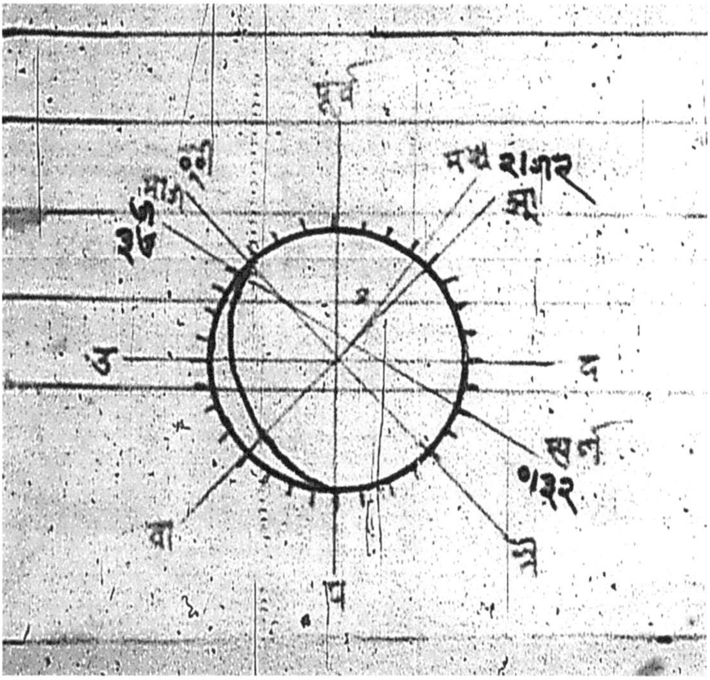

This too is a carefully drawn diagram. Again, the circle has evidently been constructed with a compass. The cardinal directions are ruled and labeled with their first letter abbreviations, as are the intercardinal directions, with the north point situated at the beginning of the fourth quadrant. Three main phases have been indicated: *sparśa* "first contact" 0|32, *madhya* "mideclipse" 2|59, and *mokṣa* "release" 5|36. Each of these phases has been marked off with reference to the appropriate angle of inclination: first contact, two notches west of the south point, mideclipse, five notches east of the south point, and release, three notches north of the east point.

Unlike the previous lunar eclipse diagram, the drawn lines associated with these angles of inclination do not go through the center. Rather a line is ruled through the points of first contact and release, and the mideclipse line has been extended through to the center of the circle, but not beyond. The significance of this is unclear. There is also an additional arc drawn on the interior lefthand side of the circle, creating somewhat of a crescent shape. The purpose (whether intentional or not) of this additional graphical feature is unclear. Given the thickness of the line, it seems to have been a deliberate act. Perhaps related is that fact that modern records indicate that this was an annular eclipse. It may be this arc was intended to indicate this. Overall there is a marked consistency of execution in both of these graphical representations by the scribe.

6.3.1.2 A Pair of Eclipse Diagrams Executed by Two Different Scribes Pertaining to the Same Datable Lunar Eclipse

Having examined the consistency in graphical representation across two examples by the same scribe, we now turn to the same example rendered by two separate scribes. This is a particularly significant example as it derives from a commentary which is both datable and the author is identified, as well as the scribe! The commentary has been composed by Viśvanātha, Mallāri's brother. Accordingly, this commentary may give extra special insight into Mallāri's original text, given the familial connections.

The worked example concerns a lunar eclipse which is dated precisely as: Saṃvat 1818, Śaka 1683, vaiśākha(-month), bright fortnight, 15(th *tithi*). This corresponds to Tuesday, 19 May 1761. An excerpt from the commentary proceeds as follows:

> In *saṃvat* 1818, *śaka* 1683, Vaiśākha, bright fortnight, 15th (lunar day). ... Now, in the case of an eclipse, the computation of the time of first contact and so on is at the end of the *tithi*—thus, it is mideclipse. Now, in a lunar eclipse, the corrected instant of full moon is 54|39 (*ghaṭīs*). The time of mideclipse is 54|39. The length of the day is 33|21. From sunset, the time of mideclipse is 21|17. The half-duration from first contact is 4|24. Subtraction is to be done. From sunset, the time of first contact is 16|53. Mideclipse is 21|17. The first half-duration of totality is 1|43. Subtraction is the be done. The time of the beginning of totality is 19|34. The time of mideclipse is 21|17. The second half-duration of totality is 1|51. Addition is to be done. 23|8 is the time of the end of totality. Mideclipse is 21|17. The second half-duration is 4|32. Addition is to be done. 25|49 is the time of release. ... Thus the lunar eclipse.

The example opens with the date and time of contemporary syzygy and uses this, along with the intervals of half-duration, to establish the key instants of eclipse. The text alludes to Mallāri's base text, referring to it by retrieving data from tables, but the process by which data is extracted from the tables has not been made explicit. There is no direct mention in the text as to how this data is to be represented on a graphical diagram.

MS Oxford The first graphical representation of this particular eclipse is included in MS Oxford CSS I 90 d. 800(7).[6] The colophon reveals this was copied by a scribe, named as Rāmajī Vyāsa. The copy is dated to Thursday 15th lunar day, bright fortnight of Mārgaśīrṣa month in Saṃvat 1821, which is equivalent to 12 December 1799.

[6] See Pingree 1970–1994, CESS IV: 365.

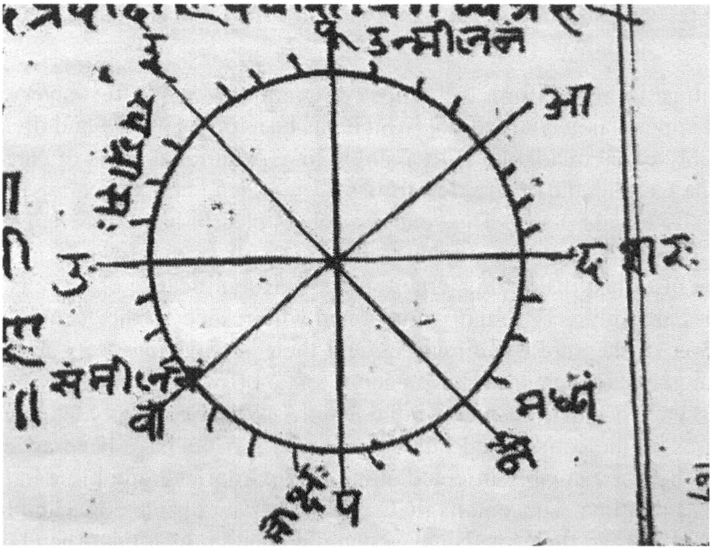

The diagram in this Oxford manuscript has been carefully executed. It is likely a compass was used to the construct the main circle. The cardinal directions have been marked and labeled (using their first letter abbreviations) and have been joined up with ruled perpendicular straight-line segments, as have the intercardinal directions. The north point is oriented at the beginning of the fourth quadrant. Each quadrant has been divided into eight using small marks which appears to have been done by sight (rather than measured and marked of). Next the phases of the eclipses have been identified. This particular lunar eclipse is a total eclipse, and thus has a period of totality. Accordingly the five distinct phases of the eclipse are noted in full (except for the first one). These are: *sammīlana* beginning of totality, *madhya* mideclipse, *unmīlana* end of totality, *mokṣa* release. It appears the omission of *sparśa* or first contact is a mistake.

In addition, two technical aspects of an eclipse have been noted. The first is *śara* or lunar latitude. The second is *kha-grāsa* or "sky obscuration." This is a technical term related to excess of the obscuration over the diameter of the eclipsed disk during totality. Notably, none of these eclipses phases have been quantified by an accompanying numerical value. The only indication is where and how the technical words have been positioned. The first or last letters of the words appear to have been aligned with a notch; perhaps this is meant to indicate a rough quantitative amount? Possibly, the scribe deemed it unnecessary to duplicate this data on the diagram as it appears in the prose text.

MS Benares The second graphical rendering of this particular lunar eclipse has also been included in MS Benares (1963) 35485 = Benares (1904) 1387. The colophon reveals this manuscript was copied in 1764 CE.[7] The name and details of the scribe are unknown.

In contrast to MS Oxford, the eclipse diagram in MS Benares is very roughly drawn. It appears in a boarded box which has been double ruled and divided into four roughly equal quadrants with a single line. Additional signs of carelessness include the fact that the diagram itself has been drawn in the opposite sense to the text, literally upside down! The central circle is clearly hand drawn. The largest circle has been drawn with a double line. Two smaller roughly concentric circles have been inscribed inside this larger circle. The significance of these is unclear. The circles are only very roughly positioned with respect to the "center," making the division of the cardinal directions and their subdivisions very uneven. The circular arcs in each quadrant have been marked off with an arbitrary number of roughly sketched long lines of variable lengths. Words corresponding to the distinct phases of the eclipse and other significant aspects have been added to the diagram. These are notably oriented along the appropriate radial line in the same direction as that line. This entails that some words are upside down and harder to read. Extra data and their labels have been included in the top left hand corner. It is unclear whether the placement of this text or labels have quantitative significance, or whether this is, in essence, a "caption" to the diagram with key aspects, along with their data listed.

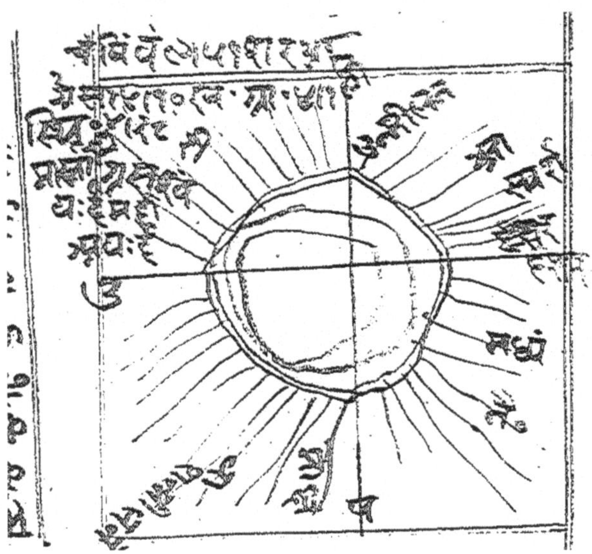

This text includes the follow labels and data: The meaning and significance of additional text tentatively:

[7] Ibid.

Lunar disk (*bimba*)	9 | 51
Latitude (*śara*)	2? | 8
Obscuration (*grāsa*)	14 | 10
Sky Obscuration (*kha · grā ·*)	4 | 1
Half-duration (*sthi ·*)	4 | 28
Mideclipse (*madhyagrasa*	2(?)

read as *yaḥ marda* and *ayaḥ rda* right below is unclear.

The label "end of totality" is positioned along the first mark south of the east point; "first contact" along the fifth mark south of the east point; "mideclipse" along the third mark west of the south point; "release" one mark north of the west point, and "beginning of totality" five marks north of the west point.

6.4 Reflections on the Features of Diagrammatic Execution

Overall, there is a fair bit of consistency across scribal practice relating to diagrammatic exposition of the *Parvadvayasādhana* when it comes to the eclipse *parilekha*. Features that were common to all the examples above include:

cardinal directions (lettered abbreviations)
intercardinal directions (letter abbreviations)
quadrants divided into eight
mideclipse (*madhya*)
impact (*sparśa*) and release (*mokṣa*)
beginning of totality (*sammīlana*) and end of totality (*unmīlana*)

Additional details, such as latitude and magnitude appeared to be relevant to some scribes and less important (and thus not included) by others. Cardinal directions were always invoked and abbreviated by their initial letters:

North	Uttarā, Udīcī
East	Pūrvā, Prācī
West	Paścimā, Pratīcī
South	Dakṣiṇā, avācī
North-East	Aiśānī
South-East	Āgneyī
North-West	Vāyavyā
South-West	Nairṛtā

All graphical representations include a central circle divided into octants with the cardinal and intercardinal directions ruled as perpendicular line segments. Each quadrant is divided into eight sections by small notch marks. Features are then noted against this backdrop; data is either registered explicitly along with the technical terms, or simply placed in the approximate location.

While the texts might have invoked practitioners to employ tools, such as a compass and ruler, to enhance the precision of the diagram, apparently this was not always followed. In some cases, a rough sketch obviously good enough. No other circle disks, other than concentric ones are drawn on the diagram. None show any attempt to include the eclipsed body at distinct phases of the eclipse nor include any shading to indicate the shadowing effect of the eclipser.

What is immediately clear when comparing textual instructions of procedures to execute diagrams is that at least in the case of these examples, the diagrams are much simpler and basic than those that might be produced if the set of textual instructions that appear in other works were followed exactly. The textual sources suggest a highly intricate diagram with multiple phases represented and distinct aspects of the eclipse rendered. This is far from the outputs of what we have seen in the above evidence.

In this particular case, in both the original text of Mallāri and of the various commentaries, actual descriptions pertaining to graphical procedures are not spelled out. There are no explicit instructions on how to render these in the text, yet there is consistency across textual sources of diagrammatic practice. This is significant and shows there was obviously a larger well-established accepted and known tradition. The fact that these *parilekha* were largely consistent across scribal practice separated by at least a century reveals a stability of tradition when it comes to eclipse diagrams.

These sorts of diagrams also embody subtleties when it comes to the representation of space and time, two essential features critical to the depiction of an astronomical event like an eclipse. At a fundamental level, quantification appears to have been achieved in two distinct ways: explicit (i.e., time of impact, by including a numerical value in the diagram), or on a scale (i.e., inclination, by a mark corresponding to the appropriate notch). Similarly, space, and the indication of representing a dynamic physical process has been achieved by means of invoking the cardinal directions on the diagram as well as locating key phases of the eclipse in distinct positions on a single backdrop. The more modern approaches to depicting movement in a static diagram, but drawing separate disks in distinct places, such as those done by NASA, although alluded to in instructional textual passages, don't appear in the known corpus of Sanskrit *parilekha* diagrams.

This lack of visual detail also raises further questions as to the purpose and scope of these diagrams. Without the carefully derived parameters and numerical data, diagrams didn't serve as vehicles to be used to answer detailed questions about timing and exact features or a precise astronomical representation. Rather they seemed to have fallen into a more public facing artifact to account for some of the specific features or to be a token image with just enough specificity for a particular event to single them out for a more generalized eclipse diagram.

More broadly eclipse diagrams (like lunar crescent diagrams) are a very special sort of diagram. They are not a typical generalized abstracted figure that we usually associate with mathematical diagrams intended to emblematize a mathematical relationship or stand alongside a demonstration of a technical fact or relation. Rather they are the instantiation of an actual physical occurrence of a natural phenomenon, and furthermore, a precise timebound instance of a particular event with its own specificities. In contrast to the timeless instantiations of the Euclidean diagram, these visual artifacts are in essence, a *unicum*, a unique example or a one-off diagram. They stand in direct contrast to other sets of procedures and graphical examples in the Sanskrit astral science tradition, such as the trigonometric sine diagram. These eclipse *parilekha* are a visual imprint of a specific local astronomical configuration, for validating against reality, or relying on it if direct observation is not possible. Indeed, as Mallāri's text indicated, all the patron king was interested in was a direct answer to his question in which direction will the eclipse occur? The diagrams we see in the text are perfectly able to answer that question in a visually appealing way. Indeed, this simplicity appears somewhat at odds with Brahmagupta's appeal to these outputs representing complete mastery and expertise in the astral sciences. More research and sources will help us enrich our understanding of the profession and social context of Sanskrit science.

References

Benares. 1963. 35485 = Benares (1904) 1387 Copied in Sam 1821 = AD 1764. See Pingree 1970–1994.

Bhāskara. *Karaṇakesarī*. See Montelle and Plofker 2013.

Mallāri. 1588. *Parvadvayasādhana* [Computation of the Two Syzygies].

Minkowski, Christopher. 2006. Why Should We Read the Maṅgala Verses. In *Śāstrārambha: Inquiries into the Preamble in Sanskrit*, ed. Walter Slaje, 1–16. Wiesbaden: Otto Harrassowitz.

Montelle, Clemency, and Kim Plofker. 2013. *Karaṇakesarī* of Bhāskara: A 17th-Century Table Text for Computing Eclipses. *History of Science in South Asia* 1:1–63.

NASA. Eclipse Web Site. https://eclipse.gsfc.nasa.gov/SEsearch/SEsearchmap.php?Ecl=17450402

NASA. Five Millennium Canon of Lunar Eclipses. https://eclipse.gsfc.nasa.gov/5MCLEmap/1701-1800/LE1744-10-21P.gif

Pingree, David. 1970–1994. *Census of the Exact Sciences in Sanskrit*. 5 vols. Series A. Philadelphia: American Philosophical Society.

Plofker, Kim. 2021. Mathematics and Presentation of Eclipse Diagrams in Sanskrit Astronomy. In *Body and Cosmos Studies in Early Indian Medical and Astral Sciences in Honor of Kenneth G. Zysk*, ed. Toke Lindegaard Knudsen, Jacob Schmidt-Madsen, and Sara Speyer, 304–314. Leiden: Brill.

Rana, Chandra Shumsher Jang Bahadur. MS. Chandra Shum Shere d. 751 & e.149. Bodleian Library. See Pingree 1970–1994.

Subbarayappa, B. V., and K. V. Sarma. 1985. *Indian Astronomy – A Sourcebook*. Bombay: Nehru Center.

Part IV
Europe

Chapter 7
Levi ben Gerson's Models for the Latitudes of Saturn and Jupiter

Robert Morrison

7.1 Introduction

Premodern astronomers often analyzed the longitudinal motion of the planets through the ecliptic, the path traced by the Sun around the Earth, in isolation from the motion of the planets in latitude to the north and south of the ecliptic. Although the models, i.e., configurations of orbs, for latitudes could be integrated into those for longitudes, astronomers presented them separately. Historians of astronomy in Islamic societies and in Jewish cultures, whether they study theoretical models or tables, have researched planetary longitudes much more than planetary latitudes. This imbalance may be because planetary latitudes did not have applications whereas longitudes had many.

There are some exceptions to that trend. One is an article by A. I. Sabra in which he edited Ibn al-Haytham's (d. ca. 1040)[1] attempt to improve on Ptolemy's (ca. 100–70) latitude models from the *Almagest* (Sabra 1979). Another exception is George Saliba's study in which he found that Naṣīr al-Dīn al-Ṭūsī (d. 1274) proposed a rudimentary version of his innovative Ṭūsī-couple, in his 1247 recension of the *Almagest*, to account for the planets' motions in latitude (Saliba 1987).[2] Ṭūsī's application of the Ṭūsī-couple to latitudes came only a year after the first incorporation of the Ṭūsī-couple into models for the planets' motion in longitude (Ragep

[1] All dates are CE.

[2] On the Ṭūsī-couple, see, e.g., Ragep 1993, 194–221.

R. Morrison (✉)
Department of Religion and Program in Middle Eastern and North African Studies, Bowdoin College, Brunswick, ME, USA
e-mail: rgmorris@bowdoin.edu

F. Rochberg (ed.), *Studies in Premodern Sciences in Memory of Noel M. Swerdlow*, Archimedes 74, https://doi.org/10.1007/978-3-032-04040-4_7

2020). Still, Saliba ranked Islamic models for planetary latitudes beneath innovations in theories for planetary longitudes (Saliba 1996, 125).

Subsequently, Noel Swerdlow studied the models for planetary latitudes in Ptolemy's *Almagest*, *Handy Tables*, and *Planetary Hypotheses* and found that the models in the *Planetary Hypotheses* were Ptolemy's best (Swerdlow 2005). In making that assessment, Swerdlow focused on the consonance of the models with observations, not on the details of the physical movers theorized by Ptolemy. Drawing on the available secondary literature, Swerdlow added that the latitude theory of the *Planetary Hypotheses*, despite its relative superiority, was without influence (Swerdlow 2005, 67).[3] That was a fair conclusion given what Swerdlow could have known, as he did not read Arabic or Hebrew. Andrea Murschel was the first to explore the physical models for planetary latitudes in the *Planetary Hypotheses* only a decade earlier (Murschel 1995, 44–50). Subsequently, Sajjad Nikfahm-Khubravan, using Arabic manuscripts that Swerdlow could not read, has shown in irrefutable detail how the models for the planets' motion in latitudes found in the *Planetary Hypotheses* were highly influential for scholars of astronomy in premodern Islamic societies (Nikfahm-Khubravan 2022).

This chapter contains a preliminary survey of the theories of Levi ben Gerson (1288–1344), aka Gersonides, for the motions of Saturn and Jupiter in latitude. Levi was one of the most significant premodern Jewish scholars and has been judged to be the most creative premodern Jewish astronomer (Goldstein 1992). He invented an instrument known as the Jacob Staff with which the user measured the altitude of an object relative to the observer and the angular diameter of a celestial body. Although Levi drew on the latitude theory of the *Almagest*, he also incorporated elements of the latitude models found in the *Planetary Hypotheses* into his own models. Medieval Jewish astronomers like Levi who did not read Arabic learned of and built on the Islamic heritage through translations from Arabic into Hebrew. The *Planetary Hypotheses* was translated into Hebrew in 1317, during Levi's lifetime. Bernard Goldstein has presented convincing evidence that Levi had access to a Hebrew translation of the Arabic version of the *Planetary Hypotheses* and that Levi was the only Jewish astronomer to quote extensively from Book Two of the *Planetary Hypotheses* (Goldstein 2002, 18). Hence, Levi's work is part of the history of the reception of the latitude models found in the *Planetary Hypotheses*.

7.2 Earlier Scholarship on Gersonides's *Astronomy*

Levi's magnum opus, the six-book *Wars of the Lord*, was translated into English by Seymour Feldman in the 1980s and 1990s (Ben Gershom 1984–1999). Part one of Book Five of *Wars*, which consists of 136 chapters, contains Levi's *Astronomy*. This

[3] Swerdlow commented, "The latitude theory, for all its ingenuity, seems to have been without influence, for it appears that no one made use of Ptolemy's final, corrected latitude theory."

section of *Wars* circulated independently in manuscript, though many manuscripts are incomplete. For that reason, Feldman chose not to translate it. In the rest of this chapter, I will call this text the *Astronomy*. Goldstein published the tables from the *Astronomy* with a commentary (Goldstein 1974b) and then edited, translated and commented upon the first 20 chapters of the *Astronomy* (Goldstein 1985).

Despite these excellent books by Goldstein, and numerous informative articles by him and others (e.g., Mancha and Freudenthal 2005) on Gersonides's astronomy, there is work left to do on his latitude models. In chapter 122 of *Wars*, Gersonides listed some of his own observations of planetary latitudes, from the 1330s, though there is no evidence that any of these observations motivated his models for planetary latitudes (ben Gerson 1397, 235b; Goldstein 2002, 22). These models for planetary latitudes, which are in chapters 123–25 of the *Astronomy*, have not been fully studied. In chapter 123, Levi investigated aspects of the physics of the orbs' motions. He determined that the poles of orbs were metaphorical; that is, there were not physical protuberances from the orbs (Goldstein 2002, 24). He was also concerned with the question of whether motions were transmitted from the encompassing orb to the encompassed orb, or from the encompassed orb to the encompassing orb. Elsewhere, in chapter 29 of the *Astronomy*, Levi concluded "that the surrounding orb is moved by the orb it encloses."[4] Chapters 124 and 125 contain the details of Levi's models for planetary latitudes. In chapter 126, Levi discussed the discrepancies between observations of the planets' motions in latitude and the latitudes predicted by Ptolemy's tables. In chapter 127, Gersonides promised tables for planetary latitudes, though these tables are not found in surviving manuscripts of the *Astronomy*. Goldstein (2002) has published an edition, translation, and study of chapters 123 and 126, and of the prose portion of chapter 127. In this publication, I focus on chapters 124 and 125.

There are several surviving Hebrew manuscripts of Levi's *Astronomy*, but the chapters with Levi's latitude models are found only in two. One, Paris BnF hébreu 724, is the oldest Hebrew witness; the copy was executed in 1397. According to the colophon, the original text was completed in 1328, but Goldstein has found observations in the text dated as late as 1339 (Goldstein 1974b, 74–75). Paris BnF hébreu 724 was eventually purchased by Caleb Afendopolo (d. 1499), a Jewish scholar in Istanbul, from Rabbi David b. Judah Messer Leon (d. 1526). I have also used Paris BnF hébreu 725, a 1510 copy produced in Istanbul (Goldstein 1974b, 75–76). Goldstein has determined that these two copies were not from the same branch of the manuscript tree. I have not consulted the Latin manuscripts of Levi's *Astronomy* to write this chapter, as Goldstein has found that most of the material missing from the available Hebrew manuscripts is also omitted in them (Goldstein 1974b, 79). In my draft edition and translation, from which I quote in this chapter, Paris BnF hébreu 724 is the base manuscript.

---

[4] Goldstein 2012, 20. Goldstein concluded that Levi found support for this interpretation in the *Planetary Hypotheses*.

Goldstein commented (2002, 16) that Gersonides's models for planetary latitudes were "difficult to interpret," and I agree. Yet even the ambiguities cannot obscure how his models for planetary latitudes reflect the same creativity and willingness to criticize Ptolemy that scholars have noted in other areas of his astronomy. According to Goldstein (1974b, 28), gaps in the manuscript record "make it quite difficult to reconstruct Levi's planetary theory," specifically his models for planetary longitudes. Because phenomena of planetary latitudes were correlated with phenomena of planetary longitudes, greater understanding of Gersonides's models for latitudes may lend insight into his models for planetary longitudes. There are certainly theoretical parallels between Levi's models for longitude and latitude. For example, in chapter 43 of the *Astronomy*, Levi determined that epicycles did not exist (Mancha and Freudenthal 2005). One reason Levi excluded epicycles was because he presumed, as Aristotle had, that planets were fixed in their epicycles. Levi reasoned that since we see always the same side of the Moon, the Moon cannot be moving on an epicycle. He proposed a lunar model that did not include an epicycle (Goldstein 1972, 1974a, b, 53–74). By extension, the models for planetary longitudes also had to exclude epicycles and, indeed, Goldstein has found no trace of epicycles in what has survived of Levi's discussions of planetary theory. We will see that epicycles are absent from Gersonides's models for planetary latitudes as well.

7.3 Ptolemy's Models for Planetary Latitudes

Because Levi took Ptolemy's latitude models as a point of departure, a brief review of Ptolemy's latitude models will help (Fig. 7.1).

The figure shows a cutaway view of the parecliptic (which stands in for the ecliptic in a model for a planet), eccentric, and epicyclic orbs. The north poles and planes of the revolution of each orb are shown as well. There are two Ptolemaic parameters that are relevant to Gersonides's models for the latitudes of Saturn and Jupiter. The first is the inclination of the plane of the motion of the planet's eccentric orb (which produced the planet's mean motion) to the plane of the motion of the parecliptic. The other parameter is the inclination of the equator of the epicycle, which produced the anomaly (variations from the mean motion), to the plane of the motion of the eccentric. These parameters varied between the models of Saturn and Jupiter. In the figure, the epicycle is at the apogee of the eccentric and the planet is at the perigee of the epicycle, meaning that the planet is at its maximum northern latitude. The first diameter of the epicycle, which is not shown in the figure, runs from the apogee of the epicycle to the perigee.

To explain observed variations in the inclinations of the first diameter of the epicycle, Ptolemy proposed in the *Almagest* that the endpoints of the first diameter of the epicycle of the upper planets revolve on small circles, the radius of which was the maximum deviation of the epicyclic perigee from a plane parallel to the plane of

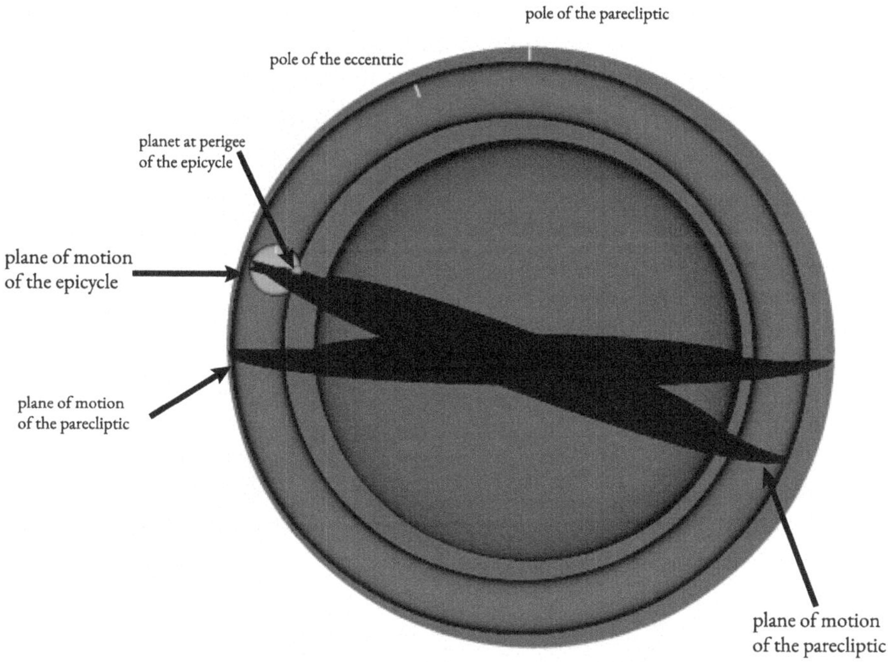

Fig. 7.1 A physicalized reconstruction of Ptolemy's model for an upper planet

the parecliptic (Toomer 1984, 599).[5] The period of revolution on the circle was the period of the motion in longitude. Ptolemy did not explain how these circles were moved and did not investigate the attendant displacement in longitude occasioned by the motion of the endpoints of the epicycles' diameters. These small circles are not shown in the figure. Ptolemy acknowledged shortcomings in his theory (Toomer 1984, 600–1) and proposed an improved latitude theory in the *Planetary Hypotheses*.[6]

In the *Planetary Hypotheses*, the equator of the epicycle was always parallel to the plane of the ecliptic. In order to maintain this orientation throughout the epicycle's motion on the eccentric, Ptolemy theorized that the endpoints of the first diameter of the epicycle were moved by concentric epicycles with inclined poles (Fig. 7.2).[7]

Imagine that each of the two orbs (shown in a cutaway view) revolved with uniform angular velocity α, equal to the velocity of the epicycle center on the eccentric,

[5] Here, in *Almagest* XIII.2, Ptolemy specified that the circle was attached to the endpoint of the first diameter that is closer to the earth.

[6] See Saliba 1987 on how Ṭūsī used the Ṭūsī-couple to replace these circles.

[7] See Nikfahm-Khubravan (2022), 112 for an explanation of why the inclined epicycles are necessary to maintain the inclination of the epicycle and 104–12 for a detailed presentation of Ptolemy's models for the upper planets in the *Planetary Hypotheses*.

Fig. 7.2 Concentric epicycles revolving in different directions

in opposite directions. The outer orb revolved opposite the sequence of zodiacal signs. The planes of the motion of each orb are shown as disks that protrude from the cutaway orbs. The endpoints of the first diameter of the inner orb traced a hippopede, a narrow figure eight that is close to an oscillation on a great circle arc, from A to B to C, and back. This hypothesis of concentric orbs with inclined poles to produce a hippopede is sometimes known as a Eudoxan Couple after the reconstruction of the astronomy of Exodus (d. 355 BCE) by Schiaparelli.[8]

In the model for the motion of the upper planets in the *Planetary Hypotheses*, the inner orb was the original epicycle, and it revolved with velocity $\alpha + \beta$, which was the sum of the motion on the eccentric (α) and the anomaly, i.e., the planet's motion on the epicycle (β) (Fig. 7.3).

[8] I found the term "Eudoxan Couple" in Ragep 1993, 2:451. Ragep included citations to the relevant secondary literature.

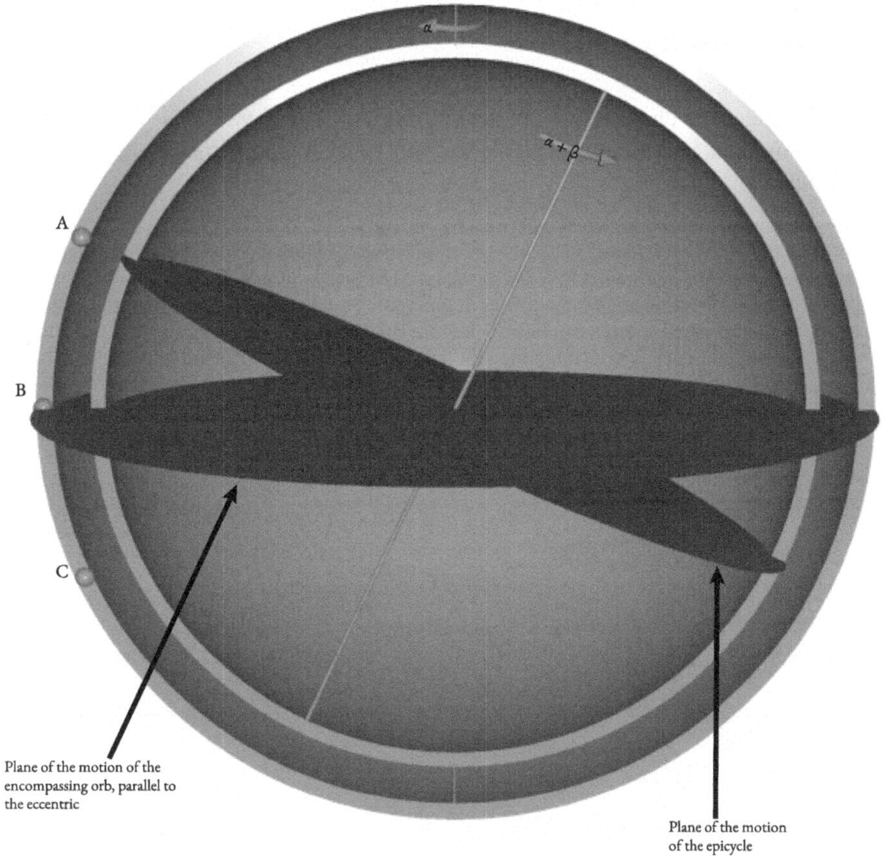

Fig. 7.3 The model for planetary latitudes in Ptolemy's *Planetary Hypotheses*

Though the orbs did not revolve with identical velocities, the apex of the epicycle still traced a hippopede (Nikfahm-Khubravan 2022, 114). Not only did the model for latitude in the *Planetary Hypotheses* not suffer from physical inconsistencies, the mathematical precision was greater compared to that of the model found in the *Almagest* (Swerdlow 2005, 65).[9] Levi's models for the latitudes of Saturn and Jupiter were informed by the models for planetary latitudes found in both the *Almagest* and the *Planetary Hypotheses*.

[9] Swerdlow commented, "The *Planetary Hypotheses* may be considered Ptolemy's last word on latitude theory, in which he finally got nearly everything right."

7.4 The Foundations of Levi's Models for Planetary Latitudes

Levi introduced in chapter 124 of the *Astronomy* two hypotheses, i.e., building blocks of models, in which he explored the motion of a planet in latitude (Fig. 7.4). He wrote:

> We say that it is clear that when two orbs are moving about different poles, with what is between the poles being equal, and there is in the upper of them a planet, then the poles of the upper orb move in a small circle about the poles of the lower orb. The inclination of the planes of these orbs that are perpendicular to their poles is equal to the measure of the inclination of their poles, one from the other. When the planet is in the circle passing through the poles of these circles, it is at its maximum inclination either to the north or to the south. This is according to what I say, for if the inclination of the pole of the upper orb to the lower is toward the north, then the inclination of the planet is northern by the same measure, for it is 90 degrees from the pole. And the opposite is if the inclination is southern. When the planet

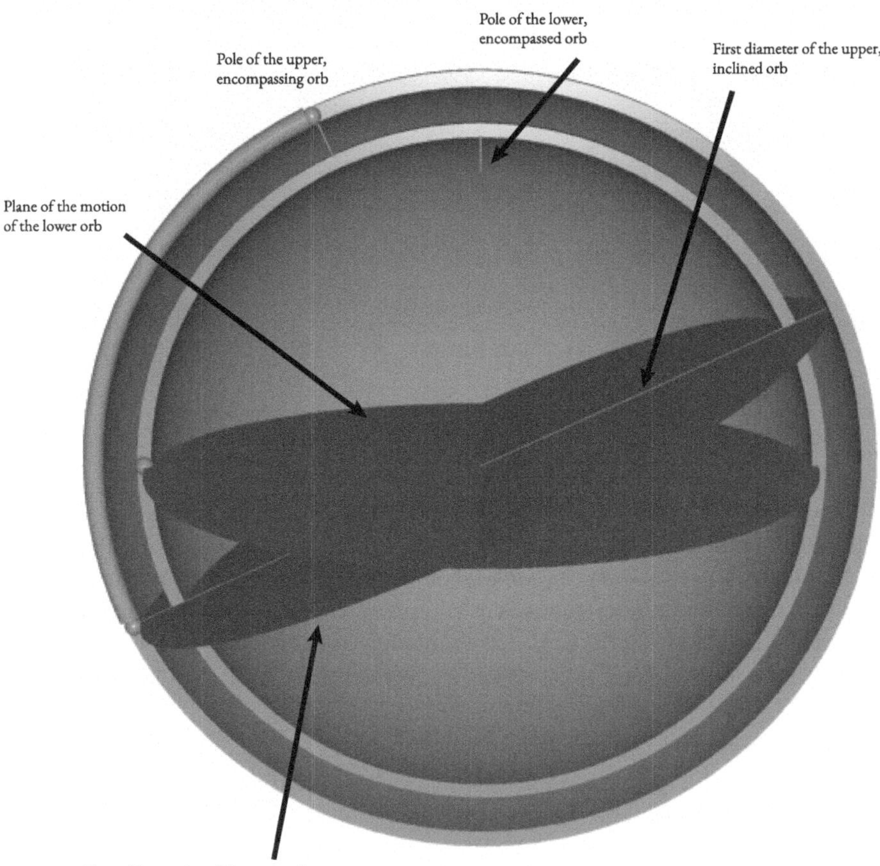

Fig. 7.4 Levi b. Gerson's first hypothesis for planetary latitudes

> is 90 degrees distant from this place, it has no inclination at all because at that time it is at the common intersection of the two planes. Due to this, it is at that time in the plane of the lower orb. (Ben Gerson 1397, 237a–b)

The figure provides a cutaway view of two concentric orbs, with the pole of the encompassing orb inclined to the pole of the encompassed orb. Levi stipulated that the pole of the encompassing orb moves about the pole of the encompassed orb. The planes of each orb are shown as well. In this figure, when the planet (B) is in the circle passing through the poles of both orbs, it is at its maximum inclination to the south of point A, which is in the reference plane, the plane of the lower orb. I have also drawn the 90° arc linking the pole of the upper, inclined orb with the planet (see Fig. 7.5).

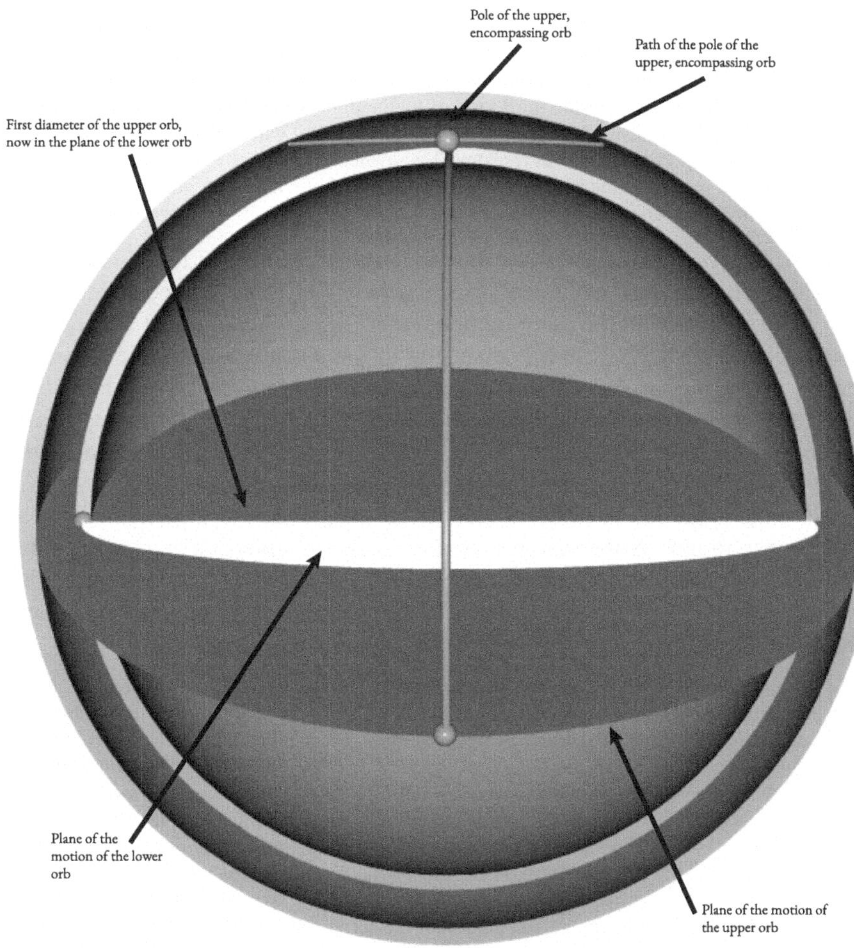

Fig. 7.5 Levi b. Gerson's first hypothesis for planetary latitudes after the pole of the upper orb has revolved 90°

Then, after the pole of the upper orb has revolved 90°, the maximum inclination of the outer orb to the inner was now right in front of the viewer. The intersection of the planes has shifted. The first diameter of the plane of the upper, inclined orb, which used to be inclined to the reference plane, is now in the reference plane. Hence, the movement of the upper, encompassing orb caused the oscillation of a diameter of that orb.

The presence of the hypothesis in the *Astronomy*, in light of Goldstein's findings of Levi's knowledge of the *Planetary Hypotheses*, is additional evidence of his familiarity with the latitude models from the *Planetary Hypotheses*. Levi's hypothesis resembled a statement Ptolemy made in Book One of the *Planetary Hypotheses*. Ptolemy wrote:

> The same thing occurs to the orbs that are outside the earth, like what occurs to the inclination of the epicycles. We do not need to change anything—if we want to move from the first case to the second case that follows it—other than replacing the equinoctial with the ecliptic, and replacing the orb on which moves the orb which is inclined to the equinoctial—i.e., the ecliptic—with the orb on which moves the ecliptic, and replacing the ecliptic orb itself with the inclined orb itself. In the third kind of inclination, which is outside the earth, the equinoctial becomes equivalent to the fixed epicycle, [the orb] which changes the declination with respect to the equinoctial [becomes] equivalent to [the orb] which changes the inclination with respect to the [fixed] epicycle, and the ecliptic itself [becomes] equivalent to the small inclined [epicycle] itself.[10]

As seen in Fig. 7.6, Ptolemy has compared concentric epicycles to the concentric orbs, with inclined poles, of the ecliptic and the equinoctial. The equinoctial orb revolves the ecliptic with the daily motion. Hence, every point on the ecliptic will culminate, i.e., transit the local meridian. But the declination – the angle above the equator – of each successive culminating point will decrease until points below the equator culminate. Thus, the declination of the culminating point oscillated on the meridian, just as the first diameter of the upper, encompassing orb, in first Levi's hypothesis, oscillated in and out of the plane of the lower orb.

The second hypothesis that Levi presented in chapter 124 of the *Astronomy* was a creative way to modify the inclination of the plane of a planet's motion in anomaly. In the *Almagest*, Ptolemy modified this inclination by theorizing small circles moving the endpoints of the first diameter of the epicycles of the upper planets. Levi proposed:

> I say further that if there is above the aforementioned upper orb another orb, the poles of which are inclined to this middle orb, then the poles of the third orb happen to move about the poles of the lower orb. They move as well about the poles of the middle orb. Due to this, the inclination of the plane of the motion of the third orb to the plane of the motion of the second orb happens to diminish and it proceeds until it arrives at a place where this inclination disappears, in the place where there is the extreme inclination for the third orb. After this, the inclination of the plane of the motion of the third orb to the plane of the motion of the second inverts, to be to the north, at the place where it was to the south, and also the opposite. This inclination does not cease increasing until it arrives at the place of the extreme inclination of the plane of the motion of the second orb to the plane of the motion

[10] As quoted in Nikfahm-Khubravan 2022, 109–10.

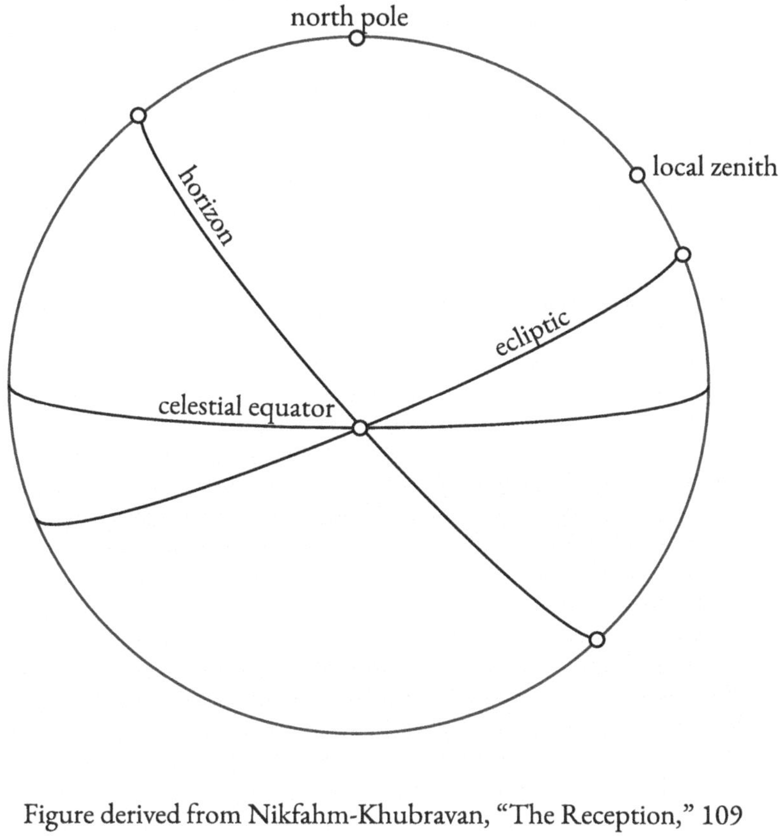

Fig. 7.6 The concentric orbs in Ptolemy's model for the motions of the fixed stars

> of the first orb. After it decreases in the previous way, this inclination inverts in the previous way, until it reaches the place where this inclination is absent in the place where the inclination is at its extreme for the third orb. (Ben Gerson 1397, 237b)

Figure 7.7 is redrawn from the manuscripts, and Fig. 7.8 is my rendition of that figure in three dimensions, with a cutaway view of the three concentric orbs. Not all of the letters from Fig. 7.7 are shown in Fig. 7.8. G is the north pole of the lower orb, the encompassed orb in the first hypothesis. D is the north pole of the encompassing orb in the first hypothesis, now the second orb. A is the north pole of the third, uppermost orb. Because the poles of the third orb were inclined to the middle orb in which the planet is found, the poles of the third orb were moved by the lower orb on the larger path, ALZ. As well, the poles of the third orb revolve about the poles of the second orb through the smaller path, AS. Hence, the inclination of the plane of the motion of the third orb is affected by the revolutions of the first *and* second orbs. Levi did not specify the inclination of the second orb to the first or the third orb to the second. In my three-dimensional rendition, the inclinations are not equal. T is a

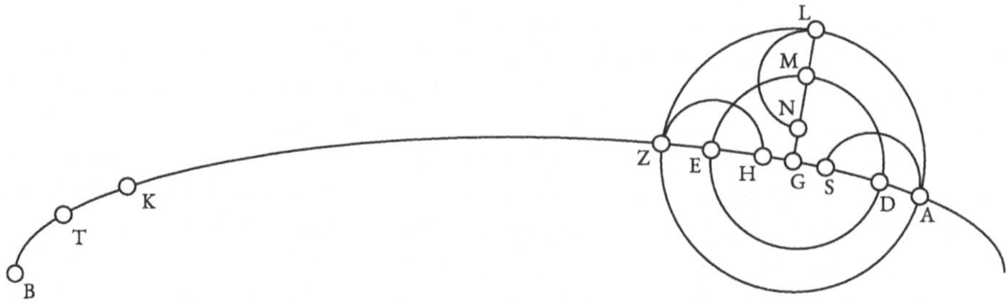

Fig. 7.7 Levi b. Gerson's Second hypothesis for planetary latitudes as shown in manuscript (BnF hébreu 724, 237b)

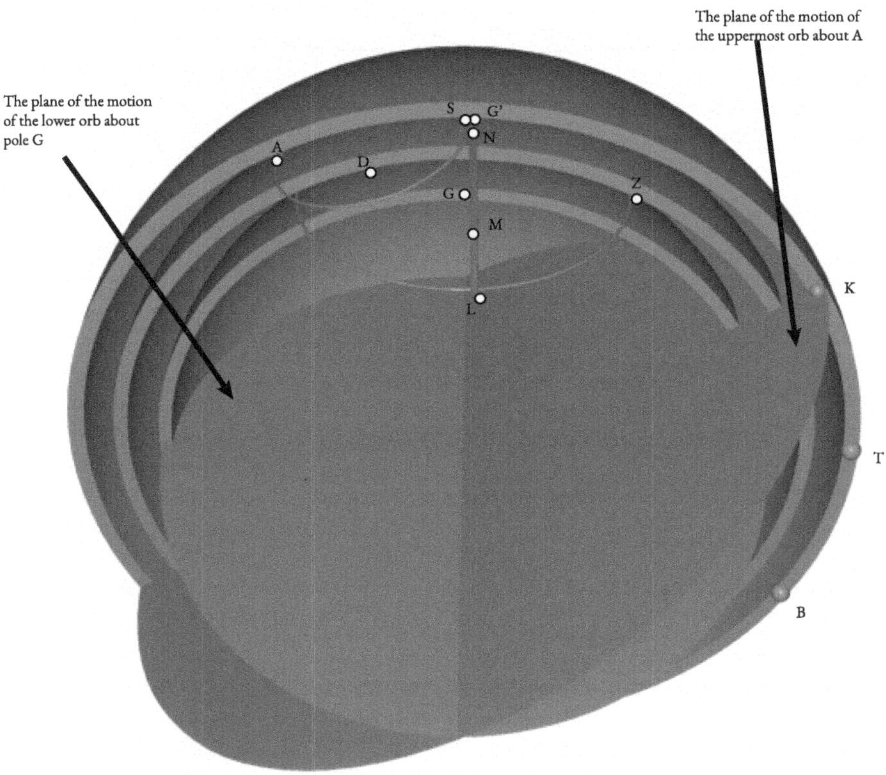

Fig. 7.8 A three-dimensional rendition of Levi b. Gerson's second hypothesis for planetary Latitudes

point 90° from G, where the plane of the equator of the first orb intersected the great circle arc, presuming the orbs had a negligible thickness, passing through poles A, D, and G. The plane of the motion of the third orb was inclined to the plane of the equator of the first orb, intersecting the first orb at K. B is the maximum southern inclination of the third orb. The plane of the motion of the second orb is not shown.

LMNG was part of a great circle arc, again presuming that the orbs had a negligible thickness, perpendicular to great circle arc BTK. To facilitate visualization, I have added point G', the projection of G onto the great circle arc. Hence, points L, M, N, and G' were all 90° from point T.

The most challenging sentence to interpret reads: "Due to this, the inclination of the plane of the motion of the third orb to the plane of the motion of the second orb happens to diminish and proceeds until it arrives at a place where this inclination disappears, in the place where there is the extreme inclination for the third orb." He has described the stage at which pole A has revolved 90° first to point L, bringing pole D to point M. Through the 180° revolution of the second orb about D (at M), pole A revolved to point N. How can the inclination of the third orb both disappear and be at its maximum? We learned from the first hypothesis that when the encompassing orb revolved 90° or 270° about the poles of the encompassed orb, the first diameters of the plane of each orb were aligned. After the combined motions of the second and third orbs, pole A has effectively revolved 270°, meaning that the inclination of the first diameter of the third orb to that of the second has disappeared. Because the poles of the first, second (at M) and third orbs (at N) were all 90° away from point T, in the reference plane, the surfaces of the second and third orbs were not inclined to each other in one respect. But the motion of pole A (of the third orb) about pole D (of the second orb), when considered in isolation from the revolution of pole D about pole G, was 180° (to S in Fig. 7.7), so in a different respect, the inclination of the third orb to the second was at its maximum.

Gersonides's statements were true only if the pole of the third orb revolved at twice the rate of the revolution of the pole of the second orb and even if the inclination of the pole of the third orb to the pole of the second was different from the inclination of the pole of the second orb to the pole of the first orb. He did not specify whether the revolutions of the second and third orbs were in the same or opposite directions.

Further on in chapter 124, Levi provided geometric demonstrations of the second hypothesis (see Fig. 7.9, a redrawing of the figure in the manuscripts).

Here Levi argued that "the inclination that is produced from the motion of pole A around pole D by 180° decreases similarly to the decrease of the first motion" (ben Gerson 1397, 238a). The overall inclination of pole A to great circle arc LMNG was produced by two components: the revolution of pole A about pole D and the revolution of pole A about pole G, through the revolution of pole D about pole G. By inspection, one can tell that if the motion of pole A around pole D was 180° and the revolution of pole A about pole G was 0° or 180°, then the impact of the revolution of pole A about pole D on the overall inclination of pole A to LMNG was at its maximum. By drawing a series of arcs, he demonstrated that the inclination produced by the 180° motion of pole A around pole D "is smaller whenever the motion approaches 90 degrees" (ben Gerson 1397, 238a). Arcs EO and HS represent the 180° revolution of pole A about pole D when pole A revolved to E and H respectively. Great circle arcs PG and ZG indicate the inclination of pole A from pole G when pole A revolved to E and H respectively. Because ZG is less than PG, Gersonides concluded that the inclination of pole A produced by the 180° revolution

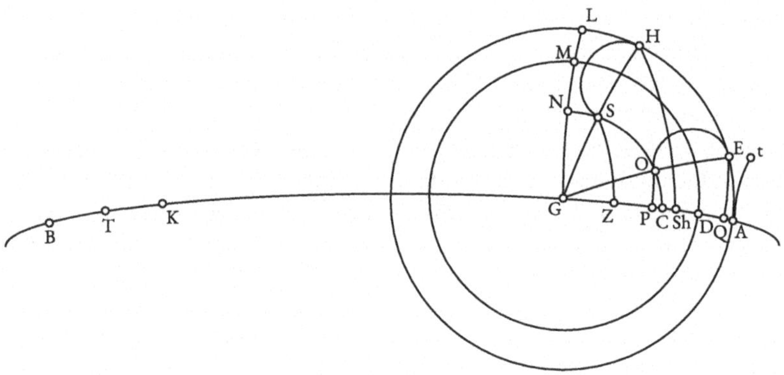

Fig. 7.9 A figure, found in manuscript, used in Levi b. Gerson's demonstration of the second hypothesis (BnF hébreu 724, 238b)

of A about D decreased the closer the revolution of A about G was to 90° (ben Gerson 1397, 238b–39a). When the revolution of pole A about pole G was 90°, the resulting inclination of pole A to LMNG was 0°.

By plotting points, I have found cases in which the combined motions of the orbs in this second hypothesis, if they were in the same direction, entailed that pole A be to the south of great circle arc LMNG before pole A revolved 90° about pole G. For instance, if the inclination of pole D to pole G equalled the inclination of pole A to pole D, pole A would be to the south of great circle arc LMNG when the revolution of A about G (facilitated by the revolution of D about G) was between 45° and 90°. If the inclination of pole D to pole G was twice the inclination of pole A to pole D, the combined motions of poles D and A, were they in the same direction, entailed that the new position of pole A be to the south of great circle arc LMNG after pole A revolved (facilitated by the revolution of D about G) approximately 60° around pole G. If the inclination of pole D to pole G was thrice the inclination of pole A to pole D, the combined motions of poles D and A, were they in the same direction, never entailed that the new position of pole A be to the south of great circle arc LMNG, when A revolved 90° about G with the revolution of pole D (Fig. 7.10).

He considered a different case, in which

> the two surfaces exist in two orbs, the first being inclined to the second by some measure. One surface approaches the other until it adheres to it and then it resumes its separation from it until they are found at their maximum possible inclination. Then they repeat this cycle like this. This is indeed possible when there is posited between them an orb that inclines to the lower orb by half the inclination. It moves the poles of the upper orb until they adhere uniformly to the poles of the lower orb, always in one direction, either to the north or to the south. (Ben Gerson 1397, 239b)

Here Levi explored how to explain cases in which an inclination should be always to the north or always to the south of a point. If the orbs revolved in opposite directions with the same angular velocity, if the lowest pole did not move, and if the inclination of the poles of the third orb to the poles of the second equalled the inclination of the poles of the second orb to the first, then the inclination produced by

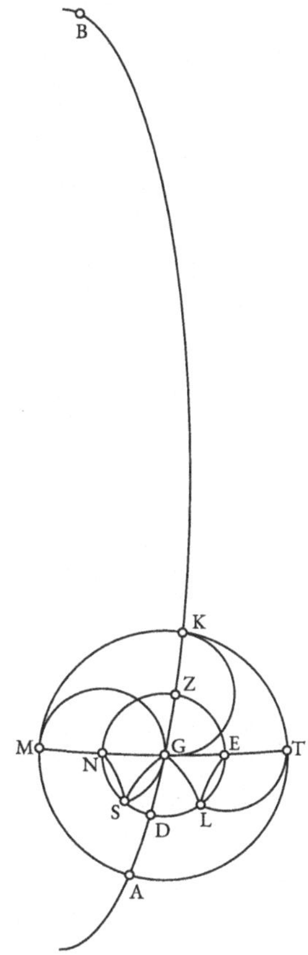

Fig. 7.10 A figure, found in manuscript, used in Levi b. Gerson's demonstration of the second hypothesis (BnF hébreu 724, 239b)

these orbs was always to a single side of the pole of the plane of motion of the first, lowest orb. One can imagine that A, the pole of the upper orb, revolved 180° about G, the pole of the lower orb. With that revolution, D, the pole of the middle orb, was carried to where is it aligned with Z, and A was carried to where it aligned with K. Then, with the revolution of the middle orb, pole A revolved about D (then aligned with Z) and moved back to being aligned with G, the pole of the lowest orb. If arc GB in the figure was 90°, then the planet would never be below B at the equator of the lowest orb. Thus, Gersonides has positioned a type of Eudoxan Couple to effect an additional inclination in a single direction with respect to a third pole (ben Gerson 1397, 241a–b).

Finally, in chapter 124, Levi noted that motions that occurred about parallel axes occurred in a single plane. He explained that "if these poles, that they are in this way not aligned, the inclination of the surface of one of them to the other equals half of the excess of the distance between the poles on one side over the distance between the poles on the other side" (ben Gerson 1397, 239a). Although Levi incorporated elements of homocentric astronomy, such as the concentric orbs of the Eudoxan Couple into his latitude models, the orbs in Levi's latitude models were not all homocentric. In a model composed solely of homocentric orbs, this discussion of parallel poles would be inconceivable.

7.5 Levi's Models for the Latitudes of Saturn and Jupiter

In chapter 125 of the *Astronomy*, the second hypothesis from chapter 124 became the foundation for Levi's models for the motion in latitude of Saturn and Jupiter. He commented that Ptolemy's models for the latitudes of Saturn and Jupiter were similar to each other, though the parameters differed (ben Gerson 1397, 240a). Because Levi rejected epicycles, his models differed from those found in both the *Almagest* and the *Planetary Hypotheses*. I will present the model for the motion of Saturn in the most detail. Levi stated that with Saturn:

> The path of the mean motion is presumed to be about poles inclined from the poles of the ecliptic orb by approximately 2;30° in that the mean distance of Saturn from the earth is sixty parts. The lower path of the motion of the anomaly is always about poles parallel to the poles of the mean motion, without any inclination at all. It moves two points aligned with the poles of the lower path of the motion of the anomaly in a small circle, with them being inclined from the poles of the upper path of the motion of the anomaly by the likes of half a degree on the measure that the mean distance of Saturn is 60 degrees. (Ben Gerson 1397, 240b)

Despite the complexity of chapter 125, neither manuscript includes many figures for that chapter. I will point out multiple interpretations, where appropriate, of the models. Levi theorized that the path of the mean motion of Saturn was about poles inclined to the poles of the ecliptic orb by approximately 2;30°. According to what Ptolemy wrote in the *Almagest* and the *Planetary Hypotheses*, the inclination of the eccentric, akin to the path of the mean motion in Gersonides's model, to the ecliptic was 2;30° (Toomer 1984, 613). Because Levi wrote, using the plural of "poles," that (both) poles of the path of the mean motion were inclined to (both) poles of the orb of the ecliptic, the observer, who was certainly at the center of the ecliptic orb might, in turn, be located at the center of the path of the mean motion. Thus, the inclination of the poles of the path of the mean motion to the poles of the ecliptic orb could be analogous to the inclination of the poles of the orbs of a Eudoxan Couple.

There is a more likely interpretation of Levi's words because Ptolemy did not make the center of the path of the mean motion coincident with the center of the ecliptic (Toomer 1984, 613; see Fig. 7.11).

In the preceding quote, Levi referred to the mean distance of Saturn and, a few paragraphs later, he mentioned the apogee of Saturn, the point where Saturn was farthest from the observer (ben Gerson 1397, 240b). Were the orb of the ecliptic and the orb with the poles of the path of the mean motion concentric, there would be no apogee and Saturn would have a single distance from the Earth, not a mean distance. Thus, it is probable that the orb with the path of the mean motion was eccentric to the orb of the ecliptic in Gersonides's model. 2;30° is the angle at the intersection of the axes of the ecliptic and the eccentric as shown in the figure.

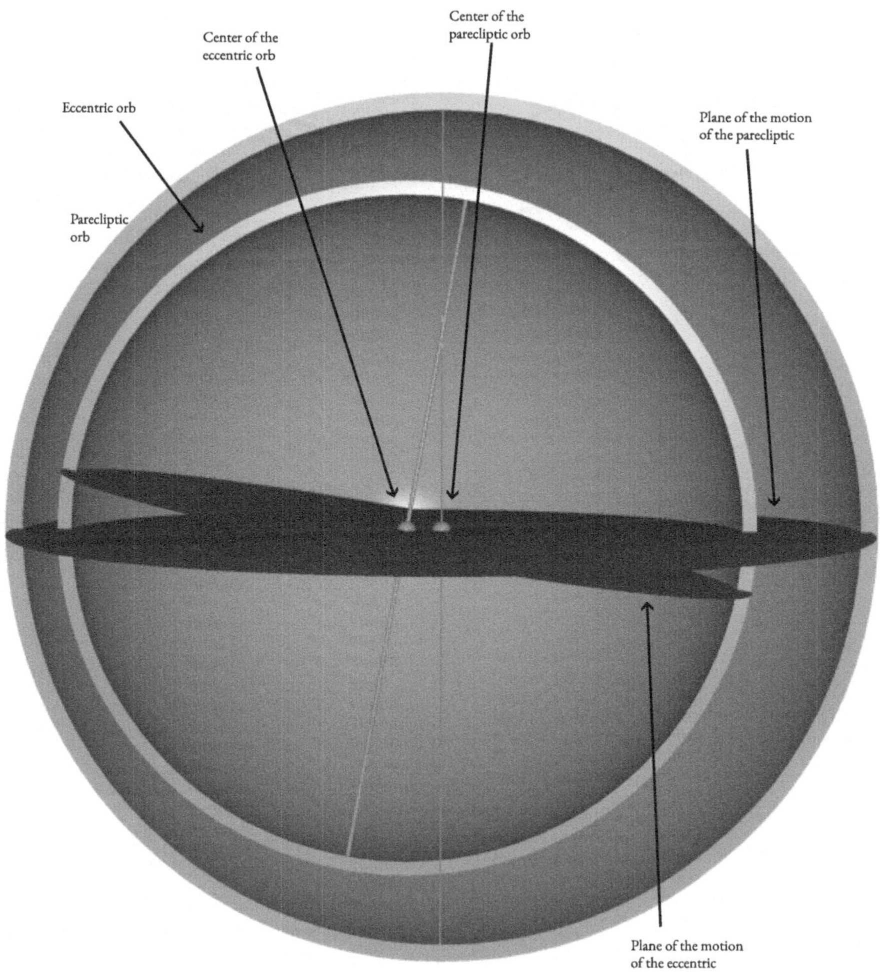

Fig. 7.11 My interpretation of Levi b. Gerson's statements in this chapter about the mean motion of Saturn

Then Levi mentioned the location of the poles of the path of the anomaly. The anomaly had a lower path and an upper path, each with its own set of poles (see Fig. 7.12). The motion in anomaly was about the pole of the lower path of the anomaly. The motion of the pole of the upper path of the anomaly varied the inclination of the path of the anomaly.

The poles of the lower path were facing or parallel (Heb, *n^{e}ḵoḥi*) to the poles of the path of the mean motion. Thus, it is possible that the poles of the lower path of the anomaly were parallel to and thus separated from the poles of the path of the mean motion. The poles of the lower path of the anomaly would be, by extension, inclined to the poles of the ecliptic orb. The centers of the lower path of the anomaly and of the mean motion were also separate from the center of the ecliptic. The path

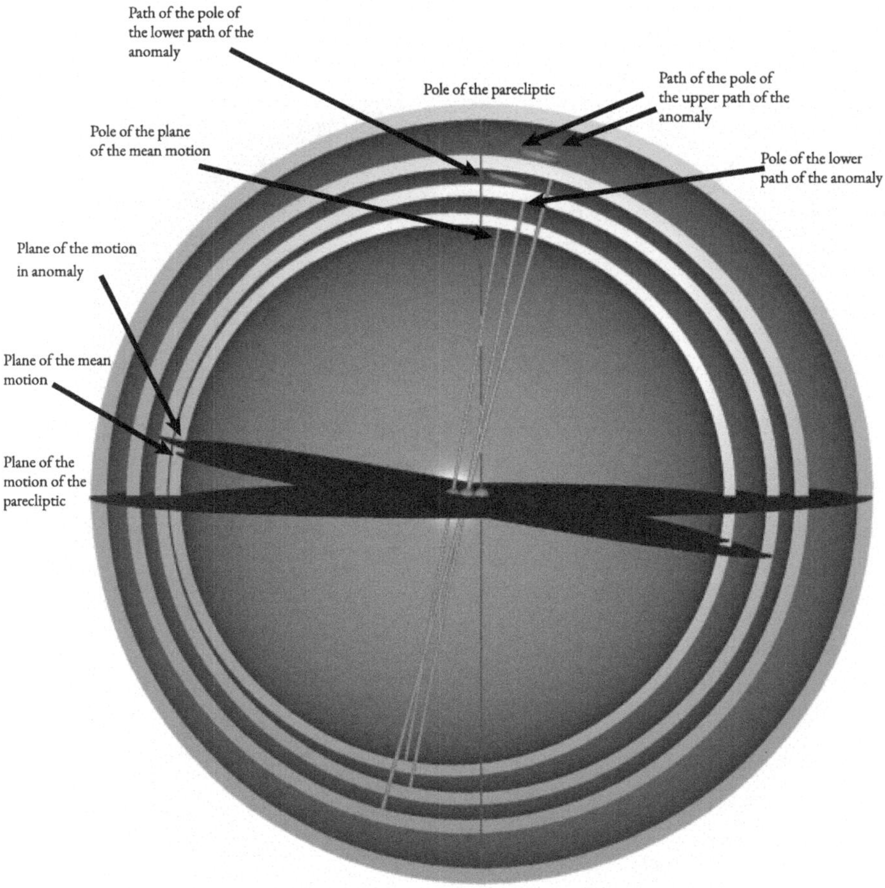

Fig. 7.12 One interpretation of Levi b. Gerson's model for the varied inclinations of the motion of Saturn in anomaly

of the anomaly is shown, in the figure, to be inclined to the path of the mean motion due to the motion of the pole of the upper path of the anomaly (see below). Gersonides wrote that the orb of the mean motion moves the poles of the lower path of the anomaly in a small circle. That revolution would affect further the inclination of the plane of the motion in anomaly to the plane of the mean motion.

Another interpretation is possible (see Fig. 7.13).

If the poles were facing, then the poles of the lower path of the anomaly would be coaxial with the poles of the path of the mean motion. That would mean that the paths of the anomaly and the mean motion shared the same center. In *Almgest* XII.1, Ptolemy theorized a double-eccentric model in which the eccenter revolved around the center of the ecliptic with the motion in anomaly (Toomer 1984, 540); Pedersen 2011, 339–40). It is possible that Levi has proposed a variation of this model. The orb containing the poles of the upper path of the anomaly was concentric with and

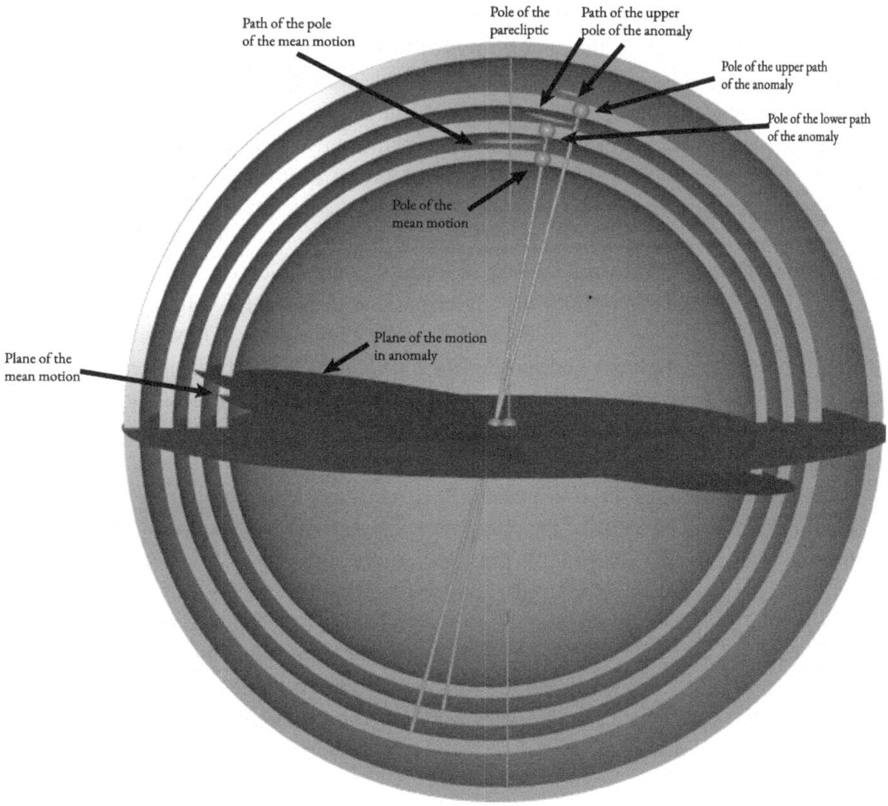

Fig. 7.13 Another interpretation of Levi b. Gerson's model for the varied inclinations of the motion of Saturn in anomaly

encompassed the orb containing the poles of the lower path of the anomaly. Thus, the orbs of the mean motion and of the upper and lower paths of the anomaly were concentric. Gersonides did not explain how the mean motion was not transmitted to the poles of the lower path of the anomaly. In Fig. 7.13, the path of the anomaly is inclined to the path of the mean motion due to the position of the pole of the upper path of the anomaly.

Levi explained how the motion of the pole of the upper path of the anomaly altered the inclination of the path of the anomaly to the ecliptic:

> From the motion of the pole of the upper path of the motion of the anomaly around the pole of the motion of the path of the lower anomaly, there occurs what is observed of the inclination due to the motion of the anomaly, when we combine this with the motion of the pole of the path of the motion of the upper anomaly around a point bisecting the distance that is between it and between the pole of the path of the motion of the lower anomaly. This motion is completed twice in the time period of the motion of the longitude, as we mentioned previously. (Ben Gerson 1397, 240b)

In this paragraph, we read that the pole of the upper path of the anomaly has two motions, just as pole A in the second hypothesis in chapter 124 had two motions. The orb with the poles of the lower path of the anomaly moves the poles of the upper path in a small circle with radius 0;30° at the rate of the motion in longitude about the poles of the lower path of motion in anomaly. Levi specified also that the poles of the upper path of the anomaly move about a point bisecting the distance between the poles of the upper and lower paths of the anomaly. Hence, the poles of the upper path move on a small circle with radius 0;15°. That motion is completed twice in the period of Saturn's motion in longitude (ben Gerson 1397, 240b).[11] The combined motions of the pole of the upper path of the anomaly altered the inclination of the path of the anomaly to the path of the mean motion. These changes in the inclination of the path of the anomaly replace the oscillation of the apogee and perigee of the epicycle in the latitude model for Saturn found in the *Almagest*. Levi did not posit an additional mover for the second motion of the pole of the upper path of the anomaly.

Although Gersonides did not specify the directions of the revolutions of the poles of the upper and lower paths of the anomaly for Saturn and Jupiter, if the revolutions of the two orbs moving the pole of the upper path of the anomaly were in opposite directions, then the pole of the upper path of the anomaly and, hence, a point on the path of the anomaly would oscillate almost on a great circle arc (Saliba and Kennedy 1991). In the model for the latitude of Mars, Levi stated that the motions were in opposite directions. Levi would be the first astronomer outside of an Islamic society to integrate a Ṭūsī-couple into an astronomical model. There is evidence that the Ṭūsī-couple was known to Jewish scholars in Europe in the fourteenth century (Ragep 2017, 177).[12]

It is also possible that the revolutions were in the same direction, though I noted earlier in the chapter the attendant imprecisions in the case in which the inclination of the pole of one orb was twice the inclination of the pole of the other orb. We have seen in the discussion of chapter 124 that Levi theorized a Eudoxan Couple in order to impart an additional inclination in a single direction. But if Levi understood the Ṭūsī-couple, then there would be no reason to propose a Eudoxan Couple at all. A Ṭūsī-couple yielded an oscillation on a path closer to a great circle arc.[13] In fact, even a Eudoxan Couple also would be a more precise way to impart an oscillation to the plane of the motion of the anomaly than what Gersonides theorized for Saturn were the two orbs carrying the poles of the upper and lower paths of the anomaly to revolve in the same direction. While we cannot be certain about the relative directions of the revolution of the orbs in the second hypothesis, we should consider the

[11] Levi did not explain, in chap. 125, that, according to Ptolemy, the planet's motion in longitude was uniform about a point, the equant, that was removed from the center of the orb that caused the motion in longitude.

[12] Avner de Burgos (ca. 1270–1340), a convert from Judaism to Christianity who was a contemporary of Levi, proved a version of the Ṭūsī-couple in the context of an intervention into the quies media debate. On Avner, see Langermann 1996.

[13] For more analysis, see Ibn Naḥmias and Morrison 2016, 370–75.

possibility that Gersonides had not fully analyzed his latitude models mathematically when he composed these chapters.

Based on the parameters supplied by Gersonides, the inclination of the first diameter of the plane of the motion of the anomaly to the plane of the mean motion varied from two to three degrees; when the inclination was 2:30°, the plane of the motion of the motion of the anomaly was parallel to the plane of the ecliptic, which was always true for the latitude model for Saturn in the *Planetary Hypotheses*. These parameters match those found in the *Almagest*, in that the greatest observed latitude of Saturn was about three degrees and the least was about two degrees (Swerdlow 2005, 45; Toomer 1984, 632).[14] In Fig. 7.13, the inclination of the plane of the motion in the anomaly to the plane of the mean motion is 3°. Levi's model for the latitude of Jupiter was structurally the same as the model for the latitude of Saturn, with the exception that the inclination of the pole of the path of mean motion and the pole of the lower path of the anomaly to the ecliptic orb was 1;30°. The inclination of the pole of the upper path of the anomaly to the pole of the lower path of the anomaly was 30 min (ben Gerson 1397, 241a). Again, these parameters match those from the *Almagest* as the greatest observed latitude of Jupiter was about two degrees and the least was about one degree.[15] Ptolemy did not provide tables or maximum and minimum latitudes in the *Planetary Hypotheses* so Gersonides's dependence on the values from the *Almagest* is understandable. Astronomers in Islamic societies borrowed tables for planetary latitudes from the *Almagest* (Van Brummelen 2006, 354).

Levi did not feel constrained by Ptolemy's observations (ben Gerson 1397, 244b; Goldstein 2002, 25–27). For example, after reporting that Ptolemy found a single inclination for the Moon, Levi commented, "We, however, have found some variation in its inclination except we have not redoubled the observation of this as would be suitable" (ben Gerson 1397, 244a). Levi furnished his own value for the lunar latitude: approximately 4;30° (ben Gerson 1397, 244a), and in chapter 126, Levi alleged that Ptolemy's mistaken parameter for the lunar latitude might have distorted other measurements (Goldstein 2002, 26).[16] Other shortcomings of Ptolemy's latitude models were due to errors in the available observations. Gersonides mentioned that his parameter for the inclination of the ecliptic to the celestial equator, 23;33°, differed from the Ptolemaic value by 18 min. Hence, Ptolemy's observations for the altitude of a planet when it culminated could be erroneous by the same amount. Levi saw no need to concern himself with the question of whether his models predicted Ptolemy's observations.

---

[14] See Swerdlow 2005, 66 for his computations of the maximum and minimum observed latitudes of Saturn and Jupiter in the *Planetary Hypotheses*.

[15] Swerdlow 2005, 45; Toomer 1984, 632.

[16] The rest of the paragraph draws on this page of Goldstein's translation.

7.6 Conclusion

Levi's models for the planets' motion in latitude are distinctive in that he transferred some of the building blocks of Ptolemy's models from the *Planetary Hypotheses* to models from which epicycles were excluded. Levi's latitude models are also distinctive because he theorized that one of the concentric orbs in the latitude models for Jupiter and Saturn revolved at twice the angular velocity of the other. Though Levi's models for latitude may have provided insight into his models for planetary longitudes, there are limitations to my reconstruction. He considered elsewhere in the *Astronomy* multiple longitude models with multiple locations for the equant point (Goldstein 1974b, 32–52).

As Levi did not mention the equant point in chapter 125, I am wary of reconstructing Levi's longitude models just from the latitude models. I look forward to studying the models for Mars and the lower planets and to learning more about Levi's creativity and the extent of his dependence on the works of his predecessors.

References

Feldman, Seymour, trans. 1984–1999. *Levi ben Gershom: The Wars of the Lord.* 3 vols. Philadelphia: Jewish Publication Society of America.

Gerson, Levi ben [as Gersonides]. 1397. *Ma'mar ḥamishi, ḥeileq ri'shon mei-Milḥamot ha-Shem.* Paris: BnF hébreu.

Gerson, Levi ben [as Gersonides]. Wars. See Feldman 1984–1999.

Goldstein, Bernard R. 1972. Levi ben Gerson's Lunar Model. *Centaurus* 16:257–284.

Goldstein, Bernard R. 1974a. *The Astronomical Tables of Levi ben Gerson.* New Haven: Transactions of the Connecticut Academy of Arts and Sciences.

Goldstein, Bernard R. 1974b. Levi ben Gerson's Preliminary Lunar Model. *Centaurus* 18:275–288.

Goldstein, Bernard R. 1985. *The Astronomy of Levi ben Gerson (1288–1344).* New York: Springer.

Goldstein, Bernard R. 1992. Levi ben Gerson's Contributions to Astronomy. In *Studies on Gersonides*, ed. Gad Freudenthal, 3–19. Leiden: Brill.

Goldstein, Bernard R. 2002. Levi ben Gerson's Preliminary Remarks for a Theory of Planetary Latitudes. *Aleph* 2:15–30.

Goldstein, Bernard R. 2012. Levi ben Gerson on the Principles of Cosmology. *Aleph* 12 (1): 17–36.

Ibn Naḥmias, Joseph, and Robert G. Morrison. 2016. *Astronomy in al-Andalus: The Light of the World.* Berkeley: University of California Press.

Langermann, Y. Tzvi. 1996. Medieval Hebrew Texts on the Quadrature of the Lune. *Historia Mathematica* 1:31–53.

Mancha, José Luis, and Gad Freudenthal. 2005. Levi ben Gershom's Criticism of Ptolemy's Astronomy. *Aleph* 5:35–46, 48–167.

Murschel, Andrea. 1995. The Structure and Function of Ptolemy's Physical Hypotheses of Planetary Motion. *Journal for the History of Astronomy* 26:33–61.

Nikfahm-Khubravan, Sajjad. 2022. *The Reception of Ptolemy's Latitude Theories in Islamic Astronomy*. Ph.D. dissertation, McGill University.

Pedersen, Olaf. 2011. *A Survey of the Almagest.* Edited and with commentary by Alexander Jones. New York: Springer.

Ptolemy. *Planetary Hypotheses.* See Swerdlow 2005.

Ragep, F. Jamil. 1993. *Naṣīr al-Dīn al-Ṭūsī's* Memoir on Astronomy (al-Tadhkira fī 'ilm al-hay'a). 2 vols. New York: Springer.
Ragep, F. Jamil. 2017. From Tūn to Toruń: The Twists and Turns of the Ṭūsī-Couple. In *Before Copernicus: The Cultures and Contexts of Scientific Learning in the Fifteenth Century*, ed. Rivka Feldhay and F. Jamil Ragep, 161–197. Montreal: McGill-Queens University Press.
Ragep, F. Jamil. 2020. The Origins of the Ṭūsī-Couple Revisited. https://archive.nyu.edu/bitstream/2451/61288/59/14.%20Ragep.pdf
Sabra, A. I. 1979. Ibn al-Haytham's Treatise: Solution of Difficulties Concerning the Movement of Iltifāf. *Journal for the History of Arabic Science* 3:388–422.
Saliba, George. 1987. The Role of the Almagest Commentaries in Medieval Arabic Astronomy: A Preliminary Survey of Ṭūsī's Redaction of Ptolemy's Almagest. *Archives internationales d'histoire des sciences* 37:3–20.
Saliba, George. 1996. Arabic Planetary Theories after the Eleventh Century AD. In *Encyclopedia of the History of Arabic Science*, eds. Roshdi Rashed and Regis Morelon, 1:58–127. 3 vols. London: Routledge.
Saliba, George, and E. S. Kennedy. 1991. The Spherical Case of the Ṭūsī Couple. *Arabic Sciences and Philosophy* 1:285–291.
Swerdlow, Noel. 2005. Ptolemy's Theories of the Latitude of the Planets in the Almagest, Handy Tables, and Planetary Hypotheses. In *Wrong for the Right Reasons*, ed. J. Z. Buchwald and A. Franklin, 41–71. Dordrecht: Springer Netherlands.
Toomer, G. J. 1984. *Ptolemy's Almagest*. London: Duckworth.
Van Brummelen, Glen. 2006. Taking Latitude with Ptolemy: Jamshīd al-Kāshī's Novel Geometric Model of the Motions of the Inferior Planets. *Archive for History of Exact Sciences* 60:353–377.

Chapter 8
Calculating the Length of Life in Sixteenth-Century Europe: Arabic Astrology and Humanist Reforms

Margaret Gaida

Astrologers in sixteenth-century Europe were tasked with making a variety of different types of predictions. They used the practice of annual revolutions to predict weather or other potential natural disasters for the upcoming year. The practice known as elections helped them to determine the most propitious days and times for undertaking journeys, going to war, getting married, or laying the foundation for a new building. And they engaged in the standard astrological practice of nativities, or casting horoscopes predicting the details of an individual's life. According to the wisdom of the second-century CE astronomer-astrologer Claudius Ptolemy, knowing the future enabled one to adequately prepare one's soul for the calamitous or the fortunate. Calculating the length of a client's life was part of the standard procedure for nativities, although by the sixteenth century, the precise method for this procedure was far from straightforward. Much of the confusion surrounding this calculation was related to two technical terms which had been transliterated from Arabic-Latin translations of the twelfth and thirteenth centuries: the *hyleg* and *alcocoden*.

Indeed, as Juste (2016) has pointed out, astrological texts of Arabic origin formed the majority of astrological manuscripts available to Latin scholars. I have argued elsewhere (Gaida 2018) that Arabic texts were considered an authoritative source of astrological knowledge for centuries in Europe. However, humanist-led reforms of astronomy and astrology in the sixteenth century involved critiques not just of the language of Latin translations of Arabic astrological texts, but also of *practices* perceived as "Arabic" in origin.

The question of the extent to which astronomer-astrologers of the sixteenth century evaluated perceived "Arabic" ideas against the Greek tradition is central to

M. Gaida (✉)
Department of History, Occidental College, Los Angeles, CA, USA
e-mail: gaida@oxy.edu

F. Rochberg (ed.), *Studies in Premodern Sciences in Memory of Noel M. Swerdlow*, Archimedes 74, https://doi.org/10.1007/978-3-032-04040-4_8

understanding the tension between the need for rigorous astronomical and astrological reform according to the emerging empiricism of the sixteenth and seventeenth centuries, and the commitment to authoritative traditions of the past that also informed this development. Several sixteenth-century astronomers and humanist scholars were interested in turning back to original Greek sources and casting doubt on an authoritative Arabic tradition that had persisted for centuries. But "casting doubt" isn't quite the right expression. Many humanist authors were downright derisive toward the Arabic tradition. Pico della Mirandola (1946–1952), for example, writes explicitly against the Arabs in his *Disputationes adversus astrologiam divinatricem* when he condemned their superstitious "false stories," which the Latins followed. Dag Nikolaus Hasse (2018) documents this tension in the fields of philosophy, medicine and astrology and rightly acknowledges that there is often no obvious "winner" in many cases where Greek and Arabic authorities are put to the test, in terms of which theories were adopted, by whom, and when.

This paper traces the development of the calculation of the length of life from its origins in classical texts, through Arabic and medieval Latin authors, to the reception and treatment of the topic by the sixteenth-century astronomer-astrologer Valentin Naibod (1523–93 CE). By placing Naibod's discussion of the length of life within the humanist context of anti-Arab prejudice and efforts toward astrological reform, this paper aims to provide a case study for understanding the tensions that developed between science and humanism in the sixteenth century, and especially concerns governing the question of authority versus the role of empiricism and observation in establishing truths about the natural world.[1]

Before turning to classical astrology, let's first consider why Naibod presents such an important case study. Naibod is best known for being one of Nicolaus Copernicus's early supporters. He was the first astronomer to popularize an image of the Sun-centered cosmos, in his 1573 textbook for gymnasium students, 30 years after Copernicus proposed this cosmological arrangement in his *De revolutionibus orbium coelstium* in 1543. Naibod's unfortunate fate was also immortalized in a tale in which he predicted his own death. As recounted by Tommaso Campanella in 1629, in consulting his own horoscope, Naibod predicted that he was approaching his date of death. Fearing the worst, he stocked up on food and drink, locked himself in his house, closed all the doors and windows, and planned to wait several days until he was astrologically out of the woods. A band of passing robbers, assuming no one was inside due to the closed house, broke in and stabbed him to death. Now, Campanella recounts the tale as part of a discussion of whether we have the capability to escape our own fate.[2] But, it is also part of his larger concern with Arabic authority, as illustrated by the extended title of his book, which claims to purge

[1] This paper is inspired by Swerdlow's seminal 1993 paper on science and humanism, along with his more recent (as yet unpublished) work on the fifteenth-century astronomer-astrologer Regiomontanus's innovations on the topic of calculating the length of life.

[2] Campanella considered this subject important enough to be worthy of its own treatise, *De siderali fato vitando* (*How To Avoid the Fate Dictated by the Stars*), which he published alongside his longer astrological work *Astrologia libri VI* in 1629.

astrology of all the superstitions of the Arabs and Jews.[3] Both Naibod and his successor Campanella were part of a much broader movement to reform and restore astrology to its classical heritage, while embracing other astronomical reforms and challenges to traditional authorities such as Aristotle.

Naibod was one of the first astrologers to begin discussing the Copernican hypothesis and its application to astrological topics, as illustrated by his 1560 commentary on the most popular astrological text in premodern Europe, Alcabitius's *Introduction to Astrology*. This commentary is also where we find Naibod's discussion of the calculation of the length of life, which he apparently used to some success. Naibod also wrote a commentary on Ptolemy's *Tetrabiblos*, which also discusses the length of life calculation. One might wonder why there was such antagonism toward the Arabic tradition, given its medieval role as a highly authoritative source.

Despite continued attention to Arabic texts in the sixteenth century, such as Naibod's commentary on Alcabitius, the return to Ptolemaic ideas and criticism of concepts perceived as Arabic grew especially potent in the sixteenth century. There's a special reason for this in the case of astrology. More so than other disciplines, astrology in premodern Europe had an especially strong Arabic tinge. In addition to the question of numbers of manuscripts and prevalence of texts by Arabic astrological authors, a good portion of the technical vocabulary associated with making calculations such as the length of life had been transliterated from Arabic to Latin, rather than translated. In the manuscript tradition, these terms are often copied in red in the margins, which serves to "highlight" them, and there's an overwhelming number of alternate spellings of Arabic transliterated terms.

From the standpoint of the humanist authors of the sixteenth century, the effort to get at the original meaning of a text is confounded by a technical vocabulary where certain terms seem to have no fixed referent or the referent changes according to different authors. For astronomical and astrological reformers also concerned with the accuracy of their predictions, the linguistic fluidity of terms across centuries of use also demanded a more rigorous examination of the Arabic technical vocabulary. As we shall see, calculating the length of life provides an insightful case study for how Naibod and other sixteenth-century authors grappled with these stakes.

8.1 Calculating the Length of Life

The primary means for the calculation of the length of life begins with the horoscope, and the calculation of the house cusps. Astrologers divided the sky into houses along the ecliptic, with the divisions known as cusps. The astrological houses are what enabled the astrologer to make predictions about the various factors of the

[3] *Astrologia libri VI, in quibus Astrologia, omni superstitione Arabum et Iudaeorum eliminate, physiologice tractatur.*

individual's life, and so the first house was devoted to the life of an individual, the second to wealth and estates, the third to brothers and sisters, and so on. There were several methods for calculating the cusps, but the one attributed to the tenth-century author Alcabitius, also referred to as the "Standard Method," was most popular in premodern Europe. After the calculation of the cusps, the astrologer would usually use a table of planetary positions, and assign each of the planets to its place in the horoscope chart. For their interpretations, astrologers were interested in the position of the planet in both the house and the zodiacal sign, and the planetary relationships with each other according to their angular relationship (called aspects), and their positions in the signs.

Before interpretations, however, astrologers first needed to calculate the length of life. As Ptolemy puts it (3.10, 271): "The consideration of the length of life takes the leading place among inquiries about events following birth, for, as the ancient says, it is ridiculous to attach particular predictions to one who, by the constitution of the years of his life, will never attain at all to the time of the predicted events." In other words, why should one go to the trouble of interpreting the chart for the full breadth of a life if the subject won't live through it. In simplest terms, the progression or direction (that is, the length of life) is basically the length in degrees along an arc drawn between two points across the ecliptic, i.e., the angular distance between a starting point P and an ending point P′, with each degree corresponding to 1 year of life.

While this seems straightforward from a mathematical perspective, the methods for determining the starting point and ending point are more complicated. There are also what seem to be endless variations on the method itself. To simplify things for our purposes, I'd like to focus more on the technical language used, especially for the arc and the two points (the starting and end point) involved in the calculation, beginning with Ptolemy.

In the *Tetrabiblos*, Ptolemy (3.10, 271) dismisses some of the other methods on calculating the length of life and presents a method "most pleasing to us, and, besides, in harmony with nature," based on a starting point, often the position of a planet, which he calls the ἄφεσις (sending out, releasing) or *aphesis*. The *aphetic* point[4] is determined according to astrological principles which Ptolemy describes at length. First, Ptolemy describes the astrological houses which are suitable places for the *aphetic* point, and then gives their order of preference. These places are, in short, the 1st, 7th, 9th, 10th, and 11th houses. He then considers the positions of the Sun, Moon, ascendant, Lot of Fortune, and their rulers. Depending on whether the birth is diurnal or nocturnal, Ptolemy gives an order of preference for each of these. If none of these positions is suitable for the *aphetic* point, Ptolemy then considers the conjunction of Sun and Moon before the birth (the syzygy, i.e., New Moon or Full Moon). One must then determine a corresponding end point or "destroyer," the *anaeretic* point, again by a set of astrological principles. The *aphetic* point is

---

[4]To facilitate the linguistic discussion here, "aphetic point" refers to the starting point of the calculation of the arc.

"directed" to the *anaeretic* point, signifying the death of the subject.[5] This method became the basis for subsequent treatments of the topic in Arabic astrology, and is thus worth exploring in some detail. Ptolemy describes a method for this calculation which Casulleras (2009) refers to as the Position Semi-Circle Method, and an approximation for this method has been called the hour-line method. Casulleras and Hogendijk (2012) have described the technical details for some of these methods.

8.2 Arabic Astrology

In the ninth and tenth centuries, especially in flourishing urban environments such as Baghdad and Damascus, several new and important Arabic astrological treatises were composed by authors who synthesized Greek, Syriac, Pahlavi, and Sanskrit sources. One of these authors, al-Qabīṣī (d. 967 CE), or Alcabitius as he became known in the West, composed an introductory treatise that included a chapter on the technical terms of the astrologers. This chapter addresses several different calculations, including one for the length of life. Al-Qabīṣī favors the hour-line method for this calculation. He first defines the *tasyīr*:

> Pertaining to this is the *tasyīr* (prorogation). This is that you move an indicator to a position on the ecliptic, and you <want to> know what is <the distance> between the two in equatorial degrees—that is, what is rotated of the time-degrees of the equator with regard to the position from which it is moved until it is in the position to which it is moved. One takes a year for each degree.[6]

To calculate the *tasyīr* arc, he gives a list of conditions for the position of *aphetic* point, here translated as "indicator" (*dalīl*), starting with the simplest and moving to the more complex cases, and then instructs the reader to add or subtract the corresponding rising times in right or oblique ascension to the degree of the ascendant or the indicator, depending on the conditions.

The simplest examples are for when the indicator is in the cardines (1st, 4th, 7th, and 10th houses), and al-Qabīṣī describes these circumstances first. For the ascendant, al-Qabīṣī writes: "If you want this and the indicator which you want to move to a degree of the ecliptic is in the ascendant, you subtract the rising times of the degree of that which you want to move from the rising times of the degree to which you want to move it. What remains are the degrees of the motion (*tasyīr*)."[7] Depending on where the indicator lies (in the ascendant, the midheaven, the 7th house, or 10th house), there are different instructions for arithmetic calculations involving the rising times in right or oblique ascension of various points (the indicator, degree of midheaven, degree of the ascendant, etc.). The result of following the

---

[5] The notion of "direction" is reflected in subsequent treatments of the topic, in Arabic (*tasyīr*) and Latin (*directio*). The modern term is "prorogation."

[6] For a critical edition of the Arabic and Latin texts, along with an English translation, see BYY.

[7] BYY 4:[11], 121–22.

addition or subtraction of these rising times in right or oblique ascension will be the *tasyīr*.

The more complicated situations are described when the "indicator" or the point to which you want to move it *are not* in the cardines. For this calculation, al-Qabīṣī provides guidance for calculations in the eastern and western hemispheres, which are quite a bit more complex. The basic operation is to convert the indicator and the degree to which it is moved to seasonal hours by subtracting the rising times in right ascension from both, then dividing by horary magnitude, or the hours of the day in which the degree of the indicator is. What is interesting to note, however, is the language that al-Qabīṣī is using for the *aphetic* and *anaeretic* points. He uses the general term "indicator"/*dalīl* for *aphetic* point and is quite vague about the *anaeretic* point, usually referring to the "degree to which it is moved."

The eleventh-century astronomer-astrologer Kushyar ibn Labban provides the same method, the hour lines method, although with more detail, examples, and two *tasyīr* tables.[8] Kushyar refers to the two points (the *aphetic* and *anaeretic*) as just degrees (*darajat*). What I want to emphasize here is that in their descriptions of the *tasyīr*, neither al-Qabīṣī nor Kushyar refer to arguably the two most important and common terms for calculating the length of life, the *haylāj* and *kadhudhāh*. In almost all sources, the *haylāj* is obviously the *aphetic* point. But the case is much more complex for the *kadhudhāh*. Kushyar and al-Qabīṣī discuss both terms, but not always in ways that are directly linked to their description of the calculation of the *tasyīr*. Kushyar explicitly states that they are important for calculating the length of life, but in al-Qabīṣī this link is more implied. The description of the *haylāj* and *kadkhudhāh* occur several pages prior to the *tasyīr* in al-Qabīṣī's text, and if you weren't already familiar with the topics, there is no obvious link between them. This lack of internal referencing comes into play later when European astronomer-astrologers attempted to understand the links between terms. So, we will look more closely at what is said about the *haylāj* and *kadkhudhāh*.

8.2.1 Al-Qabīṣī's and Kushyar ibn Labban's Discussion of Technical Terms

There is much ambiguity in the sources regarding the relationship between the *tasyīr*, the *haylāj* and *kadkhudhāh*. While the term *tasyīr* is often reliably employed in describing the method for calculating the length of life, and descriptions of the mathematical process behind its determination is clear, the precise role of the *haylāj* and *kadkhudhāh* is more difficult to pin down. In al-Qabīṣī's description of the *haylāj*, for example, the astrologer examines the planetary positions and goes through a process of elimination in order to determine which celestial element is most suitable, according to the house in which it appears. Al-Qabīṣī first makes a

[8] Yano and Viladrich 1991.

distinction between diurnal births and nocturnal births. For diurnal births, the Sun is suitable for the *haylāj* if it's in the 10th or 11th house, or before the ascendant by five degrees or less, in a masculine or feminine sign. It would also be suitable if it's in the 7th, 8th, or 9th house in a masculine sign, but not if it's in this house in a feminine sign. For the Moon, it is suitable for the *haylāj* if it's in the 1st, 2nd, 3rd, 7th, or 8th house in a masculine or feminine sign, or in a feminine sign in the 10th, 11th, 4th, 5th house or five degrees before the ascendant. If neither the Sun nor Moon meets these conditions, then you consider the degree of conjunction or opposition preceding the birth, then the degree of the Lot of Fortune, then the degree of the ascendant, in that order, and only when these points are in the cardines or their succedents (houses directly following the cardines).[9] There is also a general rule regarding whether the *haylāj* aspects one of the rulers of the five shares. This further qualification means that the astrologer would also need to have knowledge of the projection of the rays, although that technique is not described in the *Introduction*.

Kushyar's description of the *haylāj* differs from both Ptolemy's and al-Qabīṣī's, which is interesting since they are all using the same method for the *tasyīr* calculation. Following Ptolemy, Kushyar first lists the houses which are suitable places for the *haylāj*. However, his list of houses is somewhat different from Ptolemy's, and much less stringent than al-Qabīṣī's. These are: the 4th, 1st, 11th, 7th, 9th, and the degrees of the term of the ascendant. Kushyar then lists the Sun, Moon, Lot of Fortune, and ascendant for diurnal births, and the Moon, Sun, ascendant, and Lot of Fortune for nocturnal births.[10] Kushyar also includes the conditions for the *haylāj* several sections before his description of the *tasyīr*, and, like his predecessor al-Qabīṣī, does not refer back to the *haylāj* in this description. Several modern scholars have noted the complexities and difficulties with sorting out these terms and conditions. John North (1988) remarks: "The choice is made following a series of rules that are so intricate and seemingly pointless that at least a part of their function must have been to keep the dilettante at bay."[11] I argue that paying attention to these seemingly arbitrary details may give some clues about transmission. Presumably, for both al-Qabīṣī and Kushyar, the *haylāj* is intended to be the *aphetic* point. But this is far from clear in the text. With that in mind, we will consider the description of the *kadkhudhāh*.

The case is equally unclear for the *kadkhudhāh*. This point is chosen based on a set of factors relevant to the *haylāj*. Al-Qabīṣī's text reads:

> Another is the *kadkhudhāh*. It is the indicator of the length of life. When you want to know this and you already know the *haylāj* by the method which we have described above, you look at the lord of the place of the *haylāj* or its exaltation or its term or its triplicity or its decan, and whichever of them has the most leadership in the degree of the *haylāj* and is aspecting the *haylāj* is the first choice for the *kadkhudhāh*ship.[12]

---

[9] BYY 4:[4], 111–17.

[10] Yano 1997, 3: 173–79.

[11] North 1988, 215.

[12] BYY 115.

In this case, al-Qabīṣī has already explained how to determine which astrological share (lord, exaltation, term, triplicity, or decan) has the most leadership in a degree. This simple arithmetic calculation is demonstrated in the first chapter of the *Introduction*.[13] An aspect is defined by the positional relationship between the planets as defined by degrees (be it trine, quartile, sextile, opposition, or conjunction). Whether or not the *haylāj* or *kadkhudhāh* are aspecting each other is determined from the knowledge of the casting of the rays, another related and complex astrological calculation which is not described in the *Introduction*. If the *kadkhudhāh* and *haylāj* are not aspecting each other, then the astrologer must find another suitable point for the *haylāj*. There are a few additional conditions to take into consideration, which we will not mention here. Al-Qabīṣī also admits the possibility that the *haylāj* and *kadkhudhāh* can be the same point. However, al-Qabīṣī does not explain the role of the *kadkhudhāh* in determining the length of life.

At first, Kushyar gives a similar description to al-Qabīṣī's version of the *kadkhudhāh* for a point he calls *wālin*, which is taken into consideration along with the *haylāj*. This is perhaps similar to al-Qabīṣī's description of a third term, *al-mubtazz*, which he says some authors use in place of the *kadkhudhāh*. However, Kushyar also provides a different method for calculating the length of life which includes the *kadkhudhāh*. In Kushyar's account, the *kadkhudhāh* determines the length of life based solely on its position and associated numbers of great years, middle years, and small years. This calculation for the length of life does not mention the *tasyīr* at all. Kushyar also provides a table with the values for great, middle, and least years for each of the planets.[14] These values can then be adjusted depending on the planet's position in the chart. This different method does not appear at all in al-Qabīṣī's *Introduction*.

This discussion raises the question of the precise relationship between the *haylāj* and *kadkhudhāh*. The determination of the *haylāj* and *kadkhudhāh* are not complicated once the degrees of the house cusps are known, as long as one follows the rules for their selection. However, it is not clear at all how the calculation of the *tasyīr* using the *haylāj* as the indicator relates to the other calculation of the *kadkhudhāh* mentioned by Kushyar. Especially in the case of al-Qabīṣī's account, I want to emphasize that al-Qabīṣī does not describe how the length of life is determined from these parameters. He does not say, for example, that the length of life can be determined directly from the *kadkhudhāh*, as Kushyar does. Nor does he refer to the *haylāj* in his description of the *tasyīr*. Indeed, in his description of the two points in the *tasyīr* calculation, al-Qabīṣī consistently uses "indicator" and "point to be moved toward." In Kushyar's *Introduction*, he refers to the *haylāj* and *qatila* in his general description of the length of life, which implies the use of the *haylāj* as aphetic and *qatila* as *anaeretic* points.

In considering these technical terms in addition to techniques, we can see the variety in description of just two astrological terms, even where technical methods

[13] For an example determining the ruling planet for the house of property, see BYY 1:[77], 61.

[14] Yano 1997, 3: 177–79.

are the same. This discussion raises the question of whether there was a logic to the selection of the *haylāj* and *kadkhudhāh*, or if it was, as North put it, "pointless." Were Arabic authors interested in preserving and following tradition? Or were they interested in inventing their own techniques? Al-Qabīṣī's synthesis of introductory material certainly provides an account that's in line with the Ptolemaic tradition, but is presented in a way that separates the calculation of the *aphetic* point from the calculation of the progression. The same is true of Kushyar, who also adds an additional consideration of great, middle, and least years, along with two tables of calculations of progressions. In the *Tetrabiblos*, Ptolemy also gives a physical justification for his choices. There is no justification, however, in al-Qabīṣī's or Kushyar's introductions. These important distinctions lay the groundwork for the entrance of the Arabic and Greek astrological traditions into Europe, through the work of Ptolemy and al-Qabīṣī.

8.2.2 Translation Confusion, Then and Now

Johannes Hispalensis translated al-Qabīṣī's *Introduction* into Latin in the 1230s, and rather than finding suitable Latin equivalents for several technical terms, he chose to transliterate them. And so, *haylāj* became *hyleg*, *kadkhudhāh* became *alcocoden*, and al-Qabīṣī became Alcabitius. As far as we know, Johannes Hispalensis's translations were the first appearances of *hyleg* and *alcocoden* in the Latin astrological tradition. *Tasyīr* was translated as *directio*, but several other texts retained a transliterated version, *atazir*. Several astrological works were translated and assimilated within about one hundred years, including the works of Masha'allah and Abū Maʿshar, Sahl ibn Bishr, Haly ibn Ridwan's commentary on the *Tetrabiblos*, and 'Alī ibn Abi al-Rijāl's *Kitab al-Bari'*. Latin scholars were faced with comparing authors and selecting the best explanations of terms, or best values. For example, in early manuscripts of Alcabitius's *Introduction*, we see corrections of tables based on values that appear in Albumasar's *Abbreviation to the Introduction*, in addition to cross-references in the margins of manuscripts (Burnett et al. 1994, 2004). It was perhaps an unfortunate accident that the clearest expressions and discussions of the *tasyīr*, *haylāj*, and *kadkhudhāh* in Kushyar ibn Labban and the clearest description of *tasyīr* in the eleventh-century astronomer-astrologer al-Biruni were not translated into Latin. And so Latin scholars were faced with a complicated puzzle of foreign terms and meanings, which took some time to reconstruct.

Texts in the thirteenth and early fourteenth century acknowledge the interdependence of the *hyleg* and *alcocoden*, but again, their precise roles in the calculation of the length of life as it relates to the *directio* or progression are less clear. A Latin annotator has written in the margin of one early fourteenth-century manuscript an association of the *hyleg* with the *uxor* or wife and the *alcocoden* with the *vir* or husband. This distinction can be traced back to both Messahella's *Book on Eclipses* and Omar Tiberiades' *Book on Nativities*. Both references appear in the works of Abraham ibn Ezra, whose thirteenth-century Hebrew astrological treatises were translated into Latin. Again, though, this association is more philosophical than technical. The confusion was likely exacerbated, too, by the high proportion of spelling variants for these terms in the manuscript tradition. Marginal annotations also illustrate some of these alternative spellings, or various Latin translations of the meanings of terms: "id est locus vite," for *hyleg* and "quod est significator vite" in the case of *alcocoden*.[15]

Confusion over the meaning of terms is reflected in modern historians' scholarship as well. Returning to John North, he describes the calculation of progressions in his study of the English version of the text of Alcabitius's *Introduction.* North's description reflects the confused fourteenth-century state of affairs in Latin astrology. In his reconstruction of the theory of progressions, North writes:

> According to Ptolemy, the aphet (the planet, prorogator, alcochoden) may cast the life of the native from the aphetic place (prerogative place, *hyleg*) in either direction. … Reckoning in the sense of daily motion, it all seems fairly natural: the destructive (anaeretic) planet or aspect is carried in that sense until it reaches the aphetic place, *hyleg*. (The destructive planet or aspect is sometimes called the "following place," and the aphetic place the "preceding place") Working in the other sense, the motion representing the length of life is a hypothetical one, and it is the aphet, *alcochoden* in later parlance, that is regarded as being carried to the destructive thing, in this case invariably the western horizon.[16]

Here, we see the use of the *alcocoden* as part of the *tasyīr* calculation, when according to Kushyar, it is an independent calculation one might make in addition to the *tasyīr*. It is also confounded here with the *aphet* or *prorogator*, which is sometimes referred to as the *hyleg*. In this description, and many others, it's the *aphetic* point that's referred to as the *hyleg*. North writes, "Seen as a whole, in fact, the procedure was wrapped up in so convoluted a way that one might be excused for wondering whether it was ever meant to be understood."[17]

It is perhaps this confusion that encouraged John of Saxony to write a commentary on Alcabitius's *Introduction* in the 1320s. John of Saxony's commentary displays a remarkable understanding of the available sources, and includes worked out examples for the *hyleg*, *alcocoden*, and *directio*. In the case of the *alcocoden*, he gives several examples illustrating the different cases provided by Alcabitius. However, he also makes the distinction between the *hyleg* and the *alcocoden* that is missing in Alcabitius's text, namely, the notion that the calculation of the *alcocoden* is related to a separate method involving great, middle, and least years.

Drawing on the work of Haly Abenragel, John provides a lengthy account of how to determine great years, middle years, and least years according to the *alcocoden*. Referring to the "covert messages" and "hidden secrets" reported by Haly, John says:

> When the *alcocoden* is in a degree in the tenth house it gives the greatest years, and when it is in a degree in the eleventh house it gives the middle years, and when it is set aside from one of those two houses and goes to any aspect however many the degrees are from one to

---

[15] Gaida 2017 discusses several of these annotations and other marginalia.

[16] North 1988, 224.

[17] North 1988, 225.

the other, you divide through those years which are between the greater years and the middle years.[18]

John recounts several other cases drawn from Haly, including when the *alcocoden* is combust (under the Sun's rays). Consulting the printed text of Haly's *De iudiciis astrorum* from 1485, we find broad agreement in Haly's discussion of the great, middle, and least years, but no specific reference to the correspondence mentioned by John (i.e., the 10th house with greatest, etc.).[19] It is possible that this appeared in the earlier manuscript tradition of Haly's work and got left out of the printed edition, but this point needs verification.

While John's explanation is certainly more specific than what we find in the text of Alcabitius's *Introduction*, especially through illustrating that the *alcocoden* is a separate calculation, we are still missing an explanation of the relationship between the *alcocoden* and the *directio* calculation. In addition, John's commentary only circulated in approximately 30 manuscripts, whereas the number of extant manuscripts of Alcabitius's *Introduction* is over 230, and thus this information would likely have reached fewer readers. Latin scholars were themselves responsible for seeking out sources in addition to Alcabitius in which this information was available.

By the middle of the fifteenth century, efforts at astrological reform were well underway.

Humanists interested in a return to Ptolemy very carefully read the *Tetrabiblos* in order to ascertain which techniques were in the original Greek and which were Arabic inventions. The mathematician, astronomer, and astrologer Regiomontanus published the first tables of progressions in Europe, his substantial *Tabulae directionum et profectionum* (1490) in which he provides a complete reworking of several basic astrological techniques and an obvious preference for Latin vocabulary in his title. He proposes to use the Position Semi-Circle Method as a means of calculating progressions, Ptolemy's preference, rather than Alcabitius's adoption of hour lines. In addition, Regiomontanus updated the "Standard Method" of the calculation of the house cusps attributed to Alcabitius, to the "Equatorial (fixed boundaries)" method. Several scholars have identified a potential Arabic source for this method as well, the eleventh-century Andalusian astronomer Ibn Mu'adh al-Jayyani. Lastly, there is no obvious reference yet found to the terms *hyleg* and *alcocoden* in Regiomontanus's text, just to the *directio* and *profectio* (which is the *anaeretic* point).

Despite Regiomontanus's humanist preference toward the Latin language, he retained a deep respect for Arabic authors, and cites Alcabitius explicitly in his *Oration on the Dignity and Utility of the Mathematical Sciences*. In ridiculing those who make no real effort to master astronomy and astrology, he writes: "For you are truly worthy to be a laughing stock, now scarcely skimming over the first rudiments of John of Sacrobosco, then being introduced to a few preliminary parts of Alcabitius (and not even his fourth section where the difficulty of directions and rays is attended

[18] Alcabitius 1485, 127v–28r.

[19] Haly Abenragel 1485, f. 56–60.

to)."[20] Regiomontanus was the first Latin author to extend the calculation of directions and topic of the length of life well beyond his Arabic predecessors, and so he had certainly mastered the calculation on his own terms. That said, there is still some ambiguity about terms among other authors, particularly as they continued to turn to the *Tetrabiblos* for clarification.

Around the same time, at the end of the fifteenth century, Pico della Mirandola was quick to point out that there is also no reference to the *alcocoden* in the *Tetrabiblos*. Other Arabic authors, including Haly Abenragel, *had* referred to Ptolemy in discussions of the *alcocoden*, but this was likely due to pseudo-Ptolemaic works such as the *Centiloquium*. As Dag Hasse has recently shown, even authors who were disdainful of the Arabs were unaware that certain aspects of "Arabic" astrology could be found in Greek authors other than Ptolemy, such as Dorotheus. The *alcocoden* is a perfect example of this, and Girolamo Cardano and Tomasso Campanella also engaged in debates about its inclusion in calculating the length of life.[21]

By the beginning of the sixteenth century, humanist critiques directed explicitly at the Arabic tradition were in full force. Albertus Pigghius denounced prognosticators ignorant of mathematics who were dependent on the "abandoned fables" of Albumasar and the Arabs.[22] According to the historian Germana Ernst, Girolamo Cardano, the famed physician and astrologer, "maintained that a 'return to Ptolemy' was necessary to restore dignity and rigor to a discipline seriously corrupted by the 'follies' of the Arabs' manuals, which [he] criticized for offering rules as manifold and minute as they were unfounded and useless."[23] Cardano also mentions Alcabitius explicitly in his condemnation of the Arabs, writing that "astrologers must free themselves from all the ballast of the various 'Albumasars, Abenragels, Alchabitiuses, Abubatres, Zaheles, Messahalas, and Bethenes.'"[24]

Dag Hasse has illustrated that the process of debates over techniques did lead to some clarification on the relationship between terms. In his own account, he writes, "For determining the *alcocoden*, the *hyleg*—which for some authors is a point, for others a planet—needs to be found first. If the *alcocoden*, the planet ruling over the *hyleg*, has a powerful position in the chart, the full amount of its planetary years will deliver the maximum lifespan of the individual; if it is less powerfully placed, it will grant fewer years."[25] As we have seen from our analysis, this is a vast oversimplification of both terms. While Hasse admits that the technique is enormously complicated and that it received some lengthy treatments in the Latin handbooks of Guido

---

[20] This is Swerdlow's translation of the *Oration*, forthcoming as the introduction to *The Renaissance of Astronomy*.

[21] Hasse 2018, 262–65.

[22] Pigghius 1518, 5v.

[23] Ernst 2001, 47.

[24] Quoted in Ernst 2001, 47.

[25] Hasse 2018, 263.

Bonatti and Leopold of Austria, this is the closest published modern scholarly account of the relationship as it appears in early modern Europe.

By the sixteenth century, methods for calculating the length of life were well-established in Latin astrological texts, and much work remains to be done in deciphering the precise relationship between the *hyleg* and the *alcocoden* in each case. The printing of important medieval astrological works by Erhard Ratdolt, such as Haly Abenragel's *De iudiciis astrorum*, and Alcabitius's *Introduction*, no doubt facilitated the comparison and collation of important astrological terms. The editorial revisions of the text in subsequent editions, which continued to be published well into the sixteenth century, sometimes aimed to clarify doctrine, but not always. For example, there is at least a clear relation made between the calculation of progressions and the *hyleg* in the 1551 edition of Haly's *De iudiciis astrorum*: "And if neither of these are suitable [for the *hyleg*], the nativity will not have a *hyleg*. But it is necessary for you to return to directing through progression [*atazir*] the place of the *hyleg* to the unfortunate place, and to its rays, and there coming to its end place."[26] Several pages following are also devoted to the *alcocoden* and also clarify its relationship in that context. However, the work of the editor in presenting an exposition of both the *hyleg* and *alcocoden* found in the medieval Latin renditions of Haly's text certainly illustrate a sophisticated understanding of the doctrine, not the confusion evident in earlier centuries. This is also evident in Valentine Naibod's commentary on Alcabitius's *Introduction*, printed in 1560.

In this case, we see further evidence of humanist influence, as Naibod selects the Greek *aphesis* in place of the transliterated *hyleg* and explains the differences between Ptolemy's account and those of the "Arabs." Naibod gives a clear explanation of the *aphetic* point and the *anaeritic* point. In his discussion of the *alcocoden*, much like Cardano, he is dismissive of the account because it "differs from Ptolemy," but still references a scheme that has the greatest years corresponding to the cardines, the middle years corresponding to the succedents, and the least years corresponding to the cadent places, which also differs from Ptolemy, as well as from Kushyar, John of Saxony, and Haly Abenragel. This again raises the question of how these particular choices appeared in his text. This may be a simplified version of the great, middle, and least years which appears in an Arabic text not yet considered, and thus also demands further investigation, and inquiry into how Naibod discerned which techniques were "Arabic." The tension between Arabic authority and technical terms and a "return to Ptolemy" is evident just in Naibod's attention to Alcabitius in the form of his commentary, which is by far the lengthiest and most substantial commentary written on Alcabitius's *Introduction*. Whether or not Naibod believed Alcabitius to be an authority on astrology, his attention to the text itself certainly indicates his firm interest.

By way of conclusion, I return to the link between Naibod and Copernicus, mentioned at the outset. Copernicus himself demonstrates the tension between ancient authority and reform, through his commitment to the uniform circular motion of

---

[26] Haly Abenragel 1551, 147.

Aristotle and proposal of a Sun-centered cosmos. Naibod, on the other hand, spends a fair amount of ink devoted to how heliocentrism impacts astrological theory in his commentary on Alcabitius, and the issue of predictive accuracy. This is particularly poignant given Naibod's unfortunate end. But despite Naibod's "progressiveness" in his consideration of Copernican theory, his commitment to the authority of Ptolemy looms large, especially in his discussion of the length of life.

But the characterization I have just given leaves out the most essential element to the story: the role of the Arabic tradition in creating this tension. As critics of Ptolemy and innovators in their own right, Arabic astronomer-astrologers adapted and developed complex astronomical and astrological techniques and methods. After the transmission of some of this tradition into Europe, some European astronomer-astrologers, such as John of Saxony and Regiomontanus, were able to develop and augment these ideas. However, the complexities associated with transliterated terms and differences of technique led to the muddied astrological waters that humanist authors sought to clarify. To be clear here as well, the Arabs were by no means the "mud"—as the humanist reformers would have us believe—it was the complex process of transmission, assimilation, and comparison of sources that stimulated reforms and a requestioning of authority.

I hope also to have made clear the importance of comparative work that takes place across centuries. As sixteenth-century European authors were still reading the ancient works of Ptolemy, ninth-, tenth-, and eleventh-century Arabic authors, and the treatises and commentaries composed in the medieval Latin tradition, as historians we must be sensitive to the evolution of ideas across time and geographical space. This will contribute to further insights into how Europeans viewed themselves with respect to Arabic intellectual traditions from the twelfth to the sixteenth centuries, a question which is still far from resolved.

References

Abenragel, Haly. 1485. *De iudiciis astrorum*. Venice: Ratdolt.

Abenragel, Haly. 1551. *De iudiciis astrorum*. Venice: Ratdolt.

Alcabitius. 1485. *Libellus Ysagogicus Abdilazi*. Venice: Ratdolt.

Burnett, Charles, Keiji Yamamoto, and Michio Yano, ed. and trans. 1994. *Abū Maʿshar: The Abbreviation of the Introduction to Astrology, Together with the Renaissance Medieval Latin Translation of Adelard of Bath*. Leiden: Brill.

Burnett, Charles, Keiji Yamamoto, and Michio Yano, ed. and trans. 2004. *Al-Qabisi (Alcabitius): The Introduction to Astrology: Editions of the Arabic and Latin Texts and an English Translation*. London: Warburg Institute.

Campanella, Tommaso. 1629. *Astrologicorum libri VI in quibus astrologia, omni superstitione Arabum et Iudaeorum eliminate, physiologice tractatur*. Lyons: Iacobi, Andreae & Matthaei Prost.

Casulleras, Josep. 2009. Mathematical Astrology in the Medieval Islamic West. *Zeitschrift für Geschichte der Arabisch-Islamischen Wissenschaften* 18:241–268.

Casulleras, Josep, and Jan Hogendijk. 2012. Progressions, Rays and Houses in Medieval Islamic Astrology: A Mathematical Classification. *Suhayl* 11:33–102.

Ernst, Germana. 2001. 'Veritatis amor dulcissimus': Aspects of Cardano's Astrology. In *Secrets of Nature: Astrology and Alchemy in Early Modern Europe*, ed. Anthony Grafton and William Newman, 39–68. Cambridge, MA: MIT Press.

Gaida, Margaret. 2017. *Encounters with Alcabitius: Reading Arabic Astrology in Premodern Europe*. PhD dissertation, University of Oklahoma.

Gaida, Margaret. 2018. Reading between the Lines: Attitudes towards Arabic Astrology in the Marginalia of Latin Manuscripts. In *Knowledge in Translation: Global Patterns of Scientific Exchange, 1000–1800 CE*, ed. Patrick Manning and Abby Collier, 244–261. Pittsburgh: University of Pittsburgh Press.

Hasse, Dag Nikolaus. 2018. *Success and Suppression: Arabic Sciences and Philosophy in the Renaissance*. Boston: Harvard University Press.

Juste, David. 2016. The Impact of Arabic Sources on European Astrology: Some Facts and Numbers. *Micrologus* 24:173–194.

Naibod, Valentin. 1560. *Enarratio elementorum astrologiae*. Cologne: Arnold Birkman.

North, John D. 1988. *Chaucer's Universe*. Oxford: Oxford University Press.

Pico della Mirandola, Giovanni. 1946–1952. *Disputationes contra astrologiam divinatricem*, 11.2. Edited by Eugenio Garin. 2 vols. Florence: Vallecchi.

Pigghius, Albertus. 1518. *Adversus prognosticatorum vulgus, qui annuas predictiones edunt, et se astrologos mentiuntur, astrologiae defensio*. Paris: Henricus Stephanus.

Regiomontanus. 1490. *Tabulae directionum et profectionum*. Augsburg: Erhard Ratdolt.

Swerdlow, Noel. 1993. Montucla's Legacy: The History of the Exact Sciences. *Journal of the History of Ideas* 54 (2): 299–328.

Swerdlow, Noel. Forthcoming. *The Renaissance of Astronomy: Regiomontanus, Copernicus, Tycho, Kepler, Galileo*. New York: Springer.

Yano, Michio. 1997. *Kushyar Ibn Labban's Introduction to Astrology*. Tokyo: Institute for the Study of Languages and Cultures of Asia and Africa.

Yano, Michio, and M. Viladrich. 1991. Tasyīr Computation of Kūshyār ibn Labbān. *Historia Scientarum* 41:1–16.

Chapter 9
Albertus Magnus's First Lunar Model

Betsey Price

By the thirteenth century, all the ingredients necessary for medieval Latin scholars to exercise an evaluative response to texts of mathematical astronomy, increasingly available and varied, were present: adequate numeracy, and use of astronomical tables, as well as preceding authorities' manuscript texts and illustrations. Thus, it is asked how the exemplary thinker Albertus Magnus (1200–1280) addressed ideas and sources on astronomy. Visual imagery or modeling is key to understanding the approach of his predecessors and his own, as demonstrated in 1243–1244 in Albertus's earliest writings on lunar phase phenomena. As can be seen through parametric modeling of his sources' ideas and his own, Albertus, in understanding his authorities, modified their analytical accounts to his own ends.

9.1 Introduction

Latin medieval thinkers have not figured among the most revered contributors to the advancement of the history of astronomy. Albertus Magnus, the Dominican scholar (1200–1280)[1] whom many believe was the leading figure in thirteenth-century Latin learning, had, however, explicitly foreseen natural scientific and

[1] A painting of Albertus Magnus can be found in the Chapterhouse of the Seminary, formerly the Dominican convent, of San Nicolò, Treviso. His portrait is part of a series depicting 40 illustrious Dominicans.

B. Price (✉)
Department of History, Glendon College, York University, Toronto, ON, Canada
e-mail: bprice@glendon.yorku.ca

F. Rochberg (ed.), *Studies in Premodern Sciences in Memory of Noel M. Swerdlow*, Archimedes 74, https://doi.org/10.1007/978-3-032-04040-4_9

mathematical texts to be important portions of his own projected corpus of commentaries, and many works related to astronomy and astrology as well as other natural scientific writings were attributed to him in the earliest catalogs of his corpus of works. While, most likely simply for lack of time, a *De astronomia* Albertus projected to write[2] never came into being, it was clearly part of his intention to write on "all the mathematics"[3] and thus for the work to become a part of his whole "personal intellectual and cultural project" (Bazán 2005, 586–87, n. 7). The august historian of astronomy Noel Swerdlow aligned himself, however, as expressed in one of his last lectures, with Renaissance thinkers, as, for example, Regiomontanus (1474), who when introducing an edition of Euclid's *Elements* noted, about the medieval Campanus of Novara: "the opinions of Campanus ought to be removed" (Swerdlow 2019). To those who knew Noel well, it was clear that for Campanus, he felt should largely be read "all medieval Latin authors commenting on mathematics and astronomy."

Swerdlow articulated the criticism that the qualities reflecting an understanding of astronomy were simply missing from thinkers within the Latin Middle Ages: their texts are simple, without quantifiable details; their ideas are not presented as "real mathematical astronomy"; and their statements cannot be understood in mathematical terms. As will be discussed here, however, even Albertus Magnus's earliest writings on astronomy undermine these negative critiques of medieval European astronomy. Putting Noel's obvious concordance with Renaissance assessments aside, he did point to the explicating of Albertus's interest in astrology as most appropriate and potentially fruitful given the richness in research material available to that end.[4] Noel's perspective will be highlighted here, albeit within the technical context of medieval astronomy, showing that Albertus had the necessary regard for numerical specification and precise quantitative references for both astronomy and astrology. In order to appreciate in what way both the tools and the necessity for technical discernment were presented to Albertus as well as to his contemporaries, the following provides some background.

[2] See Albertus Magnus (1251; ed. Hossfeld 1987), *Physica* I.1.1 (ed. Colon. 4:3.38–41). The notation for this reference and all subsequent ones relying on the Aschendorff Verlag publications are cited as part of the so-called Cologne edition (editio Coloniensis), reflecting the now current way of evoking the new Albertus Magnus *Opera Omnia* project supervised by the Albertus-Magnus-Institut of Bonn (Geyer ed. (1243-1279) 1951-).

[3] E.g., see the cautious assertion of Craemer-Ruegenberg 1980, rpt. 2005, 41–42, who, while also questioning whether Albertus completed his own ambitious desire to write on "all the mathematics," nonetheless asserts: "On astronomy, namely 'astrologia', Albertus must, however, have written, according to his own testimonies and the older catalogues of works." ["Uber die Astronomie, bzw "Astrologie" muss Albert jedoch—seinen Selbstzeugnissen und alteren Werkskatalogen nach—geschrieben haben"]. See also Magnus 1262-63; ed. Geyer 1958.

[4] Some of Swerdlow's most personal, remarkably kind, and cajoling comments to me reflect this: "I don't know what to suggest except to … concentrate on other things in Albert's astronomy and astrology. Surely there is enough there, especially if you pick up more on the astrology or natural philosophy of the heavens that considers the relations of the heavens and the sub-lunar regions and living things, including humans, medicine, and whatever else is there."

9.2 European Medievals's Methodological Toolbox

In general, as medieval tools for analysis of astronomical ideas, one could point to three: (1) an epistemological approach, adopted from Aristotle's *Posterior Analytics* (Aristotle ca 350; ed. Minio-Paluello and Dod 1968), organized as the so-called classification of the sciences, (2) observation, instruments, and astronomical tables, and (3) manuscript writings and graphics.

9.2.1 Classification of the Sciences

Understanding the thirteenth-century Latin writers' intellectual context must include their "classification of the sciences," that is, their schematic divisions of all knowledge. E.g., one popular example of such a classification scheme found in Dominicus Gundisalinus (after 1140; ed. 1903), *De divisione philosophiae.* Albertus's interest in astronomy and astrology was fully structured within it. He submitted *Astronomia*, a theoretical division of mathematics, synonymous to *scientia astrorum*,[5] to a division into three parts:[6] *astrologia*[1], *astrologia*[2], and *astronomia* (Price 2012, 418) (Fig. 9.1).

While the focus of *astrologia*[1] and *astrologia*[2] is primarily on the quantitative abstractions from matter, i.e., form, arrangement, size, location, and speed, their subject is recognized as mobile but only in terms of local motion, not in terms of qualitative motion or "change" (Albertus Magnus (1272-71; ed. Borgnet 1890), *Analytica posteriora* 2.2.11 [ed. Borgnet 2:192b]). *astrologia*[2] Albertus called mixed or conjoined with physics (Magnus, 1262-1263; ed. Geyer 1964, *Metaphysica*

[5] "Sed oportet propriam de hoc alibi esse considera quia ipsa quae de astris est mathematica, videtur esse de corpore, quod est determinatae naturae mobilis, sicut de caelo et stellis. Sed quoad unam eius partem, [astrologiam], quae de quantitatibus caelestium, licet consideret id quod movetur, non tamen considerat in eo principia motus, sed potius mensuras quantitates eorum quae moventur, et motuum ipsorum. Et ideo mathematica est illa consideratio. Altera autem pars ipsius, [astronomia], quae considerat effectum astrorum in inferioribus, miscet se physicae, sicut diximus in II Physicorum" ("But there needs to be proper consideration of this elsewhere since that which about the stars is mathematical seems to be about the body, that is the mobile thing, as the heavens and the stars, of a determined nature. But as for the one part of it [astrologiam], which is about the quantities of celestial things, although it may consider that which is moved, nevertheless it does not consider the principles of motion in it, but rather the measurements of the quantities of those things which are moved and of their motions. And therefore, that consideration is mathematical. The other part of it [astronomia], however, which considers the effect of the stars on inferior things, mixes itself with physics, as we said in Physics II") Albertus Magnus (1263-1264, ed. Geyer 1964), *Metaphysica* VI.1.2 (ed. Colon. 16/2:304.69–80).

[6] Far from being confusing, Albertus had simply adopted the labelling of greatest use and clarity among his contemporaries. Cf, e.g., Roger Bacon, 1267–1270; ed. 1913 (Steele 1913, 6).

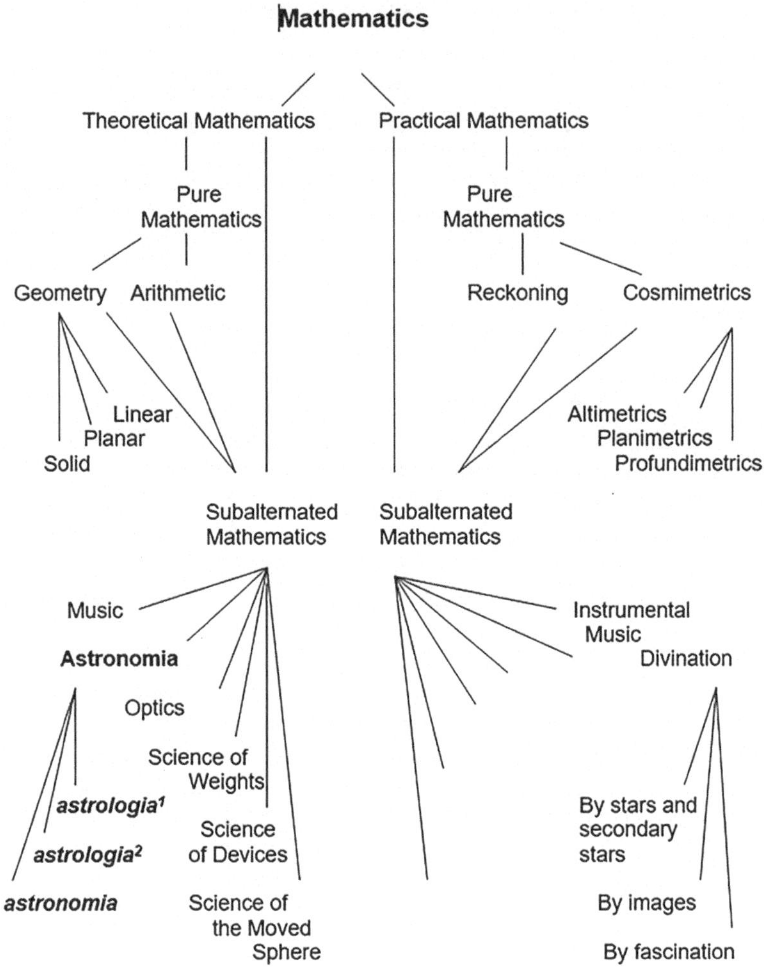

Fig. 9.1 Schematic classification of the mathematical sciences by Albertus Magnus

1.5.4 [ed. Colon. 16:73.34–43]). *astrologia*[2] is less abstract (conceptually) than *astrologia*[1] in respect of its subject matter. No part, however, of *Astronomia*, a mathematical science by reason of its principles (Albertus Magnus 1251; ed. Hossfeld 1971 and Moraux 1965), *De caelo* 2.3.11 [ed. Colon. 5/1:167.83–85]), makes use of the principles of natural philosophy. There is a third part, *astronomia*—in addition to the division of *Astronomia* that is *astrologia*, into its two parts—which concerns celestial virtue (Albertus Magnus, 1258-62, *De animalibus* XX.2.2 [ed. Stadler 1920 16:1310.41–42]) and the effects of stars on inferior or terrestrial things. Physical signs determine first and foremost the necessity for this third division.

Albertus mentions many sources for *astrologia*, but more often than not their authoritative conclusions are noted without stating how they were obtained. While

Albertus identifies the use of observational instruments explicitly, as will be seen below, he is not specific about what information is derivable from them. To his mind, the practitioners of *astrologia*[1] employ exclusively, "demonstratio quia per causam remotam" [demonstration of a fact by its remote cause],[7] while those exercising *astrologia*[2] use "demonstratio quia per effectum" [demonstration of a fact by its effect] to arrive at authoritative statements (Albertus Magnus (1250; ed. Simon 1972a), *De divinis nominibus* 4 [ed. Colon. 37/1:150.20–23]) and sometimes even to conjecture for the attribution of "facts." Still to Albertus, the unprovable opinions of the *moderni* appear in marked contrast to the long-standing definitions of the ancient *astrologi*, which have achieved the timelessness of demonstrated proofs (Albertus Magnus, 1262-1263; ed. Geyer 1964, *Metaphysica* 11.2.24 [ed. Colon. 16:513.94–95] and 1243-44; ed. Geyer 1974, *(De natura boni* 2.3.2.2.3.A.1.1 [ed. Colon. 25/1:49.78–79])).

9.2.2 Technical Tools

9.2.2.1 Observation (with or without Instruments) and Astronomical Tables

Since, in the history of science, use of technical tools by astronomers, particularly premodern astronomers, seems to have become a kind of litmus test for detecting serious understanding of things astronomical, it is important to assess their role in Albertus's astronomy. Albertus has many times over been recognized as an astute observer of the natural world. In his commentary on Aristotle's *Mete orologica* (Aristotle ca 340; ed. Vuillemin-Diem 2008 and Rubino 2010), he reported that in Saxony in 1240 he had seen a comet, which "appeared as if near the North Pole and it projected its rays between the East and the South, directing toward the East" (Albertus Magnus (1252-54; ed. Hossfeld 2003), *Meteora*, 1.3.5 [ed. Colon. 6:29.50–54; trans. Thorndike 1951, 70]).[8] Albertus gave the impression that most of his contemporaries' sightings were made by means of observational instruments, both calibrated and noncalibrated. He was himself knowledgeable about astronomical instruments, referencing the armillary sphere, plane astrolabe, quadrans vetus,

---

[7] Albertus (1251; ed. Hossfeld 1987), *Physica* 2.1.8 (ed. Colon. 4:91.9–12). Aristotelian scholastics called "demonstratio quia" a general method which produces affirmations revealing the cause of an effect as having been solely producible by its identified cause. "In a demonstration *quia* we establish from what we know of the effect *that* its cause without doubt exists, and from a demonstration *quia* we know enough about the nature of the cause to know *that* it was adequate to produce this effect" (emphases added, Snyder, 1998, i).

[8] Regarding a recorded eastern sighting of the same comet, C/1240 B1 on 31 January 1240, see Murata et al. 2021. The comet does not seem to have reappeared since this one set of sightings worldwide, and so little is known about it that predictions of its return to within view of the earth are not possible. Even without noting this example, Albertus has been identified by many historians as a "poster boy" of medieval observers, as noted by Park 2011, 28, in her chapter on Latin observers between 500 and 1500.

parallactic ruler, meridian ring, and water clock (Albertus Magnus 1262-1263, ed. Geyer 1964, *Metaphysica* 11.2.22 [ed. Colon. 16:510.42–511.6])[9] and elaborating on the biblical "horologium Achaz" of Isaiah 38:7–8.[10]

Regarding the technical tools of astronomical tables, it can be found that throughout Albertus's works, he made explicit reference to at least six different types:[11] (1) geographical coordinate tables for particular places, (2) tables giving daylight length and geographical latitude, (3) eclipse tables for the Sun and for the Moon, (4) tables of accession and recession of the eighth sphere, (5) tables of planetary effects, and (6) ephemerides. Albertus noted, as the first set of tables ever mentioned in his works, having consulted tables for the access and recess motion of the eighth sphere, "Tabulae acessus et recessus octavae sphaerae."[12] The most significant use of technical astronomical skill discovered to date in the works of Albertus Magnus is his use of tables in 1250 to establish whether or not the "darkness" at the time of the crucifixion of Jesus of Nazareth was a natural or miraculously occurring event. He concluded after a complex process of elimination, using tabular data and calculation, that the lunar eclipse which obscured the Sun on the biblically noted Friday was due to divine miracle, in that not all of the requisite celestial relationships were occurring at that time in 34 CE.[13] As Tony Grafton and Noel Swerdlow (Grafton and Swerdlow 1988, 21) noted about one author's dating of the founding of Rome, "We have come a long way from simple deductions based upon phases of the moon, … Tarutius [fl. 86 BCE] evidently used, with some measure of competence, the most

[9] It has not yet been possible to establish a single one-to-one correlation between the names Albertus gives for the instruments he saw as the tools of astronomy and astronomical instruments known to have been used in the Latin West during the thirteenth century.

[10] "Achaz horologium fecerat in templo per modum cylindri vel quadrantis catheto erecto a medio centro, qui per umbram designaret horas diei secundum lineas horologii in duodecim divisi vel secundum erectionem pinnarum in extremo quadrantis, per quarum aspectum perpendiculo dependente horae diei designarentur" (Achaz made a time-telling instrument in the temple in the manner of a cylinder or a quadrant. Either it was a [cylindrical] instrument with a perpendicular erected from its midcenter which by means of a shadow designated the hours of the day according to the lines of the time piece, which had been divided into 12 or it was one, in which according to sighting pins erected on an edge of a quadrant the hours of the day are designated by means of sighting through them, a perpendicular [plumb line] hanging down). Albertus Magnus (1254-60; ed. Siepman 1953), *Postilla super Isaiam* ca. 38.8 (ed. Colon. 19:397.52–57). Versions of both these instruments, and many more, are captured in the Bibliothèque Royale de Belgique MS. IV.111, fol. 13v. French illustration of Henry Suso's *Horologium Sapientiae* (1440–1460) .

[11] For the different table typologies, see Chabás 2012.

[12] Other studies corroborate what Albertus's use of this table reveals, his engagement with the phenomenon of precession of the equinoxes in his *De IV coaequaevis* (Magnus ca 1242; ed. Borgnet 1894 and Anzulewicz 2008) (Palazzo 2023, esp. 26; see also Mercier 1976 and 1977).

[13] Price 1991. Albertus maintained that the eclipse shadow at that time extended from India to Spain (namely over the "whole earth") and noted that it would have required positions different from those recorded in the tables for apposing solar and lunar longitudinal positions, lunar latitude, and the position of the lunar-solar conjunction relatively close to the Vernal Equinox (Albertus Magnus (1250; ed. Simon 1972b), *Super Dionysium Epistulas* VII [ed. Colon. 32/2:509–10]). Influenced perhaps by the technically grounded arguments of his master, Thomas Aquinas held to the same assertion.

sophisticated astronomical techniques, available in his time." Albertus too reflected a degree of awareness, familiarity, and use of astronomical tables relatively sophisticated for his time.

9.2.2.2 Models

By extension, in the medieval mind, a type of ontological status by proxy, inherited, as it were, from its association to the scientific discussion of the existent thing, was transferred beyond number to scientific imagery, leading thus to the inclusion of a further 'technical tool' for a medieval scholar's serious understanding of things astronomical: modeling the subject of concern. Visuals of astronomical models are replete in medieval astronomical manuscripts, whether due to the author's own inclusion of them or to a later copyist's or glossator's addition. For some current scholars, simply the fact that Albertus or any medieval author used an astronomical table, meant that he had already factored in the modeling and models that lie behind that tool: "Underlying most tables, at least the most complex ones, there is a geometrical model and, thus, one or more parameters that define it."[14] Albertus, however, stretched the role of modeling beyond his use of astronomical tables. In drawing his own sketch to accompany a textual passage in his commentary on Aristotle's *De caelo et mundo* (Aristotle 350 (ed. Leggatt 1995)), he conveyed the importance of visual representations to his words for generations of copyists to come.[15]

Further, when Albertus grappled with the Ptolemaic system, he showed he had adopted specific modeling for Ptolemy's reasoning's evoking something more than a convenient geometric fiction, particularly when he interpreted the Hellenic astronomer to assume the actual existence of a ninth heavenly sphere, not to accommodate celestial bodies per se but to explain the dynamic transmission of motion within the heavens (Price 1980). Even more striking perhaps is that Albertus drew attention to a particular aspect of Al-Biṭrūjī's system (Al-Biṭrūjī ca 1190 (ed. Carmody 1952 and Goldstein 1971)), the spiral path of a planet's periodic motion and its modeled "laulab" (which he defined literally as a twisted piece of wood).[16]

[14] Chabás and Goldstein 2012, 234–35. "It is quite usual to find tables for the same purpose, based on the same model, but with different parameters. Recognizing these embedded parameters remains essential for identifying the author of a table or the tradition to which it belongs" Chabás 2012, 284.

[15] Cf. Albertus autograph (Commentary on Aristotle's) *De caelo et mundo* MS Wien OMB 273, f. 79. Certain copies of Albertus's *De caelo et mundo* place their version of his drawings in the margins. In the first and third noted here, they appear as omissions, not to be forgotten (i) MS Paris BN 6509, f. 132ra, French, 1300s; (ii); MS Paris BN 6510, f. 9rb, French, 1300s; (iii) München, Bayerische Staatsbibliothek MS Clm 28186, f. 152va, possibly southwest German, early fourteenth century (See Hauke 1986).

[16] Hossfeld 1969 believes this definition stems from Albertus himself.

9.3 Consideration of Astronomical Models

In his day,[17] thus, Albertus seems to have been quite persuaded of the importance of the visual representation of astronomical systems in his sources, just as he was absorbed by the imagery accompanying the textual descriptions of those same works. One example of this is presented here, on the theme of the Moon, stemming from an exegetical exercise Albertus undertook in his work, *De natura boni*, from the 'first period' of his writings, ca. 1228–1238 (Price 1992). Among Albertus's sources for this exercise, Martianus Capella and Macrobius, whose texts in one or more manuscript copies Albertus certainly studied, abound in visuals, and their importance cannot be denied. Writes Hiatt (2007), "For both Macrobius himself and his medieval copyists … [the graphics seem] to have functioned as an alternative to purely verbal description, indeed one that offered a superior mode of understanding." This case study of two sources Albertus used in his exegetical exercise presents itself to support the assertion that conceptual imagery and modeling, widely operative in ancient and medieval astronomy, are key to understanding how Albertus and likely, many other contemporaries, understood and discriminated among preceding astronomers' proposals and models, and to emphasize further that their analysis was philosophically and technically well grounded.[18] The aim here then is to establish that Albertus's own experience of text and image in the sources he read led him to envisage their conveying different astronomical models and their offering enough detail to permit his evaluation of them.

9.3.1 Albertus's Sources and Their Lunar Models

Martianus Capella, an early North African Roman jurist, is one of Albertus's two earliest sources on astronomy, embedded in the group he calls the *astrologi* (astrologers), *antiqui* (ancients), and *sapientes in astris* (wise ones concerning the stars). The source text has been identified as *De nuptiis Philologiae et Mercurii* ("On the Marriage of Philology and Mercury"), written between 410 CE and 439 CE. It is "the fabula of the marriage feast … [at which] Mercury offers Philology a dowry [which] consists of seven virgins, who represent the seven liberal arts, one of which is Astronomy (Book 8)" (Martianus 410-439; ed. Musso and Ghiringhelli 2019 and Cristante 2011). Book 8 was by far the most popular portion of the work, with 234

[17] Unlike ours, perhaps as observed by Jardine 2006, 472 in noting the lack of importance historians of astronomy have attributed to visual conceptualizations: "We all too easily overlook the communicative roles of letters, marginalia, images, models, and instruments."

[18] Over the course of years, albeit complemented by numerous digressions, one puzzle of my earliest study began to dominate; it was what I see as having been a major dilemma for Albertus and his contemporaries: the criteria for choosing among astronomical models.

surviving copies. It circulated as a fragment and is one source text for Albertus's passage on lunar motion in *De natura boni.*[19]

The second pertinent source author is Macrobius Ambrosius Theodosius, usually referred to simply as Macrobius, a Roman provincial who lived during the early fifth century CE. His important work in our context is the *Commentarii in Somnium Scipionis* (Theodosius, 410-439, "Commentary on Marcus Tullius Cicero's *Somnium Scipionis*"; ca. 400 CE; ed. Stahl 1952; Willis (1970) 2013). Cicero's work relates a dream experienced by Scipio Africanus (185–129 BCE), a famous Roman general, who discusses the configuration and nature of the universe with his dead grandfather in the sphere of the stars. There are over 69 extant manuscripts of Macrobius's work, which include images of astronomy. Thanks to the contributions of Bruce Eastwood and his co-authors, very many of the astronomical images in the earlier medieval manuscript copies of authors, such as Martianus Capella and Macrobius, have been identified; e.g., Eastwood 2002 and Eastwood and Grasshoff 2004. The *Commentary* was among his most popular writings, with copies of the whole work frequently identified as "de astronomia."

While both authors' works include references to lunar phases and descriptions of lunar-solar motions and phenomena, discussions of lunar phases by an author do not necessarily constitute evidence of his having an underlying independent model, in this case of lunar motion. If, however, general discussion and even visual representation do not necessarily point to a lunar model and underlying theory of lunar motion in an author's writing, what then would or could allow for Albertus's attribution of a lunar theory to Martianus and/or Macrobius, which could serve ultimately as a basis for his own, as their thirteenth-century reader?

At the time Albertus was writing his *De natura boni*, in which Martianus and Macrobius are referenced for the first time, he was a Dominican scholar, approximately 30 years old, beginning to gain greater familiarity with Latin astronomical accounts, observations, and tabulations. His initial application of source information about astronomy was to biblical exegesis, even as he likely employed other passages of these same texts to different ends in his other writings and teaching. In response to the question posed above, the data in the writings of the two authors referenced by Albertus in his exegetical exercise offer two corroborating ways of providing an answer: (1) they can be translated into the standard parameters found in early astronomical theories and their models and (2) they can be correlated to manuscript imagery accompanying the texts, images which evoke models and a theory behind them. With the aid of these two indicators, data and imagery together, construction of lunar models is feasible and revealing of how Albertus could have interpreted the texts himself as a basis of his own lunar model. It is surmised that the dominant account of phases of the Moon in medieval manuscripts, seven or eight, as shown in Fig. 9.2, a Mediterranean tradition, is in evidence here.

[19] Martianus Capella, ed. Stahl et al. 1977. This edition of the *Marriage of Philology and Mercury* is still the definitive edition/translation of Martianus Capella *De nuptiis Philologiae et Mercurii*, and even as the manuscript tradition has much variation from textual uniformity, we will adhere to this edition's divisions of textual content to identify passages under consideration.

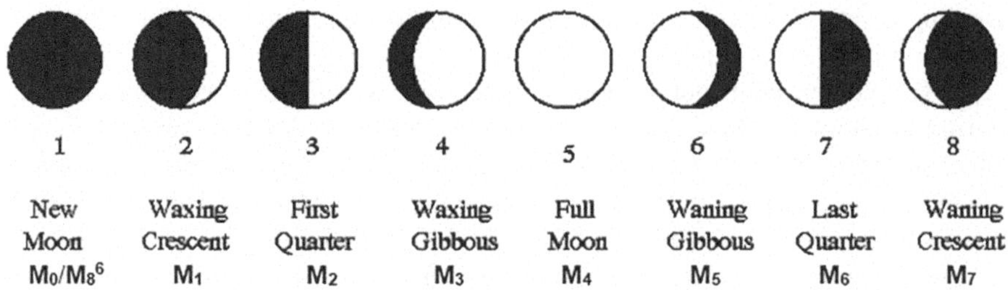

Fig. 9.2 Lunar phase markers

Much of the data from Martianus and Macrobius available to Albertus can be translated into astronomical parameters. To see their particular importance, respectively, in Tables 9.1 and 9.2, the pertinent data, stated as based on observation, are isolated from the seemingly theory-derived quantities in the texts Albertus mined. Thus, all quantities relating to lunar motion in Albertus's source texts are categorized into either values *stipulated* by other textual details *or* values deriving for all intents and purposes from *observation*, i.e., *measured*. Tables 9.1 and 9.2 present two respective sets of data, with their parameter identity (Column 3) and citation location in the source texts (Column 2).

Among noteworthy points for Martianus is that parameters expose his positing an eccentric orbit for the Sun. Martianus superimposed two periodic motions, revealed by changes in lunar angular elongations from the Sun (heliacal anomaly) and in lunar sidereal longitude (zodiacal anomaly)—with the values for the heliacal anomaly reflecting an eccentric motion of the Sun. The lunar position is determined on a geocentric circle. The value of solar eccentricity is extrapolatable from Martianus's values of equinoctial periodicity, amounting to a mean motion of 0;59°,8′25″/day between the vernal and autumnal equinoxes, and 1°17′13″30/day between the autumnal and the vernal equinox. With the data points provided by Martianus, the timing of each lunar phase could be known over an entire lunar period, but (as will be seen, in the model presented for Martianus,) the lunar phases M_1 and M_7 are to be considered not knowable theoretically, as per Martianus's reflection, that these two crescent-shaped lunar phases must be independently observed. Therein is possible indication that even with calculation, the exact timing of these two phases could not be known, at least not without the addition of hourly to daily tabulations: the daily differences of $\lambda_M - \lambda_S{}^{v}$ and $\lambda_M - \lambda_{S^{\wedge}}$ in the spring–summer months and in the fall–winter months, respectively, offer no real discrepancy as to the exact day on which phases M_1 and M_7 fall.

The data provided by Macrobius show that, like Martianus, he made an explicit distinction between the different traditionally enumerated seven/eight lunar phases, but also that, like Martianus, he was reserved about total reliance on theoretical calculation. It should be noted that nowhere does Macrobius indicate what portion of the Moon he would determine to be lit when in the gibbous phases (denoting

Table 9.1 Martianus's Data, Categorized as Theory-Derived ("Stipulated" by Theory) or Observation-Based ("Measured") Values

Values	Source Text	Parameters
STIPULATED		
M_2 90° M_3 135° M_4 180° M_5 225° M_6 315° M_0 0°	… XC partibus (864; 1-18)	λ_{M2} λ_{M3} λ_{M4} λ_{M5} λ_{M6} λ_{M0}
M_4 day 14, 15, 16 M_0 day 30	… XIII diebus (866; 12—15)	λ_{M4} λ_{M0}
13°/day	…XIII orbis (864; 18-19)	α_{vm}
	… eccentron Solis (873; 3-5)	
100 x R_E	…circularem centies (860; 6-9)	C_m
1/6C_E or 60°	…sexies minorem (859; 12-13)	d_m
354 days	… CCCLIIII diebus (866; 17-18)	$l_{m/y}$
365 days/365¼ days	… diebus XI (866;18-20)	l_s
1/6C_E or 60°	Lunam sexies … (859; 12-860)	Lunar diameter
12 x (100 x r_E or r_m)	… circulum duodecim (861; 1-2)	C_s
8° Aries	(829;…)	λ_{0v}
8° Libra	(829; …)	$\lambda_{0\wedge}$

	Source Text	Parameters
MEASURED / OBSERVED		
29 1/2 days synodic 27⅓ days sidereal	Luna circuit … (865; 1-3)	l_{msyn} l_{m*}
M_1 M_7		λ_{M1} λ_{M7}
185¼ days	CLXXXV diebus et quadrante diei (873; 21-24)	l_{sv}
180 days	CLXXX diebus (873; 21-24)	$l_{s\wedge}$
8° Aries	(824;…)	λ_{0v}
8° Libra	(824; …)	$\lambda_{0\wedge}$
360°/600	… orbe suo sexcienties (860; 1-2	d_{m-ap}
406,010 stadia	CCCCVI milia stadia et X (858 8-13)	C_E
20° or 1/18 of earth's 360° (C_E)	reperi duodeuicesimam (859; 5-8)	lunar shadow

Table 9.2 Macrobius's Data, Categorized as Theory-Derived ("Stipulated" by Theory) or Observation-Based ("Measured") Values

Values	Source Text	Parameters
STIPULATED		
28/4 = 7 days per quarter	In quarter septenos ... (Book I, vi [52]	lmsyn/4
Epoch 0° = northmost latitudinal point from node	septem diebus ... (Book I, vi [53])	$\lambda_0 - 90^\circ$
Syzygy Epoch values: 0° (M_2), 90° (☊ node, M_4) 180° (M_6) 90° (☋ node, M_8/M_0)	septem diebus ... (Book I, vi [53])	λ_{M2} λ_{M4} λ_{M6} λ_{M0} / λ_{M8}
Sidereal Epoch Values from Aries 0°: M_2 (at 90°) in 7 days M_4 (at 180°) in 14 days M_6 (at 270°) in 21 days M_8 (at 360°) in 28 days	septem diebus ... (Book I, vi [53])	λ_S at t = 7 days + 90° λ_S at t = 14 days + 180° λ_S at t = 30 days - 90° $\lambda_S = (l_s + \lambda_0)$
80,000 stadia	octoginta milia stadiorum ... diametros (terrae) (Book I, xx [20])	δ_E
Diameter x 60 = 4,800,000	Quadragies octies centena milia stadiorum (Book xx [21])	R_S

Values	Source Text	Parameters
MEASURED / OBSERVED		
30 (28 + 2) days synodic	triginta dies ... (Book I, vi [49])	lmsyn
28 days, approximate sidereal	octo et viginti prope diebus ... (Book I, vi [49])	≈lm*
27 days, 8 hours, sidereal	viginti septem dies et horas fere octo ... (Book I, vi [50])	lm*
Aries 0°	primam partem Arietis ... (Book I, vi [50])	λ_0
360 days/12 months = 30°	unum de duodecim signis integro mense metitur (Book I, vi [50])	l_s
32 days	" duos supra triginta ..." (Book I, vi [51])	l_s in Gemini
252,000 stadia	Ambitum ... stadiorum milia ducenta quinquaginta duo (Book I, xx [16-18])	C_E

them by the Greek term, *amphicyrtos*[20]), M_3 and M_5, an important detail for both theoretical and observational confirmation of the point of that specific phase. Macrobius provides two lunar frequencies, both independently measured: a sidereal revolution of 27 days, 8 hours, and a synodic revolution of 30 days (extrapolated from an epoch start at the greatest point of its northern latitudinal deviation from the ecliptic). The Sun's sidereal progression is seen as regular and independently measurable at 30°/month. A geocentric circle accounts for this motion, which ignores seasonal inequalities; there is no discussion of varying solar distance or season lengths. Heliacal anomaly is expressed in terms of the composite result of solar and multiple lunar motions. Macrobius does refer to the fact that as the Sun traverses one sign of the zodiac, Gemini, it takes it 2 days longer to complete its 30°, hence its sidereal motion has slowed (and summer would be longer than winter). There are no more parameters added to allow for positing either an eccentric or epicycle for the Sun's motion. Macrobius does integrate a periodicity of lunar latitudinal variation (without a value) into his discussion when noting eclipses, but only in passing: "when both the sun and moon are pursuing their courses along the same line at the same time, it is inevitable that one should suffer eclipse" (Book 1.XV [11]).

Returning to the question of what might have allowed for Albertus to attribute a lunar theory to Martianus and Macrobius, which would serve ultimately as a basis for one of his own, the role of manuscript illustrations needs to be highlighted. For Martianus and Macrobius, the illustrations in their manuscript copies appear to represent visually specific aspects of their respective details of lunar motion. Part of the manuscript tradition of Martianus's work, which Albertus would have experienced, includes these graphics, which appear to offer elucidation of the author's consideration of lunar nodes and eclipses,[21] lunar latitudinal deviation from the ecliptic,[22] and solar eccentricity.[23] Macrobius's manuscripts too abound in graphics which seem to elucidate his consideration of eclipse phenomena[24] and varying angular relationships between the Sun and the Moon.[25]

For both Martianus and Macrobius, it seems possible that the details in their writing about the Moon stimulated manuscript illustrators to think a possible model lay behind the author's conception of lunar motion, since connection between the very data each author provided and the illustrations can actually be made. In the Martianus

[20] As found in marginal manuscript imagery Macrobius's *Commentarii in Somnium Scipionis* [*Macrobius: Commentary on the Dream of Scipio*] in Bodleian Library MS. Selden Supra 25, f. 211v (England, twelfth century).

[21] E.g., MS BnF. Département des Manuscrits. Latin 14754, f. 209r. Chartres, France, 1151–1175.

[22] E.g., Leiden, University Library, MS Voss. lat., F48, f. 80v., France (Auxerre), ca. 850.

[23] E.g., Leiden, University Library, MS Voss. lat., F48, f. 81r., France (Auxerre), ca. 850.

[24] E.g., Copenhagen, Det Kongelige Bibliotek, ms. NKS 218 4°, f. 49v, southern France, ca. 1150.

[25] E.g., British Library Add MS 11943, f. 49r, northwestern France or England, first half of the twelfth century.

Table 9.3 Martianus's Data Reflected in Manuscript Illustrations

Values	Parameters
STIPULATED	
M_2 90°	λ_{M2}
M_3 135°	λ_{M3}
M_4 180°	λ_{M4}
M_5 225°	λ_{M5}
M_6 315°	λ_{M6}
M_0 0°	λ_{M0}
M_4 day 14, 15, 16	λ_{M4}
M_0 day 30	λ_{M0}
13°/day	α_{vm}
100 x R_E	C_m
1/6C_E or 60°	d_m
354 days	$l_{m/y}$
365 days/365¼ days	l_s
12 x (100 x r_E or r_m)	C_s
8° Aries	λ_{0v}
8° Libra	$\lambda_{0\wedge}$
Values	**Parameters**
MEASURED / OBSERVED	
29 1/2 days synodic	l_{msyn}
27⅓ days sidereal	l_{m*}
M_1	λ_{M1}
M_7	λ_{M7}
185¼ days	l_{sv}
180 days	$l_{s\wedge}$
8° Aries	λ_{0v}
8° Libra	$\lambda_{0\wedge}$
360°/600	d_{m-ap}
406,010 stadia	C_E
20° or 1/18 of earth's 360° (C_E)	lunar shadow

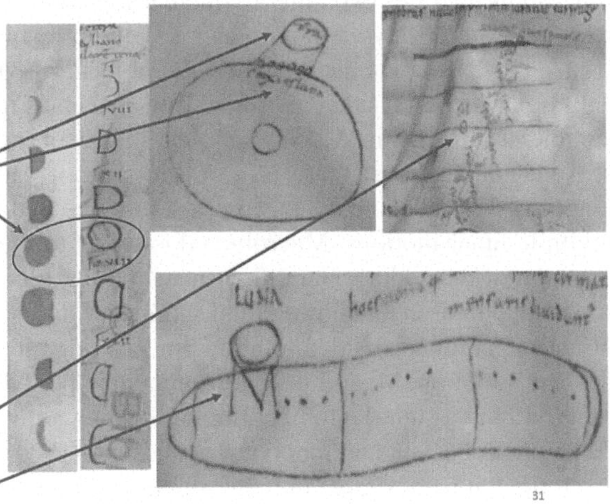

copies, connection can be made between text and illustration for the quantities defining lunar phase day counts and durations, and lunar latitude (Table 9.3).[26]

In the copies of Macrobius, it has been possible to correlate data and manuscript illustrations to his details of lunar phase day counts and duration and the solar path (Table 9.4).[27]

With texts that offer an abundance of parametric data and illustrators who had captured some of those parameters in graphic representation, how much of a stretch would it have been for the thirteenth-century Albertus, reading these texts in illustrated manuscripts, to have attributed specific lunar theories to Martianus and to Macrobius, his "*sapientes in astris*"? One could start by noting that there is sufficient parametric data provided by each author for a respective lunar model to be constructed for his source authors. In addition, however, each author has specific characteristics in his work available to Albertus to point to a model behind their ideas. In the case of Martianus there are at least three:

[26] From left to right, (a) MS Leiden B.P.L. 88, f. 163v, region of Reims, France, ninth century; (b) Vatican MS Reg.lat.1987, f. 129r, northeastern France, second quarter, ninth century; (c/e) Leiden, University Library, MS Voss. lat., F48, f. 79v, Auxerre, France, 830–40; (d) Leiden University Library, MS Voss. lat. F.48, f. 92v, Auxerre, France, 830–40.

[27] From left to right, top to bottom, (a) British Library Harley 5433, f. 70r Appendix, Northern Italy, 1446; (b/d) British Library Add MS 11943, f. 49r Appendix, northwestern France or England, first half of the twelfth century, and (c) British Library Add MS 11943, f. 27r. There is no unanimity in the manuscript drawings' interpretations regarding whether eccentricity or epicyclic motion plays a part in Macrobius's geocentric universe. (The interpretation of an eccentric lunar orbit has been opted for here.)

Table 9.4 Macrobius's Data Reflected in Manuscript Illustrations

Values	Parameter
STIPULATED	
28/4 = 7 days per quarter	lmsyn/4
Epoch 0° = northmost latitudinal point from node	λ_s - 90°
Syzygy Epoch values:	
0° (M_1).	λ_{s0}
90° (☊ node, M_4)	λ_{s90}
180° (M_5)	λ_{s180}
90° (☋ node, M_3/M_7)	λ_{s90} / λ_{s270}
Sidereal Epoch Values from Aries 0°:	
M_2 (at 90°) in 7 days	λ_s at t = 7/8 days + 90°
M_4 (at 180°) in 14 days	λ_s at t = 14/15 days + 180°
M_6 (at 270°) in 21 days	λ_s at t = 30 days - 90°
M_8 (at 360°) in 28 days	λ_s = (l_s + λ_s)
80,000 stadia	δ_s
Diameter x 60 = 4,800,000	R_s
Values	**Parameter**
MEASURED / OBSERVED	
30 (28 + 2) days synodic	lmsyn
28 days, approximate sidereal	≈lm*
27 days, 8 hours, sidereal	lm*
Aries 0°	λ_0
360 days/12 months = 30°	l_s
32 days	l_s in Gemini
252,000 stadia	C_E

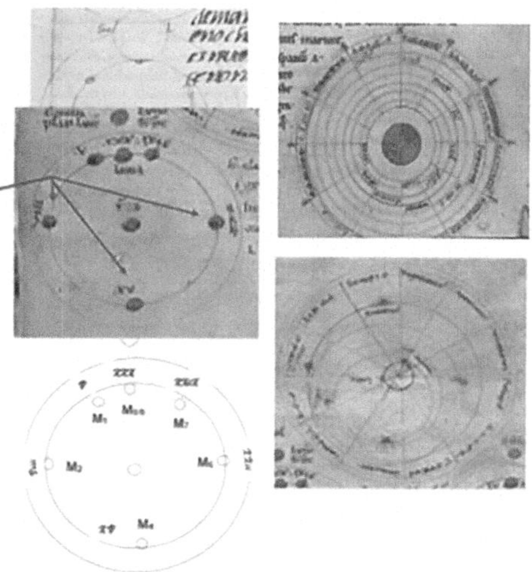

(1) His data indicate that he thought six of the traditional eight lunar phases can be predicted theoretically, i.e., known other than by visual sighting; the remaining two, M_1 and M_7, must still be observed to be recognized as occurring.

(2) He provides quantifications, which evoke the varying angular relationships between the Sun and Moon against the backdrop of the fixed stars.

(3) He considers that lunar phases are the result of the combined (geocentric) motion of the Moon and the (eccentric) motion of the Sun and that this leads to specific periodicities (sidereal, tropical, etc.).

In the case of Macrobius, there are two other features. (1) The timing and location of only five of the lunar phases can be known other than by visual sighting, which excludes from theorical locating the phases M_3, M_5, and M_7, although Macrobius does note that M_3 and M_5 could be known by approximation. (2) As his manuscripts, abounding in elucidating graphics, show, he considered eclipse phenomena and varying angular relationships between the Sun and the Moon.

As already shown, there is sufficient quantifying data in their writings to allow for a model for each author of late Roman antiquity to be constructed from the parameters they provided (Figs. 9.3 and 9.4).

These proposed results affirm that two different lunar models can easily be seen to exist behind the descriptions and data of Albertus's two authoritative sources. In the end, the reconstruction of lunar models for each of the Albertus's source authors and the contrasting of them with a reconstruction of his own in the context of his exegetical passage is intended to be the most persuasive argument for adopting the historical interpretation in this study.

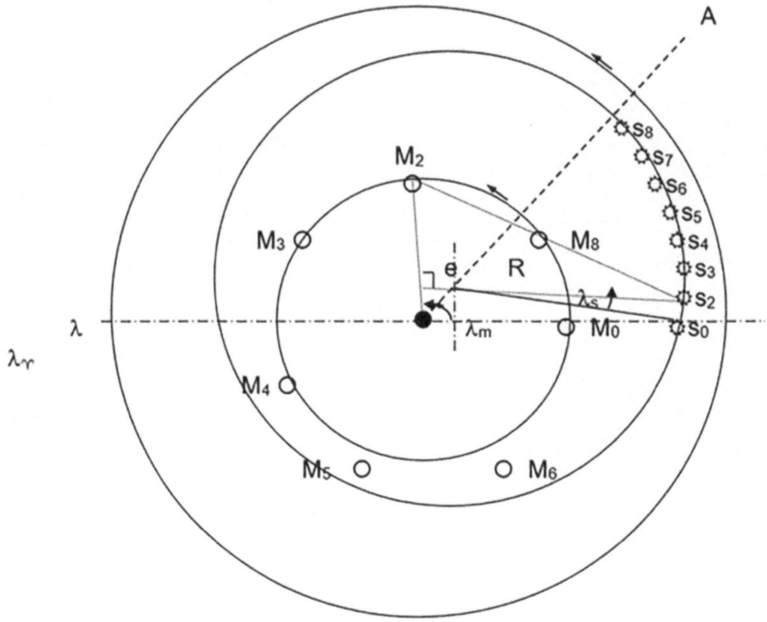

Fig. 9.3 Martianus's lunar details modeled

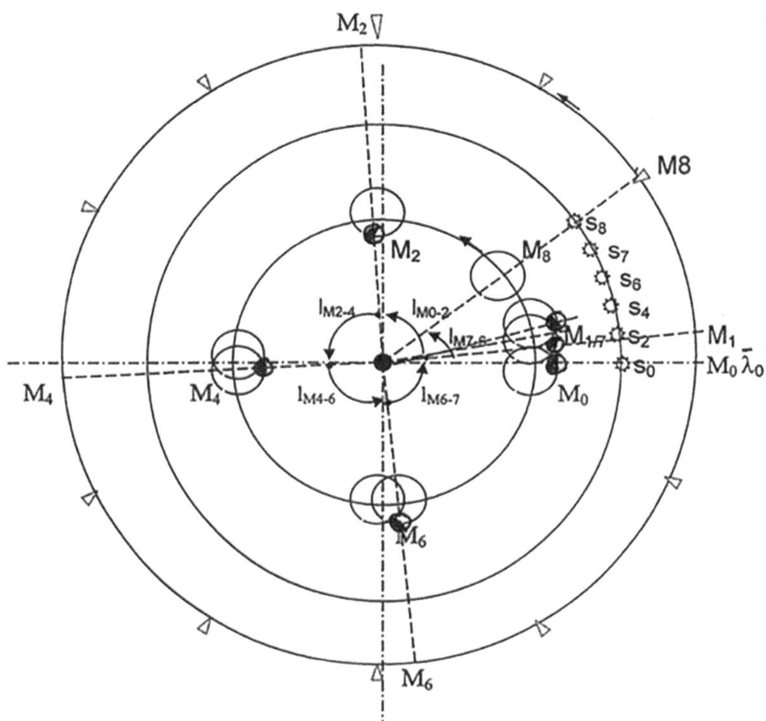

Fig. 9.4 Macrobius's lunar details modeled

9.3.2 *Albertus's Own Lunar Model*

The remaining question yet to be answered and in light of the argument for a thirteenth-century scholastic's perception of models in his sources, the most purposeful one, is what did Albertus make of his sources? One way of addressing such a question is in the direction of analysis adopted by other historians for deciphering an author's thoughts,[28] namely, to ask what is it that the author wishes to "prove"? To analyze Albertus in this light, understanding the context of his use of the sources Martianus and Macrobius is important. They were assistants in his exegesis of an Old Testament passage, Ecclesiasticus 43:6–8. Although connected to the main concern of Albertus's treatise on the nature of "good," which led him into the exegetical exercise, his digression into astronomy is brief and isolated within the work. One of the five details Albertus provides about the Moon in his discussion draws close attention, however, since it leads to an awareness that Albertus too had a theory of lunar phases, embedded within *De natura boni*. This would be the earliest of Albertus's astronomical theories.

Although Albertus's details include the two essentials for contemporary lunar theory— (1) a conception of the relationship of the periodic motions of the celestial bodies relative to the fixed stars (and thus to one another) and (2) an explanation for phase phenomena—his data are noticeably more sparse than those of his sources. Albertus identifies lunar periodicity (monthly, calendrically), lunar phases (noted as changes in appearance, color), and eclipse phenomena, writing "the proximity of the moon to the sun determines the moon's light: when it is close, there is none; when far, it is full; when close, sometimes an eclipse occurs." He observes that specific amounts of the Sun's light on the lunar surface are markers on its monthly path and that each of these was given a name by the "ancients," indicating the lit amount. He seems to have realized that the times for the seven phase markers could not be determined without a theory of relative lunar-solar positions, since he recognized that only four changes in the quantity of the lit lunar surface are striking enough to be designated by observation alone: *monoides* (M_0/M_8), the first *dichotomos* (M_2), *panselinos* (M_4), and the second *dichotomos* (M_6). By inference, only theory could determine the points of the two other phase markers, *amphicrytos* (M_3 and M_5) (Fig. 9.5).

Albertus gives no numerical value for the length of time required by the Moon to complete its own course, and hence no month length. His statement, that the year and month are commensurate, seems to indicate, however, that it is the synodic month of a 30-day length he had in mind. Albertus also does not otherwise provide values for solar motion. The mean location of the Moon in its orbital course is thus the value for Albertus, as for his two source authors, stipulated by the angle between

[28] The analysis of Carman 2017 comes to mind, in which he let that question guide his incredibly insightful decryption of a puzzling illustration of the text of Martianus in the margins of Leiden University Library, MS Voss. lat. f. 48, f. 89 and f. 93r (Auxerre, France, 830–40). His conclusions as well are very pertinent to this study.

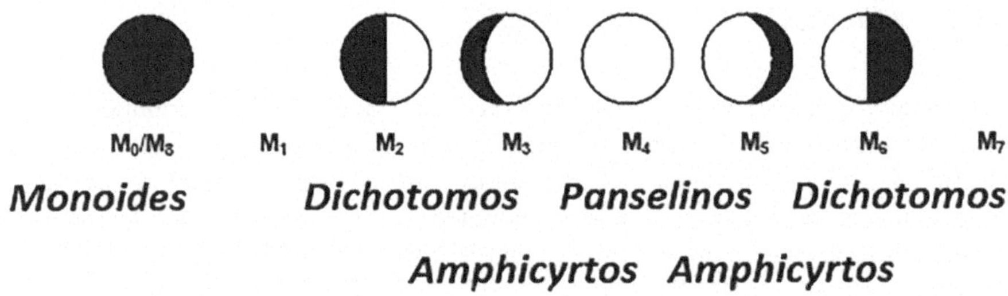

Fig. 9.5 Lunar phases, determinable theoretically as to time and place, as explained by Albertus

the Sun and Moon. Albertus relied solely on mean longitudes. Lunar latitude is not included in Albertus's theory (Table 9.5; Fig. 9.6).

If all the essential parameters of Albertus's theory and subsequently his reconstructed model are compared with his sources' parameter tables and models, the following conclusions emerge. The differences between his theory and model are striking. Albertus's comparative lack of detail could be read as an indication of less accuracy and less sophistication. Before it is discounted altogether, however, we return to the guiding question of analysis: so, what did the author wish to "prove" or establish? In Albertus, comprehensive contemplation of the seven phases of the Moon and of the length of the "month" is sacrificed due to the main aim of his discussion: to establish the lunar cycle as analogous to the solar course.

> 6 et luna in omnibus in tempore suo ostensio temporis et signum aevi 7 a luna signum diei festi luminare quod minuitur in consummatione 8 mensis secundum nomen eius est crescens admirabiliter in consummation
>
> 6 And the moon, unto all things, in its time, is a signal of weather and a sign of the epoch 7 From the moon comes the sign of a feast day to be luminous, as it diminishes/wanes in its fullness. 8 The new moon, according to its name, is growing wonderfully into fullness.

As the Vulgate biblical text Albertus designated for his exegetical exercise says, the Moon too is a sign of the "seasons." The Moon's course is divided into four parts, whose quantities are determined by their identification with one of the four phases. This correlation is further supported for Albertus by examples from other *antiqui astrologi*, including Firmicus Maternus. Albertus's theory of lunar seasons needs definite markers to distinguish one part of the lunar "year" from another, but only four are required, as there are only four solar seasons. Among the seven lunar phases, the most easily observable are the likeliest candidates, providing they are separated into balanced intervals, as are the annual solar seasons: the first sighting of the Moon, M_1, is a traditional beginning point of a lunar period; the phases of M_2, M_4, and M_6, whose occurrences can be quite accurately determined, would mark culminations of the first season, spring, then summer, and fall; winter would extend from the fading light of M_6 through the darkness after M_7, until abruptly halted with the beginning of a new cycle, with another M_1.

Like the Sun, the Moon was thought by Albertus and his sources to have observable effects on the Earth. Albertus explains that its effects are caused by the quantity

Table 9.5 Albertus's Data, Categorized as to Theoretically Derived ("Stipulated" by Theory) or Observation-Based ("Measured") Values

Values	Description of the value	Parameter
Stipulated (Albertus)		
30 days x 12 = year		$l_m \times 12 = l_s$
M_2	... XC partibus (864; 1-18)	$\lambda_s + 90^\circ = \lambda_{M1}$
M_4		$\lambda_s + 180^\circ = \lambda_{M1}$
M_6		$\lambda_s + 90^\circ = \lambda_{M1}$

Values	Description of the value	
Measured		
30 days		l_{M0}
M_0	... deficiens est ... (125, 78-9)	λ_{M0}
M_2	... quando media est ... (125, 83)	λ_{M2}
M_3	... plus ultra medium ... (125, 84)	λ_{M3}
M_4	... plena ... (125, 85)	λ_{M4}
M_5	... descrescens .. (126, 1)	$\lambda_{M5/6}$
M_6		
M_7	... magis deficiens (126, 1)	$\lambda_{M7/8}$
M_8		

of the lunar surface, which is lit, for this determines the amount of light the Earth receives from the Moon. A progression of effects, so analogous to that of the Sun on growing plants, led, he seems to assert, to the names of annual seasons' being attached to four of the Moon's intervals. During the Moon's first visibility and its first quarter (I), both diseased and healthy states of the human body were thought to become vivified, Albertus writes. The waxing Moon causes the natural heat of bodies to increase its retention of moisture and its burning of any excess. From the end of the first quarter through the full Moon (II), disease conditions are augmented. From full Moon through last quarter (III), these are brought to maturity. During the time, between the Moon's last quarter and its first visibility (IV), chronic infirmities

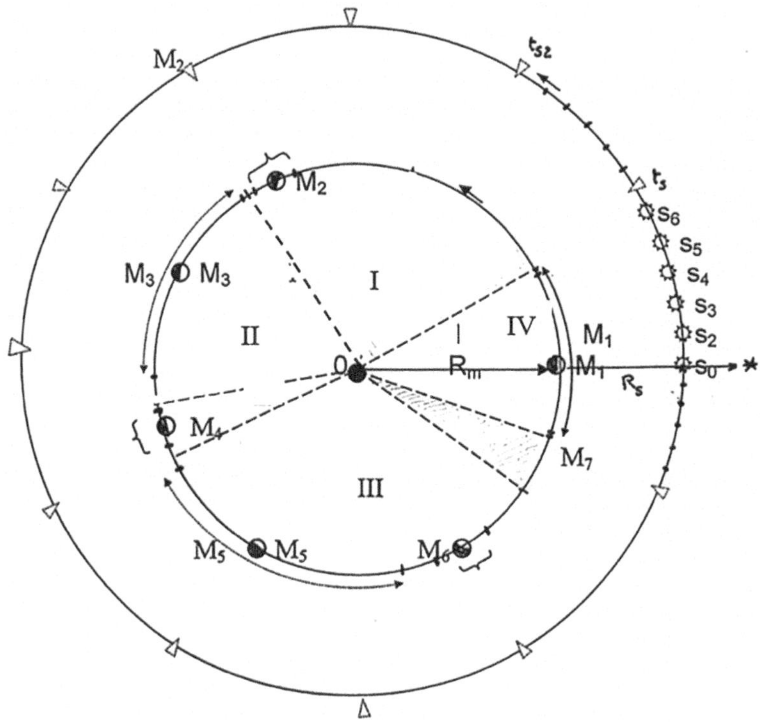

Fig. 9.6 Albertus's lunar details modeled

were thought likely to flare up. It is then that some individuals, the *lunatici*, would suffer loss of perceptual and discretionary capacities, due to the relative and ultimately complete lack of lunar light and the natural heat it contains.

Thus, the astrological theory here is clearly an extension of Albertus's astronomical one, in that it too considers the quantity of lunar surface lit to undergo a recurring cycle of gradual variation marked by the four distinctly observable points of marked change. Alteration in the quantity of lunar light during the intervals between the four marker points he describes as taking place very gradually, with the possible effect on Earth of the increasing or decreasing lunar light during the interval between markers' being treated as one in kind. Albertus also noted that the Moon's particular "season" can be determined not only by keeping track of its relation in time to the four defining observable markers, but also by observing the color of the Moon at any time. All that being said, Albertus, like Martianus and Macrobius, designates not just four lunar phases, but seven, although his model struggles to account for the longitudinal position of the first (λ_{M1}), the "old" (λ_{M7}), and the gibbous phases (λ_{M3} and λ_{M5}).

Despite his having an available mean value for lunar longitude, Albertus considered only points independently measurable by observation between phases M_1 and M_7. From lunar longitude and a theoretical correlation between the longitudinal position of Sun and Moon, he could well have determined some of these phase positions. For example, $\lambda_{M2} = \lambda_S - \lambda_M$ at t = "x" day, on which day $\lambda_S - \lambda_M = 90°$. By the mean longitudes of Sun and Moon ($\lambda_S - \lambda_M$ or $\lambda_M - \lambda_S$), at each of the four "seasonal" phases, the mean length of the seasons can be determined, and significantly, as balanced. Guided by and choosing among his source writings of the "ancient and wise ones," Albertus "modeled" his own conception of the Moon's motion over the course of a month.

It might be lamented that Albertus did not adopt more details from Martianus and Macrobius, but in the context of his exegetical discussion of the Moon in *De natura boni*, his model of lunar seasons was well conceived to his own purposes. His sources' data would at the very least have improved his accuracy of lunar longitudinal positioning, but as the purpose of his model was seen to be astrological, to determine what "season" the Moon is in and how close it is to passing into the next, his own modeling was fully adequate.

In concluding this case of a medieval scholastic's use of source writings on astronomy, it is confirmed that Albertus had an Aristotelian epistemic methodology and numeracy, and an awareness of observation, instruments and tabulated records, and geometric modeling, all the tools necessary for understanding the texts of his sources. One could, he had written, construct either a mathematical or a physical proof of a celestial phenomenon, and for him the difference was clear. It is established that the manuscript figures illustrating the scholastics' sources reflected to them that the authors they were consulting had created astronomical models in their own texts and that others reading their works could well do so too. When reading through the eyes of the medieval scholar, as Jardine has stated, these must be taken into consideration.

As shown in Albertus's own presentation of details and of models, it is evident that the most frequent observations under medieval scrutiny in astronomy were eclipses, changing planetary positions, and the relative sizes of the celestial bodies and of the spaces in between them. Led albeit by biblical exegesis, Albertus's earliest discussion of astronomical ideas was thus an analysis of lunar motion and phase phenomena, guided by Martianus Capella and Macrobius. Martianus, as Delambre wrote in 1817, "made more services to astronomy than many astronomers by far more skilled than he" (194),[29] and Macrobius was not the worst guide to periodicity. When integrating authorities into his understanding of lunar motion, Albertus's choice was thus not an exercise of choosing who was "right" and who not ; it was not even one of binary choice. He gleaned from sources the kernels of information to meet his exegetical needs, revealed in the modeling of his own details. As he was to write later in his *Commentary on the Metaphysics* about the greatest choice facing all medieval thinkers: "You ought to know that man cannot become perfected in philosophy except through knowledge of the philosophies of *both* Aristotle and Plato" (Magnus 1262-1263; ed. Geyer 1964, *Metaphysica* I.5.15 [ed. Colon. 16/1:

[29] Delambre, like some many years later is very ready to "forgive his verbiage, his blunders and his complicated language" ("pardonner son verbiage, ses bévues et son galimathias" (1817, 312). Noted by Carman 2017, 192.

89.85–87]; my emphasis), by 1238, Albertus's conception of the lunar models of *both* Martianus and Macrobius allowed him to formulate his own first astronomical model for the readers of *De natura boni*. There is much more analysis warranted to appreciate Albertus's understanding of other far more sophisticated astronomical models, the Aristotelian inspired al-Biṭrūjīan one, among them![30]

Acknowledgments Sincere thanks are extended to Francesca Rochberg, University of California, Berkeley, 2023 recipient of the Francis Bacon Award, Professor Jed Buchwald, and all those involved in hosting the 2023 Francis Bacon Conference in tribute to Noel Swerdlow. It was an honor to have spoken at the gathering in May, and it is the setting from which this contribution derives. It shares ideas which were the subject of my last scholarly exchanges with Noel Swerdlow.

References

Al-Biṭrūjī. *De motibus celorum*. ca. 1190; Latin trans. 1217. See Carmody 1952.

Al-Biṭrūjī. *On the Principles of Astronomy*. ca. 1190. See Goldstein 1971.

Anzulewicz, Henryk, ed. (ca. 1242) 2008. *De IV coaequaevis*. In *Albertus Magnus: Opera Omnia*. Edited by Bernhard Geyer. Vol. 27/1. Münster: Aschendorff Verlag.

Aristotle. ca. 350. *Posterior Analytics*. Latin translation 12th and 13th centuries. See Minio-Paluello and Dod (1953–1954) 1968.

Aristotle. ca. 340. *Meteorologica*. Latin translation ca. 1260–1270. See Vuillemin-Diem 2008.

Aristotle. ca. 340. *Meteorologica*. Latin translation before 1162. See Rubino 2010.

Bacon, Roger. 1267–1270. *Liber Secundus Communium Naturalium Fratris Rogeri de Celestibus*. See Steele 1913.

Baur, Ludwig, ed. 1903. Dominicus Gundisalinus: De divisione philosophiae. *Beiträge zur Geschichte der Philosophie des Mittelalters* 4(2–3): 3–142.

Bazán, Bernardo Carlos. 2005. Radical Aristotelianism in the Faculties of Arts. The Case of Siger of Brabant. In *Albertus Magnus und die Anfänge der Aristoteles-Rezeption im lateinischen Mittelalter: von Richardus Rufus bis zu Franciscus de Mayronis*, ed. L. Honnefelder, 585–630. Münster: Aschendorff.

Borgnet, Auguste, ed. (1254–1257) 1890. *Analytica posteriora*. In *Albertus Magnus: Opera Omnia*. Edited by Auguste Borgnet et al. Vol. 2. Paris: Ludovicum Vivès.

Borgnet, Auguste, ed. (ca 1242) 1894. *De IV coaequaevis*. In *Albertus Magnus: Opera Omnia*, Edited by Auguste Borgnet et. al. Vol. 34. Paris: Ludovicum Vivès.

Capella, Martianus. ca. 410–420. *De nuptiis Philologiae et Mercurii*. See Cristante 2011; Musso and Ghiringhelli 2019; Stahl, Johnson, and Burge 1977.

[30] In 1976, Noel Swerdlow came for a year-long visit to the University of Toronto's Institute for the History and Philosophy of Science and Technology. His course offerings and teaching cemented my interest in the history of astronomy, which study I had already launched under the watchful eye of Galileo scholar Stillman Drake and the caring guidance of historian of medieval astronomy John Abrams. After Noel's return to the University of Chicago, as is the way of the world, the rather mundane circumstances (administrative boundaries, available supervisors, etc.), which often dictate the longer trajectory, oriented me toward the writing of a thesis on the hypothetical contents of a *De Astronomia* by Albertus Magnus (Price Buchwald 1983). For Noel's teaching, insight and long years of friendship (almost 40!) I am forever grateful. May he and his work be remembered well!

Carman, Christián C. 2017. Martianus Capella's Calculation of the Size of the Moon. *Archive for the History of the Exact Sciences* 71:193–210.

Carmody, Francis J., ed. 1952. *Al-Biṭrūjī: De motibus celorum*, Critical edition of the Latin translation of Michael Scot, 1217. Berkeley: University of California Press.

Chabás, José. 2012. Characteristics and Typologies of Medieval Astronomical Tables. *Journal of the History of Astronomy* 43:269–286.

Chabás, José, and Bernard R. Goldstein. 2012. *A Survey of European Astronomical Tables in the Late Middle Ages*. Leiden: Brill.

Craemer-Ruegenberg, Ingrid. 1980. *Albertus Magnus*. Munich: C. H. Beck OHG. Reprinted Anzulewicz, Henryk, ed. 2005. Albertus Magnus. Dominikanischen Quellen und Zeuginise 7. Leipzig: St. Benno.

Cristante, Lucio, ed. (early 5th c.) 2011. *Martianus Capella: De nuptiis Philologiae et Mercurii, Libri I–II*. Translated by Luciano Lenaz, commentary by Lucio Cristante, Ireneo Filip, and Luciano Lenaz, with an unedited essay by Pietro Ferrarino. Hildesheim: Bibliotheca Weidmanniana.

Delambre, Jean-Baptiste-Joseph. 1817. *Histoire de l'astronomie ancienne*. Paris: Ve Courcier.

Eastwood, Bruce S. 2002. *The Revival of Planetary Astronomy in Carolingian and Post-Carolingian Europe*. Aldershot: Ashgate.

Eastwood, B., and G. Graßhoff. 2004. *Planetary Diagrams for Roman Astronomy in Medieval Europe, ca. 800–1500*, Transactions of the American Philosophical Society 94, part 3. Philadelphia: American Philosophical Society Press.

Geyer, Bernhard, ed. (1243–1279) 1951-. *Albertus Magnus: Opera Omnia*. 41 projected vols. Münster: Aschendorff Verlag.

Geyer, Bernhard, ed. 1958. Geometria. Die mathematische Schriften des Albertus Magnus. *Angelicum* 35: 159–75.

Geyer, Bernhard, ed. (1262-1263) 1964. *Metaphysica*. In *Albertus Magnus: Opera Omnia*, vol. 16/2. Münster: Aschendorff Verlag.

Geyer, Bernhard, ed. (1243-1244) 1974. *De natura boni*. In *Albertus Magnus: Opera Omnia*, vol. 25/1. Münster: Aschendorff Verlag.

Goldstein, Bernard R., ed. and trans. (ca 1190) 1971. *Al-Biṭrūjī: On the Principles of Astronomy*. New Haven: Yale University Press.

Grafton, Anthony, and Noel Swerdlow. 1988. Calendar Dates and Ominous Days in Ancient Historiography. *Journal of the Warburg and Courtauld Institutes* 51:14–42.

Gundisalinus, Dominicus. ca. 1145. *De divisione philosophiae*. See Baur 1903.

Hauke, Hermann. 1986. *Katalog der lateinischen Handschriften der Bayerischen Staatsbibliothek München: Clm 28111–28254*. Wiesbaden: Harrassowitz.

Hiatt, Alfred. 2007. The Map of Macrobius before 1100. *Imago Mundi* 59 (2): 149–176.

Hossfeld, Paul. 1969. Die naturwissenschaftliche/naturphilosophische Himmelslehre Alberts des Grossen (nach seinem Werk *De caelo et mundo*). *Philosophia Naturalis* 11 (3): 318–329.

Hossfeld, Paul, ed. (1251) 1971. *De caelo et mundo*. In *Albertus Magnus: Opera Omnia*. Edited by Bernhard Geyer. Vol. 5/1. Münster: Aschendorff Verlag.

Hossfeld, Paul, ed. (1251) 1987. *Physica*, Books 1–4. In *Albertus Magnus: Opera Omnia*. Edited by Bernhard Geyer. Vol. 4/1. Münster: Aschendorff Verlag.

Hossfeld, Paul, ed. (1252–1254) 2003. *Meteora*. In *Albertus Magnus: Opera Omnia*. Edited by Bernhard Geyer. Vol. 6/1. Münster: Aschendorff Verlag.

Jardine, Nicholas. 2006. Imagineering the Astronomical Revolution. *Journal of the History of Astronomy* 37:472–473.

Leggatt, S. ed. (350) 1995. *Aristotle: On the Heavens I and II*. Warminster: Aris & Phillips.

Magnus, Albertus. 1254–1271. *Analytica posteriora*. See Borgnet 1890.

Magnus, Albertus. 1258–1262. *De animalibus, 13-16*. See Stadler 1920.

Magnus, Albertus. 1243–1279. *Opera Omnia*. See Geyer 1951.

Magnus, Albertus. 1254–1260. *Postilla super Isaiam*. See Siepman 1953.

Magnus, Albertus. 1262–1263. *Geometria*. See Geyer 1958.

Magnus, Albertus. 1262–1263. *Metaphysica*. See Geyer 1964.
Magnus, Albertus. 1251. *De caelo et mundo*. See Hossfeld 1971.
Magnus, Albertus. 1250. *Super Dionysium De divinis nominibus*. See Simon 1972a.
Magnus, Albertus. 1250. *Super Dionysium Epistulas*. See Simon 1972b.
Magnus, Albertus. 1243-1244. *De natura boni.* See Geyer 1974.
Magnus, Albertus. 1251. *Physica.* See Hossfeld 1987 for Books 1-4.
Magnus, Albertus. 1252–1254. *Meteora*. See Hossfeld 2003.
Magnus, Albertus. ca 1242. *De IV coaequaevis*. See Anzulewicz 2008; Borgnet 1894.
Mercier, Raymond. 1976. Studies in the Medieval Conception of Precession. *Archives Internationales d'Histoire des Sciences* 26 (99): 197–220.
Mercier, Raymond. 1977. Studies in the Medieval Conception of Precession. *Archives Internationales d'Histoire des Sciences* 26 (100): 33–71.
Minio-Paluello, L., and B. G. Dod. (1953–1954) 1968. *Analytica posteriora*. In *Translationes Iacobi, Anonymi sive "Ioannis," Gerardi et Recensio Guillelmi de Moerbeka*, vol. 4.2–3. Bruges/Paris: Desclée De Brouwer.
Moraux, P. ed. (350) 1965. *Aristotle, Du ciel*. Paris: Société d'Édition "Les Belles Lettres".
Murata, Koji, Kohei Ichikawa, Yuri I. Fujii, Hisashi Hayakawa, Yongchao Cheng, Yukiko Kawamoto, and Hidetoshi Sano. 2021. Cometary Records Revise Eastern Mediterranean Chronology Around 1240 CE. *Publications of the Astronomical Society of Japan* 73 (1): 197–204.
Musso, Simona, and Elena Ghiringhelli, ed. (ca 410-420) 2019. *Martianus Capella: De nuptiis Philologiae et Mercurii, Libri I–II.* https://digiliblt.uniupo.it/xtf/view?docId=dlt000355/dlt000355.xml;query=De%20Nuptiis%20Philologiae%20et%20Mercurii;brand=default. digilibLT project website digilibLT. https://digiliblt.uniupo.it
Palazzo, Alessandro. 2023. The *Annus Magnus* in Albert the Great's Parisian Theological Works: *De IV Coaequaevis* and *Commentarii in Libros Sententiarum*. *Vivarium* 61:26–58.
Park, Katharine. 2011. Observation in the Margins, 500–1500. In *Histories of Scientific Observation*, ed. Lorraine Daston and Elizabeth Lunbeck, 15–44. Chicago: University of Chicago Press.
Price, Betsey B. 1980. The Physical Astronomy and Astrology of Albertus Magnus. In *Albertus Magnus and the Sciences*, ed. James A. Weisheipl, 155–185. Toronto: Pontifical Institute of Medieval Studies.
Price Buchwald, Betsey B. 1983. *The Astronomy of Albertus Magnus*. PhD dissertation, University of Toronto.
Price, Betsey B. 1991. The Use of Astronomical Tables by Albertus Magnus. *Journal of the History of Astronomy* 22:221–240.
Price, Betsey B. 1992. Albert the Great. In *Medieval Philosophers*, Dictionary of Literary Biography 115, ed. Jeremiah Hackett, 31–39. Detroit: Gale Research.
Price, Betsey B. 2012. Interpreting Albertus the Great on Astronomy. In *A Companion to Albert the Great. Theology, Philosophy, and the Sciences*, Brill's Companions to the Christian Tradition 38, ed. Iven M. Resnick, 397–436. Leiden: Brill.
Rubino, E. ed. (350) 2010. *Meteorologica, liber quartus*. Translated by Henrici Aristippi. Turnhout: Brepols.
Siepman, Ferdinand, ed. (1254–1260) 1953. *Postilla super Isaiam*. In *Albertus Magnus: Opera Omnia*. Edited by Bernhard Geyer. Vol. 19. Münster: Aschendorff Verlag.
Simon, Paul, ed. (ca 1242) 1972a. *Super Dionysium De divinis nominibus*. In *Albertus Magnus: Opera Omnia*. Edited by Bernhard Geyer. Vol. 36/1. Münster: Aschendorff Verlag.
Simon, Paul, ed. (ca 1242) 1972b. *Super Dionysium Epistulas*. In *Albertus Magnus: Opera Omnia*. Edited by Bernhard Geyer. Vol. 32/2. Münster: Aschendorff Verlag.
Snyder, Steven C. 1998. *Scientific Method and the Human Soul in Aristotle's De Anima*. Notre Dame: University of Notre Dame.
Stadler, H., ed. (1258-1262) 1920. *De animalibus 13–26*, Beiträge zur Geschichte der Philosophie 16. Münster: Aschendorff Verlag.

Stahl, William Harris, ed. and trans. (410-439) 1952. *In Somnium Scipionis Commentarios*. In *Macrobius: Commentary on the Dream of Scipio*. New York: Columbia University Press.
Stahl, William Harris, Richard Johnson, and E. Burge, ed. and trans. (ca 410-420) 1977. *Martianus Capella: De nuptiis Philologiae et Mercurii*. Martianus Capella and the Seven Liberal Arts 2. New York: Columbia University Press.
Steele, Robert, ed. (1267-1270) 1913. Roger Bacon: *Liber Secundus Communium Naturalium Fratris Rogeri de Celestibus*. In *Opera hactenus inedita Rogeri Baconi*. Oxford: Clarendon.
Swerdlow, Noel. 2019. Regiomontanus' Defense of Scientific Publishing. In *Looking Back as We Move Forward. The Past, Present, and Future of the History of Science*. Pasadena: California Institute of Technology.
Theodosius, Macrobius Ambrosius. 410-439. *Commentarii in Somnium Scipionis*. See Stahl 1952; Willis (1970) 2013.
Thorndike, Lynn. 1951. Albertus Magnus on Comets: English Translation. In *Latin Treatises on Comets Between 1238 and 1368*, ed. Lynn Thorndike, 62–76. Chicago: University of Chicago Press.
Vuillemin-Diem, G., ed. 2008. (Aristotle) *Meteorologica. Translatio Guillelmi de Morbeka*. 2 vols. Turnhout: Brepols.
Willis, James, ed. (410-439) (1970) 2013. *Commentarii in Somnium Scipionis in Ambrosius Theodosius Macrobius, Opera*. Bibliotheca Scriptorum Graecorum et Romarorum Teubneriana 2. Berlin: DeGruyter.

Chapter 10
Historical Chronology Before Joseph Scaliger: Jean Bodin and His Discussion Partners

Anthony Grafton

Everyone knows that historical chronology began in 1583. In that year Joseph Scaliger published his *Opus novum de emendatione temporum*—a famously innovative, interdisciplinary work. In it he cut across languages and traditions with ease, working in Hebrew, Aramaic, Ethiopic, and Arabic sources as well as Greek and Latin ones. More important, he wielded the newest techniques of both philology and astronomy to reconstruct the calendars used across the ancient and modern worlds and establish the dates of dozens of events in ancient, medieval, and modern history.[1] True, as Scaliger himself recognized, his work was incomplete. A second edition, revised and augmented, appeared in 1598.[2] His *Thesaurus temporum*—a vast series of textual editions and commentaries, a magnificent complement to the *De emendatione*—came out in 1606.[3] Revised editions of both were published posthumously.[4] Later chronologers—from Scaliger's Lutheran disciple Seth Calvisius to his Jesuit critic Dionysius Petavius—produced treatises and handbooks that corrected his work in dozens of large and small details. Yet all of them drew on the materials he had collected and applied the methods he had developed in the first edition, even as they challenged specific arguments.[5]

[1] Scaliger 1583; for this and what follows, see Grafton 1993, 2.

[2] Scaliger 1598.

[3] Scaliger 1606.

[4] Scaliger 1629, 1658.

[5] Grafton 2019b, 154–93.

A. Grafton (✉)
Department of History, Princeton University, Princeton, NJ, USA
e-mail: grafton@princeton.edu

F. Rochberg (ed.), *Studies in Premodern Sciences in Memory of Noel M. Swerdlow*, Archimedes 74, https://doi.org/10.1007/978-3-032-04040-4_10

Though Scaliger himself corrected his original views on many particulars, he insisted that he had created a structure as novel as it was sturdy. The first edition of his work, he remarked, had served as an intelligence test for his contemporaries, which most of them had failed. Most historians since his time have agreed. Writing in the *Quarterly Review* in 1860, Mark Pattison described Scaliger's accomplishment with his customary eloquence: "By this grand effort of genius, Scaliger may be said to have created for modern times the science of chronology."[6] Even now, excellent general accounts say much the same: "in his *De emendatione temporum* (1583)," writes Rens Bod, "he defined a new timeframe for classical Antiquity, and using calendar comparisons he was able to place Graeco-Roman history in the context of Babylonian, Egyptian, Persian, and Jewish history."[7]

More than 30 years ago I tried to show that Scaliger, though astonishingly erudite and original, was one figure in an inhabited landscape. He drew vital materials and methods from a wide range of scholars, many of them comparatively little known—like the Jena scholar Paulus Crusius, who had already established a canon of epochs and eclipses and tied them to events before Scaliger went to work. More recently, Philipp Nothaft has revealed a lost medieval chapter in the story of chronological scholarship. Roger Bacon and a host of others discovered, centuries before Scaliger, that astronomy could provide a new foundation for historical chronology. Some of them put this program into practice with great skill.[8] As yet we do not know how much Scaliger and his contemporaries knew about these medieval traditions. Scaliger's many critics seized on any evidence they could find to show that his work was derivative and inconsequential. It is striking that not one of them cited these medieval precedents. A full history of historical chronology remains to be written.

In this paper, I examine the practices of chronological scholars in the years just before Scaliger compiled and published the *De emendatione temporum*: what might be called the ordinary science of chronology. We will begin with some of the chronological arguments put forward by a famous French scholar, Jean Bodin (ca. 1530–1596). Unlike most chronologers, Bodin remains a presence in multiple contemporary fields of scholarship. Political scientists remember the political theorist who at first defended the rights and powers of the French Estates and later devised the radical new doctrine of absolutism—a theory that helped to shape the new monarchy of Louis XIV. Historians of witchcraft remember the demonologist who put forward a distinctive set of theories about the mischief that witches, helped by the devil, could wreak on the bodies of their victims. Environmental historians remember Bodin's elaborate theory about how climates shape the characters of peoples. Historians of economics and natural philosophy have examined his works in those fields. More than others, students of historiography treat him as a celebrity. Bodin was the most original and imaginative member of a highly gifted group of scholars,

---

[6] Pattison 1889, I.162.

[7] Bod 2014, 169. In the other excellent recent synthesis, James Turner offers a more contextual account (Turner 2014, esp. 51–52).

[8] Nothaft 2018, 136–43, 2020.

who set out, in the middle decades of the sixteenth century, to draw up textbooks for the reading and writing of history. His *Method for Obtaining Knowledge of History with Ease* was a success from its first appearance in 1566. It went through several editions and occupied the first place in the anthology of treatises on the art of history published in Basel in 1579.[9]

Most studies of Bodin's historical work have concentrated, reasonably enough, on the interests he shared with François Baudouin and other jurists, and that played a central role in his work. They set out to integrate law with history: using a comparative method, they made clear that every system of law grew up in a particular society and state, and could in turn be used as a rich source for that nation's history. Early readers in England saw the *Methodus* as a textbook on how to read and use history; early readers on the Continent more as a source of information about historians.[10] Modern scholars have been most interested in his historical method and his political doctrines. Few have recently devoted much attention to the chapter of the *Methodus* that Bodin dedicated to chronology—a notoriously technical and contentious field, which he saw as a vital part of the art of history. In fact, though, some knowledgeable contemporary readers took him seriously as an expert.

In 1581 he arrived in England, where he and another jurist, Peter Baro, were tasked with arranging a match between a French duke and Queen Elizabeth. Though the match failed, some lively conversations took place between the visitors and their hosts. Given the chance to spend time with these French experts on history, a learned Cambridge scholar, Gabriel Harvey, directed their discussion to chronology. Harvey recorded what they talked about in his copy of Livy's Roman history, which was equipped with massive chronological tables compiled by Heinrich Glarean.[11] He wanted advice on how much credence he should give to the ancient chronicle compiled by Eusebius at the start of the fourth century and adapted into Latin a century later by Jerome. Eusebius's comparative tables established multiple connections between the histories of ancient peoples. They showed, for example, that the Hebrew prophets and the Greek Seven Sages had been contemporaries, and made clear when in the series of Greek Olympic cycles Rome was founded. Harvey did not find all of these connections convincing: "there are as yet many concerns about a reliable synchronism."[12] The advice the two Frenchmen offered sounds ambiguous, but it satisfied him:

> I benefited greatly from a comparison with two most expert Frenchmen: Jean Bodin and Petrus Baro. They credit Glarean, Funck, Mercator, Crusius with more effort and reliability

[9] Good introductions to Bodin are provided in the *Stanford Encyclopedia of Philosophy* by Mario Turchetti (2005, https://plato.stanford.edu/entries/bodin/) and by Lloyd 2017.

[10] For early readers' responses see Miglietti 2013.

[11] Livy 1555; Princeton University Library PA6452 A2 1555q ex. A digital facsimile with a transcription and translation of Harvey's marginalia is available at https://archaeologyofreading.org/viewer//#aor. On Harvey's copy of Livy and his ways as a reader, see Grafton et al. 2024. For Glarean's chronology of Livy, see Grafton and Leu 2014.

[12] Livy 1555, sig. P verso.

> than any of the ancient chronologers. Yet without detriment to each classical author's authority.[13]

Above all, the episode makes clear that Bodin continued to read about chronology after the first edition of the *Methodus* appeared. The chronologies Bodin cited here were all technical and innovative: in addition to Glarean, he named Johann Funck, who based his chronology of 1545 on the discoveries of Copernicus; Mercator, who composed his own canon of eclipses to serve as the foundation for his chronology in 1569; and Paulus Crusius, whose newly accurate and precise work on epoch dates and eclipses appeared as late as 1578. In one way or another Bodin was in dialogue with these and other writers. By comparing his work to theirs, we can gain a sense of the patterns and practices of chronological scholarship—as well as the levels at which it was practiced—before Scaliger came on the scene.

Chronology, as Bodin presented it, looked prosaic. He laid out the intervals of world history, and some of the problems they posed, in packed pages of prose that no reader can have found easy to follow. After noting that ancient and later writers differed about the date of the foundation of Solomon's Temple and the interval from that date to its fall, Bodin recorded some of the values they had put forward:

> Philo struck a mean. He also counted 3,373 from the creation to the foundation of the temple, Josephus 3,513. The former reckoned 1,717 from the flood to the overthrow of the temple, but the latter, 1913. The Jewish interpreters give considerably less for both, but Philo adopted a compromise. Then to reconcile these various authorities among themselves, let us in the first place reject Alfonso's period of 8,549 years of the world to today, that is, A.D. 1565, since he offers no support for his testimony.[14]

Passages like these suggest that Mark Pattison did not exaggerate when he proclaimed that before Scaliger, "the utmost extent of chronological skill which historians had possessed or dreamed of had been to arrange past facts in a tabular series as an aid to memory."[15]

Pattison dismissed tables too easily. They are not easy to compile, and they can reveal far more than prose. Bodin knew that complex arguments about the relations between different historical epochs and the years of different nations were far easier to follow and to verify when they were laid out in tabular form. This format was one of the salient features of Eusebius's *Chronicle*, as Bodin and Baro recognized when they advised Harvey about the trustworthiness of the *Chronicle*'s "synchronisms," the arguments that underpinned the tables, by finding points at which the chronologies of different kingdoms could be tied together. Contemporary writers whose works he knew well shared his belief that synchronistic tables were at the core of reliable chronology. When discussing the debate over the date of Christ's death, Bodin referred to the *Opusculum de emendationibus temporum* of Johannes Lucidus Samotheus (a pseudonym, meaning something like "Jean the Bright Lad from

---

[13] Ibid.

[14] Bodin 1566, 397, 1945, 330. Reynolds's translation has been silently changed in a number of passages.

[15] Pattison 1889, I.162.

France," for Giovanni Maria Tolosani, a Dominican who served as papal theologian). The book, which first appeared in 1537, offered individual lists of the rulers of dozens of lands as well as a tabular chronicle that covered the whole span of time from the Creation to 1535. Tolosani set out the histories of the different nations on a grid framed by several major dating systems. These enabled the reader to match the dates of any ruler of the Hebrews or the Assyrians, the Egyptians or the Romans to the appropriate years counted from both the Creation and the Incarnation (AM and BC), as well as, in later periods, from Olympiad 1,1, the Foundation of Rome, and other markers as well.[16] Funck cast his whole work in tabular form, using years of the world and Olympic years to serve as a fixed framework.[17] Gerardus Mercator, whose massive chronology appeared in 1569, drew up a tabular harmony of the Gospels and a canon of solar and lunar eclipses as well as a long table of rulers and events through history, framed, once again, by several systems of dating.[18] Bodin's failure to devise a mise-en-page that would clarify his arguments is revealing. His case shows that two decades before Scaliger, chronology attracted dedicated practitioners who followed—and improved on—the ancient model of Eusebius. But it also appealed to a wider range of amateurs like Bodin, who discussed chronological questions without drawing up full-scale treatments of the field, relying on the work of others. By the time Scaliger began to work on chronology, it was no longer the bare, ruined choir evoked by Pattison, if it ever had been, but a lively area of scholarship inhabited by highly competitive and quarrelsome individuals.

When a chronologer confronts discordant values for the same historical period, he needs to choose which authorities to trust. Bodin, who concluded his work with a bibliography of historians and composed a chapter on how to choose among them, was especially aware of this point. Could the Hebrew Bible serve as an exclusive basis for chronology? Some thought so: but they found themselves forced to reduce the duration of the Persian kingdom to four rulers and 54 years, as rabbinical chronologers had long before them. Bodin was inclined to use a range of sources eclectically.

The testimony of some of these ancients corresponded to and complemented that of the ancient Hebrews. Herodotus, the oldest of the surviving Greek historians, seemed to cover "almost the same princes of Egypt as did Jeremiah."[19] But he clearly knew far more Persian history than could fit in the biblical carapace. Bodin did his best to establish a canon of authorities, in which pagan authorities supplemented the Bible.

> From all the writers of the other peoples I find none older than Moses. … Herodotus is followed next by Thucydides. Then Berosus the Chaldean, 200 years younger than Herodotus. Berosus is followed by Manetho the Egyptian: then Metasthenes the Persian, in the time of

[16] Tolosani 1537.

[17] Funck 1552.

[18] Mercator 1559.

[19] Bodin 1566, 388, 1945, 322.

> Alexander the Great. Of these, though, only small fragments survive, if they were really written by these men. There is less doubt about Metasthenes and Ctesias.[20]

Some of the names Bodin brandished here are familiar to any modern reader interested in the ancient Near East. Berosus was a Chaldean, and Manetho an Egyptian, who compiled the legends and dynasties of their lands for readers in the new Greek-speaking states established after Alexander's conquests. Metasthenes, however, looks like a typographical error for Megasthenes, a Greek historian who served the Seleucids and wrote about India. And Bodin's ability to cite them with such ease seems mysterious. The works of all three survive only in fragments, which were first systematically collected in the 19th and 20th centuries. What "small fragments" of them did Bodin know in 1566, long before the rise of Karl Müller and Felix Jacoby, the great German fragment collectors?[21] Why, moreover, did he ask if these "fragments were really written" by their supposed authors?[22]

Bodin's source for these writers is not in doubt. Giovanni Nanni or Annius of Viterbo, a Dominican who served as papal theologian, published in 1498 a massive anthology of what he presented as the fragments of a dozen or so ancient writers. The listed authors included Berosus, Manetho, and Metasthenes (Bodin copied the name from Annius letter for letter), though Annius also included a genuine poem by Propertius and several more otherwise unknown texts. He embedded their work in a massive commentary: waves of closely printed interpretation break on small blocks of text in larger type. Together, the texts and the explanatory matrix that framed them became a magnificently surreal rewriting of ancient history—one in which Isis and Osiris had been the ancestors of Annius's patrons, the Borgia, and Noah's ark had landed—as some medieval traditions held—on the Gianicolo, across the Tiber from ancient Rome. Noah—who invented wine, *yayin* in Hebrew—became Janus, the god of wine, in Italy, as the name of the hill where his vessel came to rest made clear.

Annius's histories clearly contradicted those of the Greeks and Romans on many points. These discrepancies did not trouble their creator. Annius, who attacked the humanists but drew, as Giuliano Mori has shown, on their debates about historiography, found an ingenious explanation for the discrepancies.[23] The Greeks were liars—as ancient tradition had held, especially in the case of Herodotus, that teller of implausible tales. Berosus, Manetho, and Metasthenes, however, were not Greek. More important, they were not random individuals who recorded lies but priests, who wrote official histories that were publicly preserved and attested. In the fragments that Annius printed, as well as his commentaries, fake author after author belabored this point as they legitimated one another. Metasthenes, for example, wrote that "it is true that all authors who write about these kings need not be

[20] Bodin 1566, 384, 1945, 319.

[21] See Most 1997.

[22] For this "topos of critical rejection," see Borchardt 1971.

[23] Mori 2024.

accepted, but only the priests to whom is entrusted preservation of the public annals. An illustration is Berosus."[24]

A number of scholars had raised questions about these texts long before Bodin. Beatus Rhenanus, friend of Erasmus and expert on ancient and medieval history, pointed out that the Latin style, not only of all of the fragments but also of the commentary, was identical. "One of them milks the she-goat," he commented, wryly quoting one of Erasmus's *Adages*, "while the other holds out the sieve."[25] Bodin was prudent enough to accompany his references to these texts with cautious expressions of doubt about their genuineness. But he also accepted Annius's test for reliability in a historian. In fact, he quoted the passage just cited from Metasthenes to explain why he, too, saw priestly annals as superior in credibility to the narratives of individuals.[26]

In using Annius's forgeries, Bodin fell in line with the chronologers he read. Tolosani based his dynasty lists and their chronological arrangement on the work of his Dominican predecessor.[27] The Lutheran Funck, happy to find ancient pagans whose narratives matched those of the Hebrew Bible, did the same.[28] Mercator was more skeptical. Berosus—or rather Annius—argued that Nimrod was one of the great men whom Ianus sent forth, after the Flood, to populate the world. Known in pagan tradition as "the Saturn of Babylon," around the year 131 after the Flood he had begun but not completed a tower. Mercator rejected this account. The population of the world, he argued, had not been divided into different peoples until the time of Phaleg, 100 years after the Flood. The group that followed Nimrod could not have grown quickly enough in 30 years to perform the deeds ascribed to him.[29] Instead, Mercator identified the mighty Nimrod of Genesis with Ninus, the second king of the Assyrians.[30] Later on, when he used the testimony of Berosus, he made clear that he was not citing Annius's collection, but the fragments quoted by an actual ancient author, Josephus. Yet even Mercator took his chronology of Ninus from "Cato Annij" and "Maseas the Phoenician"—two Annian forgeries.[31]

In this area at least, Scaliger departed completely from earlier practices. He denounced the Annian fragments of Berosus et al. as forgeries. He became furious when he saw that scholars continued to cite them. And he devoted substantial portions of the second edition of the *De emendatione* and his *Thesaurus* to editing what he took as the genuine fragments of these and other ancient historians. In method, Scaliger differed from almost all the chronologers with whom Bodin was in

---

[24] Annius of Viterbo (1512), LXXXIV verso.

[25] Grafton 2020, 188. See also, more generally, Grafton (2019a).

[26] Bodin 1566, 48, 1945, 47. For Bodin's debt to Annius, see Grafton 1989.

[27] Tolosani, *Opusculum*, "Prooemium authoris," sig. A recto-A ij verso, at A recto: "Quare cavendum est a nonnullis authoribus Graecis qui multa mendacia veris historiis inseruerunt." A long series of quotations from Annius and his authors follows.

[28] Funck 1552, fol. 3A: "Berosus ab egressu Noah ex arca supputare orditur annos regni Babylonici."

[29] Mercator 1559, 7.

[30] Ibid., 9.

[31] Ibid.

dialogue. But in his deep interest in the ancient Near East and his commitment to integrating it with that of Greece and Rome, he was at one with them. It was not Scaliger but Annius who placed "Graeco-Roman history in the context of Babylonian, Egyptian, Persian, and Jewish history."[32] Tolosani and Funck, Mercator and Crusius saw the past in exactly the same terms.

Bodin followed Annius in one crucial respect. Some Greek and Roman writers claimed that the history of the Near Eastern kingdoms had lasted a very long time—far too long to be accommodated within the biblical span of roughly 4000 years between Creation and Incarnation. Following Annius, he dismissed them as liars:

> What Apion took from the *Timaeus* of Plato is false—that is, that the Egyptians had the history of eight thousand years put on record in their sources. Even more absurd is the tale Herodotus heard—that there had been a kingdom among the Egyptians for 13,000 years. Most absurd is what Cicero wrote in the book *On Divination*: that the Chaldeans boasted that they had for 470,000 years tested the talents of men.[33]

As Bodin put it elsewhere, concentrating his fire on the most important of his opponents:

> Since the period of Egyptian history seemed too short to Herodotus, in comparison with the reported one of eight thousand years, he wrote this: that the Egyptians boasted that before Menes, whom he called the first king of Egypt, they had the histories of 330 kings written down in sacred books, yet they did not record the name of any of these, or the deeds.[34]

Bodin knew that Aristotle and other ancients had argued that the world was eternal. He probably also knew that—as Ivano dal Prete has taught us—many thinkers in his day, from artisans to philosophers, conjectured on the basis of geological evidence that the world could not be as young as the Bible claimed.[35]

Early modern chronologers tended to be less radical than contemporary natural philosophers. Yet some were tempted to modify—if not to extend—biblical chronology. In the *Chronicle* of Eusebius and Jerome, Egyptian history begins after the Flood—with the sixteenth dynasty, which was listed next to the new Assyrian monarchy of Ninus. What, someone might wonder, had happened to the other 15? This question occurred to Mercator. He suggested that the Egyptian kingdom might have existed before the Flood, even though the Bible did not mention it. Evidently some memory of it had survived the Flood.

> The large number of dynasties recorded by Eusebius suggests that the kingdom of Egypt is very old, and was born almost with the earliest human beings. … It is likely that the Egyptians, who always took such pride in their antiquity, and were more blessed than others thanks to their ancestors, who passed on their tradition, and by their method of writing, created in time immemorial, could manage to preserve a general memory of events in their land from the Creation.[36]

---

[32] Bod 2014, 169.

[33] Bodin 1566, 384, 1945, 320.

[34] Bodin 1566, 388, 1945, 322–23.

[35] Dal Prete 2022.

[36] Mercator 1559, 5.

Following a suggestion made long before by Marsilio Ficino, Mercator digressed to offer further confirmation for his view.[37] In Plato's *Timaeus*, Egyptian priests told Solon that Athens had been founded 9000 years, and Thebes 8000, before their time (23d–e). Here too, Mercator saw a potential reference to the fact that the kingdom of Egypt had existed before the Flood. True, the duration Plato claimed for Egyptian history seemed implausible. But a deft interpretation removed that difficulty: "As to the fact that they reckon so many thousand years, it will not be easy to convict them of falsity if we count them as lunar years, that is, months, which were in use at some point."[38] If one counted 9000 lunar months backward from the time of Plato, the history of Athens would have begun in the year 160 or 170 before the Flood—a time of chaos and criminality when almost anything might have happened. Bodin composed the first version of his text just before Mercator wrote—but he was clearly responding, in his own way, to the sorts of ancient evidence that would lead Mercator to try to loosen the historical corset imposed on historical time by the Hebrew Bible.

"Fables of this sort," Bodin wrote, "can be refuted by many arguments."[39] But he placed special reliance on one particular discipline, astronomy. In his view it provided exactly the information needed to prove that ancient history was not so long as Cicero and Apion argued. "In truth," he wrote, "no argument is greater than that Claudius Ptolemy, an Egyptian, sought out the initial positions and epochs of the fixed stars from the time of Nabonassar, as though the Chaldeans had taken their first observations from that period."[40] Bodin did not fully explain what he was referring to here. But his words show that he was familiar with the tradition of ancient astronomy.

Nabonassar, as historians of astronomy know, was a Babylonian ruler. His accession date, 26 February 747 BCE, provided the baseline from which later astronomers in Babylon and Greece dated their observations. In Arabic and medieval Latin astronomical works, he was given the name of a biblical ruler, Nabuchodonosor.[41] But the Greek text of Ptolemy's *Almagest* revealed that the real name of the Chaldean king in question was Nabonassar. He too appeared in the Hebrew Bible, Bodin argued, under a different name: "Nabonassar, who was called Salmanassar in the Book of Kings, flourished in the year 3000 after the Creation, 880 years before Ptolemy, as he himself wrote in the *Almagest*." Salmanassar V was a king of Assyria, closely contemporary with Nabonassar, who subjugated ancient Israel and put down a rebellion led by the last king of Israel, Hosea. Eusebius called him king of Chaldea in his Chronicle. By assuming that Nabonassar and Salmanassar were the same person, both rulers of Chaldea, Bodin could connect biblical history with the most reliable pagan account, one that rested on king lists and astronomical observations

[37] See Grafton 1995, 15–31.

[38] Mercator 1559, 5.

[39] Bodin 1566, 388, 1945, 323.

[40] Ibid.

[41] Swerdlow and Neugebauer 1984, 1:187.

that confirmed one another—and prove that astronomical evidence supported that of the Bible.

Few of the elements of Bodin's structure were new. As Neugebauer and Swerdlow showed in a brief, packed section of *Mathematical Astronomy in Copernicus's De revolutionibus*, Copernicus argued in the manuscript of his work that Nabonassar must not be Nabuchodonosor but Salmanassar, "king of the Chaldeans," and he stated the point, more briefly, in the final version of his book.[42] Chronologers after him—including Funck, Mercator, and Crusius—regularly assumed both the identity of the two rulers and the synchronism it provided between biblical and secular history.[43] To that extent, Copernicus transformed chronological scholarship. But Copernicus himself may have drawn on an earlier source. In *De revolutionibus*, he identifies Salmanassar as "king of the Chaldeans among the historical writers." At least one historical writer had already made part of the same identification. When Tolosani reached Salmanassar, whom he set at 748 BCE in his chronology, he added a note: "The years of Nabuchodonosor begin, according to the astronomers, with Salmanassar."[44] Copernicus, working from the Greek text of Ptolemy, distinguished Nabonassar from Nabuchodonosor. But he may well have found the original suggestion to identify Nabonassar with Salmanassar in Tolosani.[45] Evidently, no iron disciplinary curtain kept philologists from infringing on the materials and methods of the astronomers, and Bodin was not the first of them to do so.

Bodin used this evidence to strive for what he saw as a vital polemical goal for Christian chronologers: to prove that astronomy—and the ancient world—were younger than many had claimed. If the Egyptians and Chaldeans had really observed the heavens for 470,000 years, "why did Ptolemy the Egyptian, a neighbor to the Chaldeans, collect the motions for only 800 years? This larger body [of observations] would have been essential for the demonstrations which he had undertaken."[46] Bodin underestimated the span of time between Nabonassar in 747 BCE and Ptolemy, who worked in the second century CE. But his general point was valid—as was his argument that later astronomers had corrected Ptolemy in essential points, as when they found that the apsidal line—the line between the apogee and the perigee—of the Sun, which he had thought immobile, moved over time.[47]

Bodin, moreover, found a further piece of astronomical evidence that confirmed his history of ancient astronomy—one that came from an author known for his

[42] Ibid., 1:183–88.

[43] Funck 1552, fol. 44A: "Vocatur autem hic rex [Salmanassar] a Ptolemeo Nabonassar"; Mercator 1559, 55: "SALMANASSAR (Ptol. Nabonaasar)."

[44] Tolosani 1537, fol. 100 recto: "Anni Nabuchodonosor secundum Astronomos sumuntur ab Salmanassar."

[45] Copernicus's and Tolosani's phrasings mirror one another, and Tolosani's book appeared in 1537, early enough for Copernicus to have used it. In any event, Tolosani, far from accepting Copernicus's theory, wrote a violent tract against it. See Garin 1975; Rosen 1975; Kempfi 1980–1981.

[46] Bodin 1566, 388–89, 1945, 323.

[47] Bodin 1566, 389, 1945, 323.

antipathy to the Christian tradition and its sources, the sixth-century philosopher Simplicius:

> I have drawn an even weightier argument from Simplicius. For although he proved a most bitter enemy of the Hebrews and the Christians, especially when he defended Proclus against the Christians, nevertheless in the commentaries on the first book of *On the Heavens*, he left witness that Aristotle had written a letter to Callisthenes, asking him to collect the antiquities and records of the Chaldeans while the others, intent on booty, plundered Babylon. Then Callisthenes wrote back that he had diligently collected the Chaldean records and had found there the history of 1903 years. This number exactly fits the sacred history of Moses and Philo, if we count back from Alexander the Great to that time in which he first reported that the sons of Noah were dispersed over the earth.[48]

Bodin, in other words, integrated astronomy with philology to establish a firm biblical framework for chronological scholarship.

His arguments were sharp and rigorous. But they were not original. More than half a century before, the eclectic philosopher Giovanni Pico della Mirandola spent the last years of his life composing a set of *Disputationes contra astrologiam divinatricem*. He argued, in granular detail, that astrologers' theories and practices, ancient and modern, lacked any foundation in nature. One of the ways in which he set out to show that the discipline lacked any foundation was to argue that its origins were relatively recent:

> When Hipparchus and Ptolemy, the founders of astronomy, produce the observations of the ancients to lay the foundations of their discipline, they produce none older than those [made] in Babylon and Egypt under King Nabuchodonosor. Hipparchus flourished in around the six-hundredth year after his reign. From him to our time no more than about sixteen hundred years have passed. Their claim to have observations from so many centuries is therefore false and mendacious.[49]

Bodin's evidence about Callisthenes also came from Pico, who had quoted the same passage as Bodin, from Simplicius's commentary on *De caelo* 2.12 (293a4):[50]

> Relevant to our project is a very manifest testimony of Porphyry's. He says that the Chaldean records about the stars that Callisthenes sent back from Babylon in the time of Aristotle and Alexander, contained observations of the stars for only 1903 years. If you count these years backward from Alexander, you will find that they run out in the times of Abraham the Patriarch or a little earlier.[51]

Scaliger, as we have seen, built his chronology in part on Paulus Crusius's earlier study of epochs and eclipses, which he did not cite. Bodin worked in a very similar way, using Pico without citing him.

True, Bodin tweaked the second piece of evidence. Pico described what Callisthenes found as "de astris annotationes." Bodin transformed these "notes on the stars" into "antiquities and records" (*antiquitates ac monimenta*). Here too, however, his practices were anything but unique. Less than 20 years after Bodin,

[48] Bodin 1566, 385, 1945, 320.

[49] Pico della Mirandola 1946–1952, 2:476.

[50] Diels 1882–1909, 506.

[51] Pico 1946–1952, 11.2, 2:476.

Scaliger cited the same passage from Simplicius in much the same way that the lawyer had. Like Bodin, he used it to show that pagan and Hebrew evidence complemented one another neatly. Like Bodin, too, he transformed the observations mentioned in the original text into the material that the Chaldeans had offered Callisthenes "when he asked them about Chaldean antiquities and origins [*de antiquitate et originibus*]." It seems entirely possible that Scaliger never examined the original text of Simplicius, but simply paraphrased what he had read in Bodin (or in one of the several sources published between Bodin's book and his that discussed the same source in the same way). In this case, as in others, a medieval prehistory beckons alluringly, since the Latin version of Simplicius had been available for centuries. Pietro d'Abano, for example, discussed Callisthenes's mission. However, he drew no chronological conclusions from the passage.[52] One conclusion is certain: Bodin anticipated Scaliger. One is probable: Scaliger drew his interpretation of the source from Bodin. At this point the distance between the humanist and the creator of technical chronology was very short.

Though Bodin mined Pico's critique of astrology for useful sources, he apparently did not wholly abandon his faith in astrology itself. One piece of astrological evidence in particular—a horoscope for the city of Rome computed by the ancient astrologer Lucius Tarrutius of Firmum (modern Fermo)—seemed to him relevant to establishing a much controverted data point: the actual foundation date of the city of Rome. Girolamo Cardano had argued that the appearance of Alkaid, the last star in the tail of the Great Bear, in the zenith of a city brought it power. He claimed that this had happened in the case of Rome. To rebut Cardano, Bodin brought in ancient evidence. The scholarly Varro, friend of Cicero and expert on the Roman past, was troubled by the variety of opinions about the era of Rome, and the lack of written sources with which to choose among them. He asked Tarrutius to practice his art retrospectively: to work out, from Romulus's characteristics and his fate, the date of his birth, and to do the same for the city of Rome itself. Tarrutius complied, establishing dates both for the birth of the city's founder and for that of the Foundation itself. His charts were transmitted by Cicero, Plutarch, Manilius, John Lydus, and other writers, with a variety of errors and inconsistencies.

Solinus, who wrote around 300 CE, described Tarrutius's horoscope for the city of Rome in detail:

> Romulus made the auspices and set the foundations of Rome at the age of 18, on the eleventh day before the calends of May [i.e., 21 April], between the second and third hours, as the highly regarded mathematician Lucius Tarrutius declared, with Jupiter in Pisces, Saturn Venus Mars Mercury in Scorpius, the Sun in Taurus, and the Moon in Libra. [1.18]

As Stephan Heilen has shown, Solinus or his source misunderstood Tarrutius's dating. He set the founding of Rome not on the traditional calendar date, 21 April, but 7 months earlier, on 4 October. This was the date that Plutarch attributed to him in his life of Romulus: 9 Pharmuthi in the fixed Egyptian calendar. By placing the Sun in Taurus, the author of Solinus's source tried to fit Tarrutius's horoscope to 21

[52] d'Abano 1985, 198.

April. But he failed to realize that making this change introduced an elementary error. The two interior planets, Mercury and Venus, retained their proper places, for 4 October, in Scorpius. But the Sun moved to Taurus, across the Zodiac from them. Mercury could never be distant by more than one zodiacal sign, or Venus by more than two, from the Sun. No competent astrologer—and Tarrutius was extremely competent—could have made this mistake.

Bodin set out to use these materials to show that Cardano was wrong. He cited Plutarch and Solinus as if their testimony could be reconciled:

> Varro, according to Plutarch, ordered L. Tarrutius Firmanus, a well-known mathematician, to fix the natal day and hour of the city and its founder, if possible. Then, by a diligent investigation of the stars he found that the foundations of the city were laid in the third year of the sixth Olympiad, to which Dionysius and Varro agreed, on 9 Pharmuthi, or 21 April, which is called the day of Palilia, about three o'clock in the afternoon, with Jupiter in Pisces, Saturn, Venus, and Mars in Scorpio, the sun traversing Taurus and the moon Libra, in the eighteenth year of Romulus.[53]

He did not comment on the positions of the inferior planets. But he did take the time to point out that the ancients never mentioned the last star in the tail of the Great Bear.

A decade later, in his *Republic*, Bodin returned to this question, which he now investigated in more detail. He ascribed his error about Mercury and Venus to his source, Tarrutius, and denounced it in no uncertain terms: "This horoscope is not only false, but also impossible in nature: for he puts Venus in opposition to the Sun, though she never departs 48 degrees from the Sun."[54] Injudiciously, he noted that a modern astrologer, Augier Ferrier, had committed the same error, "by forgetfulness," placing Venus and Mercury in opposition to the Sun.[55] This attack, far from distracting readers from Bodin's original error, won him a sharp counterblast from Ferrier.[56]

Bodin also discussed other astrologers' efforts to explicate Tarrutius's horoscope. He noted, for example, what he took to be an error in the *Tractatus astrologicus* of Luca Gaurico, which appeared in 1552. This horoscope collection began with a series of charts for cities, including a "Thema Vrbis Romanae a Romulo fundatae," which Gaurico dated 21 April 752 BCE. As his sources he named "Solinus & Lucius Terentius."[57] Though Gaurico stuck to the traditional date, Bodin complained, he followed the poet Manilius, whom he quoted, in making Libra the dominant sign.[58] Bodin was not wrong to complain about Gaurico, who simply omitted the planets and their positions from his chart—no doubt because he saw that the positions given by Solinus for Venus and Mercury were impossible. Yet it seems quite likely that Bodin found both a version of Tarrutius's horoscope of Rome and a reference to

---

[53] Bodin 2013, 276; see Heilen 2007.

[54] Bodin 1586, 397.

[55] Bodin 1586, 397–98.

[56] Lloyd 2017, 119.

[57] Gaurico 1552.

[58] Bodin 1586, 398.

Solinus in Gaurico, and that he added the positions of the planets as given by Solinus to his own account.

As this episode suggests, Bodin continued to read new chronological publications, including technically demanding ones, after the *Methodus* appeared. In the first edition of the *Methodus*, Bodin argued that the Eastern nations, near and far alike, had set the beginning of the year in the fall:

> Not the least cause of error [in computing years from the Creation] can seem to be that the Greeks began their year from the summer solstice, the Latins from the winter one; the others to the west, from the vernal equinox; the Arabs from the entry of the sun into Leo, as Firmicus teaches in book 3[1.1]. The Orientals, though, that is the Chaldeans, the Persians, the Indians, the Hebrews from the autumn, at which time the world was created by God.[59]

He cited two sources for this view: Josephus, who noted that the Jewish year had originally begun in the fall (*Jewish Antiquities* 1.15–16), and Rabbi Eliezer ben Hyrcanus, one of the second-century Tannaim (sages) of the Talmud. The Talmud was a highly contentious source. In 1553, copies of the Talmud in Italy were confiscated, and on 9 September 1553, Rosh HaShanah of the Jewish year, many of them were burnt in public in the Campo dei Fiori.[60] Bodin, however, was a determined comparatist, who treated the laws of as many nations as he could find. He made clear that he saw the Talmud as a legitimate juridical text, a parallel to the Roman *Corpus Iuris*, and that he had drawn on it with the help of Christian Hebrew scholars:

> We use the authorities of jurists and historians at the same time, so that the laws of the Persians, the Greeks, the Egyptians are considered on a par with the laws of the Romans. We have also set out to draw excellent materials from the Pandects of the Hebrews [the Talmud], especially from Sanhedrin. In this Jean Cinquarbres and Jean Mercier, royal professors of the Hebrew language, promised that I would not lack their help.[61]

Bodin argued that Eliezer saw his view as rooted in the Hebrew language: "Rabbi Eliezer on Genesis: ברשית [Bereshit], that is, 'In the beginning'; בתשרי [BeTishrei], that is, 'in the month of September': he thinks that the mystery of the month is embraced in this transposition of letters."[62] His quotation came from *Bereshit Rabba*, a massive midrash (expansive commentary) on the book of Genesis (22:40). Rabbi Eliezer took the same position, at somewhat more length, in tractate Rosh HaShanah of the Babylonian Talmud (10b).

The argument from the Hebrew language does not appear in these sources, and Bodin mangles it—partly because he misspells the word Bereshit, which lacks an aleph in his text. A Jewish tradition notes that the letters of the word בראשית ("in the beginning"), properly spelled with its א, can be rearranged to spell א בתשרי ("the first of Tishrei"). It seems likely that one of the two Hebrew professors whose advice

[59] Bodin 1566, 391, 1945, 325.

[60] See Stow 1972.

[61] Bodin 1566, sig. [*iii verso].

[62] Bodin 1566, 391, 1945, 325. For early scholars' discussion of the nature of the year at the time of the Flood, see Nothaft 2011.

Bodin acknowledged, Jean Cinqarbres and Jean Mercier, explained this point, and that Bodin remembered it incorrectly.

Not long after the *Methodus* appeared, Mercator tried to settle the date of Creation in his chronology. Noah, he argued, sent out the second dove on the 25th day of the 11th month. The bird returned with a fresh olive leaf. So the flood must have ended when the Sun was in Gemini. After all, the season when plants grew would not have changed since the Creation. The Jewish calendar is luni-solar, as Mercator knew: therefore its months do not have fixed places in the solar year. Using rough equivalences nonetheless, Mercator noted that the Hebrew month Siuan corresponds to the period when the Sun is in Gemini. He assumed that this relationship had never changed: "it is not plausible that the Holy Fathers, who felt a religious respect for the year, which had been established either by God as it was, or, prudently, by Adam, with his profound knowledge of all things, would have made any change before the Flood."[63] Accordingly, the 11th month mentioned in the Hebrew Bible, which corresponded to Gemini, must still have been Siuan—followed, in the Jewish calendar, by Tammuz and Av. If Tammuz was the twelfth and last month, then Av, the month that corresponded to Leo, "must have been the first of the months of the year, and thus the beginning of the world."[64]

As when Mercator dealt with the Flood, here too he did his best to find another ancient source that provided corroboration. Solinus—the same compiler on whom Bodin had relied—discussed the flooding of the Nile at midsummer.

> The origin of the Nile's flooding is entirely derived from the sun. The first swelling takes place when the sun is carried through Cancer. After it has passed through the thirty degrees of Cancer, when it enters Leo and arouses the rising of Sirius, all the waves are released and the whole motion of the waves breaks forth. (32.12)

He then noted that "the priests" (presumably of Egypt) had argued that this yearly event corresponded to the Creation: "The priests have judged this time to be the birthday of the world: that is, between the 20th and the 22nd of July" (32.13).[65] Since Solinus did not name the source of the priests' knowledge, Mercator argued, "it is probable that they were instructed in this by some Hebrew interpreters, for they had the writings of Moses and engaged in regular contact with the Hebrews." Only the intervention of Satan had obscured matters for later ages, "so that he might induce the peoples to accept insane idolatry."[66]

Revising the *Methodus* in 1572, Bodin added a sharp rebuttal. Mercator was wrong about the date of the Creation, he argued, and that fatal error undermined his entire work.

> Mercator is just as wrong [as those who set the Creation in the early spring]. He thinks that the sun was in Leo when the world was created. Since that foundation was badly laid, the

---

[63] Mercator 1559, sig. a recto.

[64] Ibid.

[65] Ibid.

[66] Ibid.

> other things that he drew from the motion of the stars to support the credibility of history are threatened with ruin.[67]

A fellow cartographer and antiquarian, Abraham Ortelius, called Mercator's attention to this passage. Mercator noted that he had "skimmed the book, but this passage escaped me." In typical academic fashion, he insisted at some length that Bodin's obviously inexpert book and clearly incorrect critique did not bother him in the least:

> I have examined the author carefully. I see a man who is learned and widely read, but in this text, a clearly unripe judge. He recites many opinions, many errors without refutation, with no regard for good order and method. The principles and foundations that he has laid down are often completely wrong and not at all consistent with one another. But this is of no consequence for me. So far as the beginning of the world is concerned, I wish that he had put forward another proof with more solid arguments, which would make it possible to compare my arguments with his. For what he cites on the beginning of the year from Moses and the Rabbis adds nothing here. Far less is it true that with the beginning of the world misplaced (as he claims), the way in which I applied the motion of the stars to chronological argument can be shown to be wrong. Rather, that overturns his entire method, if one painstakingly examines both in detail. I am surprised that a learned man judges matters that he has not understood or not discerned in so frivolous a way. When I have a chance I will briefly and modestly expunge his calumny.[68]

In the end Mercator did not reply to Bodin in print, but Bodin nonetheless relitigated the question in his *République* of 1576. Mercator was far more committed to the study of chronology than Bodin, and considerably more adept at it, as he saw. Yet the resemblances between the two men's methods are at least as evident as their differences. Both assumed that evidence about the Jewish calendar, drawn from ancient Jewish traditions, could reveal when the world came into existence. To that extent, Mercator and Bodin were playing the same game, by the same rules, and their debate was a case of the narcissism of small differences that so often affects scholars.

Noel Swerdlow taught us, by the example of his lifetime of work comparing Copernicus with Regiomontanus, to see that heroes need teachers, whether they encounter them in person or on paper. Scaliger, like Copernicus, needed teachers. Most of the techniques that he applied were already in use when he began to explore the field. Almost all chronologers sought information about the East as well as the West. Almost all of them attained some knowledge of ancient astronomy and integrated it into their work. Almost all of them used texts somewhat indiscriminately, sometimes citing sources as authorities which they elsewhere subjected to corrosive criticism. Almost all of them hoped that ancient texts, none of them written earlier than the first millennium BCE, and ancient calendrical practices, might help them identify the moment, millennia before, when the world began. Scaliger innovated in multiple ways, and imposed a new level of precision and coherence on the unsightly mass of materials and problems that his predecessors had assembled. But one reason that his work had such a powerful impact is that others, far less erudite and far

---

[67] Bodin 2013, 519. In this printing Bereshit receives its aleph but loses its yod. For a modern edition that traces the evolution of the text through Bodin's revisions, see Bodin 1566.

[68] Mercator to Ortelius, 26 March 1576, in van Durme 1959, 113–14.

less penetrating, had already prepared the ground. The normal science of chronology was recognizably the same discipline that he practiced and, in part, reformed.[69]

References

Annius of Viterbo. 1512. *Antiquitatum variorum volumina xvii*. Paris: Petit and Ascensius.

Bod, Rens. 2014. *New History of the Humanities: The Search for Principles and Patterns from Antiquity to the Present*. Oxford: Oxford University Press.

Bodin, Jean. 1566. *Methodus ad facilem historiarum cognitionem*. Paris: Martin Le Jeune.

Bodin, Jean. 1572. *Methodus ad facilem historiarum cognitionem*. Paris: Martin Le Jeune.

Bodin, Jean. 1586. *De republica libri sex*. Paris: Iacques du Puy.

Bodin, Jean. 1945. *Method for the Easy Comprehension of History, tr. Beatrice Reynolds*. New York. Columbia University Press.

Bodin, Jean, 2013. *Methodus ad facilem historiarum cognitionem*, ed. Sara Miglietti. Pisa: Edizioni della Normale.

Borchardt, Frank. 1971. *German Antiquity in Renaissance Myth*. Baltimore: Johns Hopkins University Press.

d'Abano, Pietro. 1985. "De eccentricis et epicyclis." Edited by Graziella Federici Vescovini. *Medioevo* 11:175–206.

Dal Prete, Ivano. 2022. *On the Edge of Eternity: The Antiquity of the Earth in Medieval and Early Modern Europe*. New York: Oxford University Press.

Diels, Hermann, ed. 1882–1909. *Commentaria in Aristotelem Graeca*. Berlin: Reimer.

Funck, Johann. 1552. *Chronologia*. Nuremberg: Wacher.

Garin, Eugenio. 1975. *Rinascite e rivoluzioni: movimenti culturali dal xiv al xviii secolo*. Bari: Laterza.

Gaurico, Luca. 1552. *Tractatus astrologicus*. Venice: Curzio Troiano Navò.

Grafton, Anthony. 1989. Editing Technical Neo-Latin Texts: Two Cases and Their Implications. In *Editing Greek and Latin Texts*, ed. John N. Grant, 163–186. New York: AMS.

Grafton, Anthony. 1993. Joseph Scaliger: A Study in the History of Classical Scholarship. In *Historical Chronology*, Vol. 2. Historical Chronology.

Grafton, Anthony. 1995. Tradition and Technique in Historical Chronology. In *Ancient History and the Antiquarian: Essays in Memory of Arnaldo Momigliano*, ed. Michael H. Crawford and Christopher R. Ligota, 15–31. London: Warburg Institute.

Grafton, Anthony. 2019a. *Forgers and Critics: Creativity and Duplicity in Western Scholarship*. Princeton: Princeton University Press.

Grafton, Anthony. 2019b. Scaliger's Chronology: Early Patterns of Reception. In *Confessionalisation and Erudition in Early Modern Europe. An Episode in the History of the Humanities*, ed. Nicholas Hardy and Dmitri Levitin, 154–193. Oxford: Oxford University Press for the British Academy.

Grafton, Anthony. 2020. *Inky Fingers: The Making of Books in Renaissance Europe*. Cambridge, MA: Harvard University Press.

Grafton, Anthony, and Urs Leu, eds. 2014. *Henricus Glareanus's (1488–1563) Chronologia of the Ancient World: A Facsimile Edition of a Heavily Annotated Copy Held in Princeton University Library*. Leiden: Brill.

Grafton, Anthony, Nicholas Popper, and William Sherman, eds. 2024. *Gabriel Harvey and the History of reading: Essays by Lisa Jardine and Others*. London/Chicago: UCL Press/University of Chicago Press.

[69] Warm thanks to Ann Blair for comments and corrections.

Harvey, Gabriel. (2019). The Archaeology of Reading. https://archaeologyofreading.org/viewer//#aor
Heilen, Stephan. (2007). Ancient Scholars on the Horoscope of Rome. *Culture and Cosmos 11*, 1–2: 43–68.
Kempfi, Andrzej. 1980–1981. Tolosani Versus Copernicus. *Organon* 16 (17): 239–254.
Livy, T. 1555. In *Decades tres cum dimidia*, ed. Celio Secundo Curione. Basel: Hervagius.
Lloyd, Howell A. 2017. *Jean Bodin, "This Pre-eminent Man of France": An Intellectual Biography*. Oxford: Oxford University Press.
Mercator, Gerardus. 1559. *Chronologia*. Cologne: Heirs of Arnold Birckmann.
Miglietti, Sara. 2013. Reading from the Margins: Some Insights into the Early Reception of Bodin's Methodus. In *The Reception of Bodin*, ed. Howell Lloyd, 193–218. Leiden/Boston: Brill.
Mori, Giuliano. 2024. *Historical Truth in Fifteenth-Century Italy: Verisimilitude and Factuality in the Humanist Debate*. Oxford: Oxford University Press.
Most, Glenn, ed. 1997. *Fragment collecting = Fragmente sammeln*. Göttingen: Vandenhoeck and Ruprecht.
Nothaft, C. Philipp E. 2011. The Calendar of the Flood: The Chronology of the Flood Narrative and the History of Astronomy in Sixteenth- and Seventeenth-Century Europe. *Journal of the Warburg and Courtauld Institutes* 74:191–211.
Nothaft, C. Philipp E. 2018. *Scandalous Error: Calendar Reform and Calendrical Astronomy in Medieval Europe*. Oxford: Oxford University Press.
Nothaft, C. Philipp E. 2020. 'With Utmost Certainty': Two Late medieval Pioneers of Technical Chronology. *Journal of Medieval History* 46:335–349.
Pattison, Mark. 1889. *Essays*, edited by Henry Nettleship. Oxford: Clarendon Press.
Pico della Mirandola, Giovanni. 1946–1952. *Disputationes contra astrologiam divinatricem*, 11.2, edited by Eugenio Garin. 2 vols. Florence: Vallecchi.
Rosen, Edward. 1975. Was Copernicus's Revolutions Approved by the Pope? *Journal of the History of Ideas* 36 (3): 531–542.
Scaliger, Joseph. 1583. *Opus novum de emendatione temporum*. Paris: Patisson.
Scaliger, Joseph. 1598. *Opus de emendatione temporum*. 2nd ed. Leiden: Plantin-Raphelengius.
Scaliger, Joseph. 1606. *Thesaurus temporum*. Leiden: Basson.
Scaliger, Joseph. 1629. *Opus de emendatione temporum*. 3rd ed. Geneva: Pierre de la Rovière.
Scaliger, Joseph. 1658. *Thesaurus temporum*. 2nd ed. Amsterdam: Jansson.
Stow, Kenneth. 1972. The Burning of the Talmud in 1553, in the Light of Sixteenth-Century Catholic Attitudes towards the Jews. *Bibliothèque d'Humanisme et Renaissance* 34 (3): 435–459.
Swerdlow, Noel, and Otto Neugebauer. 1984. *Mathematical Astronomy in Copernicus's* De revolutionibus. 2 vols. New York: Springer.
Tolosani, Giovanni Maria. 1537. *Opusculum de emendationibus temporum*. Venice: Giunta.
Turchetti, Mario. (2005). *Stanford Encyclopedia of Philosophy*. https://plato.stanford.edu/entries/bodin/
Turner, James. (2014). *Philology: The Forgorren Ancestor of theModern Humanities*. Princeton: Princeton University Press.
van Durme, M., ed. 1959. Mercator to Ortelius, 26 March 1576. In *Correspondance Mercatorienne*. Antwerp: De nederlandsche Boekhandel.

Chapter 11
Writing "Celestial History" in Early Modern Europe

Florence C. Hsia

11.1 Classifying the History of Astronomy

In June 1803, Joseph-Jérôme Lalande's *Bibliographie astronomique* finally came off the presses of the Imprimerie de la République, thanks to the support of Jean-Antoine Chaptal—chemist, Lalande's fellow academician, and, most helpfully, minister of the interior. [1] A member of the Institut national and numerous academies of science, professor of astronomy at the Collège royal in Paris, and erstwhile director of the Paris observatory, Lalande thought himself better positioned than anyone for a project of this scale and ambition, a typically immodest author's claim that, in this case, was likely true.[2] As the longtime editor of and contributor to the *Connaissance des temps*—an organ of the Paris Academy of Sciences since 1702—Lalande made the long-running annual calendar and ephemerides a showcase for astronomical progress by including notices of the latest methods, observations, computational techniques, and publications of interest to astronomers. Lalande developed these book notices into his 1803 *Bibliographie astronomique*, whose scope—at nearly 900 pages and over 2200 years of astronomy's history—far

[1] Lalande 1804, 400; 1803, 245.

[2] Lalande 1803, ii.

F. C. Hsia (✉)
Department of History, University of Wisconsin, Madison, WI, USA
e-mail: florence.hsia@wisc.edu

F. Rochberg (ed.), *Studies in Premodern Sciences in Memory of Noel M. Swerdlow*, Archimedes 74, https://doi.org/10.1007/978-3-032-04040-4_11

exceeded that of comparable works like Johann Weidler's *Bibliographia astronomica* (1755) and Johann Bernoulli's serial guide to current publications, the *Recueil pour les astronomes* (1771–1776).[3]

In the extensive index that Lalande's fellow academician, Louis Cotte, compiled to guide readers through the *Bibliographie astronomique*, less than 40 titles were classified as "History," a categorization that reveals much about shifting early modern conceptualizations of astronomy's past and its relationship to present-day concerns.[4] The rubrics Cotte selected for the index were pragmatic, not principled. Guillaume Le Gentil's 1779 account of his lengthy travels to India, for instance, appeared in Cotte's index under at least six headings, including "History" for its memoirs on Indian astronomical traditions, though not under "Venus" despite its title, as Le Gentil's plans to observe the transit of Venus across the disk of the Sun, first in 1761, then in 1769, had come to naught. "History" evidently functioned as a marker of the book's content, not its genre ("Voyage"). As Cotte remarked, "I've tried to put as much order as possible into this work that I offer to astronomers: it's not brilliant, but it is useful."[5]

Well into the eighteenth century, substantive work in the history of astronomy could still readily be found in what Nick Jardine has called "prefatory histories": overtures concerning astronomy's venerable origins, excellence, and utility that, like orations inaugurating a series of lectures, established the subject's value and legitimacy for the Renaissance reader or auditor. Cotte's index subsumed such texts within the categories appropriate to the works they accompanied: principally treatises ("Traités") and tables, two of the oldest and most durable genres of astronomical texts.[6] Cotte's headings also nodded to bibliography as a genre suitable for writing the history of astronomy, as in the works by Weidler and Bernoulli already mentioned. Biography comprised another and sometimes closely related generic option.[7] The poet and translator Edward Sherburne, for instance, concluded

[3] Lalande 1764, 228–37, 228 (Weidler); Lalande 1803, 458 (Weidler), 521 (Bernoulli). On Lalande's "additions" to the *Connaissance des temps*, see Boistel 2022, 128–29, 146, 237–44.

[4] Cotte 1803; see Cotte's "Avertissement," [919]–20, to his index and the items he listed under "History," 929.

[5] Le Gentil 1779; also indexed under "Compass variations," "Meteorology," "Refraction," and "Wind," following Lalande's brief summary of its contents. Lalande 1803, 568 (Le Gentil); Cotte 1803, 920 (quote).

[6] Jardine 1984, chap. 4. For examples of "prefatory histories" in works Cotte categorized as "Tables," see Stadius 1560, and Kepler 1627; for instances in works Cotte categorized as "Treatises," see Longomontanus 1622; Riccioli 1651; Levera 1663; Moxon 1665; Milliet de Chales 1674; Keill 1721; Cassini 1740; Le Monnier 1746; and Lalande 1764.

[7] Compare Weidler 1741, classified under "History," and Savérien 1765, vol. 5, classified under "Biography"; Cotte 1803, 929 ("Biography").

his 1675 English edition of Manilius's first-century *Astronomicon* with a brief historical narrative of astronomy's origin and progress through the reign of Augustus, paired with a long series of notices on astronomers and their works, starting with "Adam, the first Authour of Astronomy, as of all other Arts and Sciences," who had "particularly instructed *Seth* in this Astral Science, and that by writing"—and ending with various students of Tycho Brahe.[8]

In contrast, Pierre Estève—a member of the Société royale des sciences in Montpellier—explicitly set aside both the bibliographical and the biographical for his 1755 three-volume study, in which he "undertook the history of this part of our understanding, the progress of the human spirit on all which has relation to the knowledge of the material sky: this is the unique object of this work."[9] Indeed, the most dominant genre of works categorized as "History" in Lalande's *Bibliographie* constitute critical contributions to the Enlightenment genre of philosophical history.[10] Situating the history of astronomy as a key feature of civilizational development was central to George Costard's essays on how little Greek astronomers owed their Egyptian and Babylonian predecessors;[11] Antoine-Yves Goguet's expansive study of "the first steps of the human spirit";[12] Jean-Sylvain Bailly's multiple volumes on the history of astronomy from antediluvian times to the present;[13] and Jean-Étienne Montucla's history of mathematics. When death prevented Montucla from seeing a new edition through the press that would cover the eighteenth century, Lalande himself stepped in to finish off the two additional volumes, the last of which was devoted solely to astronomy.[14]

As seen through the grid of Cotte's index, the history of astronomy moved decisively from the prefatory margins to the textual center with the rise of philosophical history in the latter half of the eighteenth century. The other titles in Lalande's *Bibliographie astronomique* that Cotte classified under "History" provided source material as well as philological, bibliographical, and technical control in the service

[8] Sherburne 1675, "Appendix," 6 (quote), 71–72. See Riccioli 1651, 1: xxvi–xlvii ("Chronicon duplex astronomorum, vel astrologorum cosmographorum, aut polyhistorum, qui astronomica, vel cosmographica propius attigerunt"); Lalande 1764, 132–46 ("Des astronomes les plus célébres qui ont paru depuis un siècle"; "Table chronologique des astronomes les plus célèbres").

[9] Estève 1755, vol. 1:5 (quote); x–xi, 12.

[10] On Enlightenment approaches to the history of science as philosophical history, see Špelda 2017; Richards 2006.

[11] Costard 1746, 1–2, 156–58; Costard 1748.

[12] Goguet 1758, 1:v (quote); for astronomy, see 1:213–37, 385–87, 392–401; 2:250–58, 276–85, 397–438; 3:91–120, 261–72, 278–82, 292, 295, 302–11, 338–39. See Wolloch 2007.

[13] Bailly 1775, 1777, 1779, 1779–1782, 1787. See Raina 2003; Smith 1954, 454–66, 468–83, and 494–500.

[14] Montucla 1758; Montucla and Lalande 1799–1802, 3: v–vii ("Préface de l'Éditeur"); 336 ("Note de l'Éditeur"). See Swerdlow 1993.

of tracing the progress of the human spirit, especially with respect to astronomical traditions unfamiliar by virtue of their cultural, temporal, or geographic distance: the China Jesuit Antoine Gaubil's detailed studies of Chinese astronomy; accounts of Indian astronomy by Le Gentil, penned during his Venus transit travels, and by contributors to the *Transactions* of the Asiatic Society of Bengal (1799); Johann Albert Fabricius's erudite assessment of Ptolemy's works, surviving manuscripts, translations, and editions; and reports of astronomical matters in journals edited by the director of the ducal observatory at Gotha, Franz Xaver von Zach.[15]

Altogether absent from Cotte's category of "history" titles, however, was a genre of astronomical literature that, in the course of the 17th and 18th centuries, dramatically shifted the textual grounds on which astronomers engaged with their colleagues past, present, and future: "celestial history." Relatively few works make up the printed corpus of early modern "celestial history," though they are closely associated with some of the most important astronomers at work in the period: Tycho Brahe, renowned for his island observatory of Uraniborg off the Danish coast; John Flamsteed, the first director of the Royal Astronomy at Greenwich; and many of the leading astronomers in the Paris Academy of Sciences, from Pierre Le Monnier and Jean-Dominique Cassini (Cassini IV) to Alexandre Guy Pingré and even Lalande himself. Read together, these "celestial histories" reveal how their authors came to rethink making observations of astronomers past not just available, but also useful to astronomers present and future.

11.2 Opening the Treasury: Curtz's *Historia coelestis* (1666)

The earliest publication to bear the title of "celestial history" was an exceptional publishing project, as its grandiose dedications and frontispiece proclaimed. No fewer than four Habsburg emperors were celebrated as its patrons: Rudolph II, who had brought Tycho Brahe from Denmark to serve as imperial mathematician; Ferdinand II, to whom Johannes Kepler, Tycho's successor in imperial service, had dedicated the *Rudolphine Tables* (1627) based on Tycho's observational data; Ferdinand III, whom the volume's editor credited with safeguarding Tycho's observations in manuscript; and the reigning Holy Roman Emperor, Leopold I, under whose protection the edition appeared in print.[16]

In well over a thousand folio pages, the volume presented some 20 years of observations made by Tycho Brahe between 1582 and 1601—the first and only effort to systematically publish Tycho's empirical data in the early modern

[15] Souciet, ed., 1729–1732, vols. 2–3, by Gaubil; Fabricius 1796, 5:270–306 (on Ptolemy); Le Gentil 1779–1781, 1:206–352. Lalande cited the 1799 edition of the *Asiatic Researches, or, Transactions of the Society Instituted in Bengal* (vols. 1–5); the *Allgemeine geographische Ephemeriden* under Zach's editorship (1798–1799); and Zach's new *Monatliche Correspondenz zur Beförderung der Erd- und Himmels-Kunde* (1800–1813).

[16] Curtz [as Lucius Barettus], ed., 1666. On the frontispiece, see Remmert 2011, 152.

era—along with a wealth of observational material that its editor, the Jesuit Albert Curtz, culled from both printed and manuscript sources. Thanks to his efforts, manuscript observations of Tycho's contemporaries—Wilhelm IV, Landgrave of Hesse; Michael Maestlin, Kepler's mathematics professor—appeared in the volume as well. Curtz also broadened the temporal scope of his celestial history far beyond Tycho's lifetime, pushing both ahead to 1635 with Wilhelm Schickard's unpublished observations, and back in time to 720 BCE—the date of the oldest lunar eclipse observed at Babylon as reported by Ptolemy in the *Almagest*—all presented in chronological order.

The scale of this publishing project far outstripped in size and scope the few sets of astronomical observations that had preceded it in print, but Curtz's edition differed in kind, too. To be sure, observations of celestial—or possibly celestial—phenomena whose ontological status was actively contested or whose rarity made them resistant to predictive control (these were sometimes overlapping categories) already moved within a lively economy of print publication. What the Jesuit astronomer Christoph Borri characterized in his *Collecta astronomica* (1631) as the "new appearances observed in our time," like comets and "new" stars (versus the "common, and ancient appearances [*apparentiis*], which the Mathematicians call Phenomena") fueled the circulation of sightings by letter and in print.[17] As Adam Mosley has shown, such unusual events called for greater circumstantial context to bring them under control, whether natural philosophical or theological, astronomical or astrological. This was the case for individual phenomena—witness the flood of pamphlets that followed each of the great comets of 1577, 1618, 1665, and 1680—and also for comets in general as a major category of "new appearances observed in our time."[18] Seeking patterns in their portentous meaning, compilers brought together sightings of both comets and the "notable events" in the heavens and on Earth that followed in cometary "catalogs."[19]

Astronomers who sent "quotidian" observations to the press typically did so by putting them in harness, whether in setting the parameters of planetary models or as evidence for the predictive prowess of astronomical tables. Curtz, however, explicitly disavowed any explicit agenda for the *Historia coelestis*. Noting the difficulties faced by technical chronologers in dealing with a well-known crux—Plutarch's report of a astrological calculation for the time of Romulus's conception, which the Roman mathematician Tarutius determined to have taken place during a total solar eclipse—Curtz excused himself from all such controversies: "our purpose is … simply to relate what had been seen and observed in the sky."[20]

---

[17] Borri 1631, 73, 8.

[18] For comet print publication, see Jerratsch 2020; Gruber 2020.

[19] Mizauld 1549. This cometary catalog—likely the first such compilation—was based on lectures by Wittenberg professor Paul Eber, together with Mizauld's own additions. *Catalogus* seems to have been the preferred title term through the early seventeenth century; I have relied on Pingré 1783, 180–243. See Mosley 2013a (see 7–9 for Mizauld and Eber) and Mosley 2013b.

[20] [Curtz], ed., 1666, *3v. For motivations for print publication of observational material in this period, see Hsia 2017, 23–28. On the passage in Plutarch and a reconstruction of Tarutius's calcu-

Curtz's edition certainly distinguished itself in the imperial breadth of its coverage, both temporal and topical. Ranged over some 2300 years, the observations brought within its pages encompassed the whole of the sky and its myriad objects and phenomena: stars and planets, occultations and eclipses; comets, and even a transit of Mercury. But Curtz's celestial history also marked a sea change in access to such material. When Johann Schöner went about printing manuscript texts left by Johannes Regiomontanus, which had been purchased in 1476 by Regiomontanus's fellow observer, Bernhard Walther, and then by the Nuremberg city council in 1519, he brought together some short texts on astronomical topics by Regiomontanus and his teacher, Georg Peurbach, along with sets of extended observations by Regiomontanus—some with Peurbach—and Walther, referred to as Regiomontanus's student.[21] In his dedication to the Nuremberg city council, Schöner praised the "Treasury of observations, which we found most sacredly hidden away and carefully safeguarded in a certain casket." Records in danger of being "disregarded," Schöner's assiduous editing made them publicly available to stand as examples to students of what mathematical experts like Regiomontanus could accomplish.[22]

The material Schöner presented to readers was valuable indeed. Kepler consulted Walther's observations as published in Schöner's edition of 1544 and Willebrord Snell's edition of 1618 in researching the problem of refraction and checking the predictive results of his *Rudolphine Tables* (1627) before sending it to the press; a century later, the French academician Nicolas-Louis de Lacaille worked intensively with Walther's solar data to determine whether the obliquity of the ecliptic was decreasing, as well as to establish elements of his solar theory.[23] But "treasury" connoted more than just scientific value. The term was already in use as a title for works like Robert Estienne's *Latinae linguae Thesaurus* (1531), which, in their various editions, provided greater access to the Latin lexicographic riches that Estienne had amassed at such great effort.

Invoking his "Herculean" labors in consulting ancient texts and erudite colleagues on the definition of Latin terms and examples of their use, Estienne reminded users of the 1543 edition that its "incredible abundance" of Latin phrases had justly earned the title of "a Treasury of the Latin Language," even as he glossed the term *thesaurus* itself as a place where things are hidden (e.g., a treasury) and as something to be unlocked (e.g., a treasure room or chest).[24] In his 1538 *Dictionarium latinogallicum*, Estienne made the analogy clear in his address to readers of the work. Praising its utility for his compatriots, Estienne celebrated "exposing" what

lation, see Grafton and Swerdlow 1986.

[21] On the fate of Regiomontanus's library and Walther's observations, see Kremer 2004, 75–82; Kremer 1980.

[22] Schöner, ed., 1544, aiiiv.

[23] See Kremer 1981; Wilson 1980, 80–86, 118, 179.

[24] Estienne 1543, *.ii. v (*Hercules … labori*), .iiii. r (*incredibilem copiam, thesaurum*); 1430 (citing Cicero's *Letters to Atticus* and Virgil's *Aeneid*). For parallel formulations in other Estienne editions and the significance of both tropes, see Considine 2008, 19–24, 41–46.

had been the "hidden treasure" of Latin speech "to our language."[25] The 1555 French translation of Conrad Gesner's pseudonymous *Thesaurus* of secret remedies undertook the same task of unlocking the "riches contained" in this "Treasure," "enclosed in the Latin language for the most part," "covered" with Greek, Arabic, and "Barbarous words," and "unknown" to the French: "to open it, and discover it for them, and give them the use of it, we put it in pure French language, to be understood and practiced by all the French."[26]

"Thesaurus" captured the skill and effort that made making—and accessing—astronomical observations so difficult. The student who prepared the catalog of comet sightings from Paul Eber's lectures for publication deployed it in describing his professor's labors. So did Nicholas Mulerius in his 1617 edition of Copernicus's *De revolutionibus* (1643), to which he appended a "thesaurus of astronomical observations, collected from the writings of Nicolas Copernicus"; Mulerius described his work of compilation as "digging up" this observational monument from the mud.[27] Philips van Lansbergen likewise called the "observations of all times" that served as both the basis and confirmation of his 1632 astronomical tables a "Treasury," which he indexed along with his text to give readers better purchase on its contents, and Giovanni Battista Riccioli followed suit in his *Almagestum novum* (1651) for the massive series of solar and lunar eclipse observations he compiled—covering some 2500 years—and that he called "an enormous treasure for Astronomy, Geography, & Chronology."[28] In the lavishly illustrated description of the instruments that had equipped his observatory on the island of Hven, Tycho himself noted in his dedication to Rudolf II that he kept his observational trove safe and held close in a "treasury," but hoped—with divine mercy—to someday make them public. Doing so involved a shift in expectations about access to such material that came decades later with the support of a different Holy Roman Emperor, who made it possible for Curtz to present the *Historia coelestis* for "public use."[29]

11.3 "Celestial History" as the History of Astronomy

John Flamsteed struggled to maintain control over his life's work at the Royal Observatory at Greenwich. In this difficult period, he looked back at Curtz's "celestial history" and saw in it, as Frances Willmoth has suggested, a model for how his own observational work should appear in print.[30] Flamsteed was initially keen to

[25] "Rob. Stephanus Lectoribus," Estienne 1538, unpaginated ("latentem adhuc linguae nostrae gazam exponere").

[26] "Epistre Luminaire," [Gesner] 1555, [bb 6]r.

[27] Mizauld 1549, 210 (see note 19); Copernicus 1617, 471, 470.

[28] See "Theoricae motum coeslestium nova, & genuina," in Lansbergen 1632, 39–186 ("Observationum astronomicarum thesaurus"); Riccioli 1651, 1:361–85; 361 (quote).

[29] Brahe 1598, F2v; [Curtz], ed., 1666, dedication to Leopold I (unpaginated).

[30] Willmoth 1997, 63–64.

find his own sources of support for his "Historia Brittanica Caelestis," which he conservatively estimated would require nearly 1500 folio pages.[31] But the massive publication project was soon wrested away by Isaac Newton, who told Flamsteed that he wanted "not your calculations but your Observations only" for work on his lunar theory;[32] by a royally appointed committee of referees—including Newton—which had its own ideas about what should be printed from Flamsteed's papers, and how;[33] and by Edmund Halley, whose editorial "corrections" and truncations of the author's text Flamsteed called a "villianous outrage."[34]

Even as the production of what would become Halley's 1712 edition of Flamsteed's observations moved ahead, Flamsteed looked to Curtz's 1666 publication as a precedent for expanding the scope of his own "celestial history." He decided to include observations made by John Gascoigne and William Crabtree, two English astronomers whose work Flamsteed admired deeply, as well as the observations he had made himself before arriving at Greenwich.[35] Though Flamsteed thought his "celestial history" incomplete without these materials, they were not included in Halley's 1712 edition, only appearing in print in the 1725 edition that Flamsteed's widow and longtime assistants brought to completion after Flamsteed's death in 1719.

The 1725 edition omitted the sections of the lengthy preface in which Flamsteed described his long conflict with Newton and Halley over its publication, as well as his triumph in eventually burning 300 of the 400 printed copies of the 1712 "corrupted" edition.[36] But the preface as published did make explicit what Flamsteed saw as *historical* about "celestial history." In the opening lines, Flamsteed praised two—and only two—writers for having "formed a history of Astronomy." One was Albert Curtz, in his compilation of observations from the Babylonians through his own time. The other was the astronomer Ismael Boulliau, whose prefatory history of astronomy's "rise and progress" at the beginning of his *Astronomia philolaica* (1645) centered on the work of astronomers "whose *observations* have come down to us." This emphasis, Boulliau wrote, was designed to connect "the history of observers and observations."[37]

---

[31] Flamsteed memo (8 November 1704), in Forbes et al., eds., 1995–2002, 3:97–99 (letter 952).

[32] Newton to Flamsteed, 29 June 1695, in Forbes et al., eds., 1995–2002, 2:599 (letter 698).

[33] Clark to Newton, 11 December 1704, in Forbes et al., eds., 1995–2002, 3:114 (letter 961); "Report of the Referees" (23 January 1704/5), in Forbes et al., eds., 1995–2002, 3:130–31 (letter 977).

[34] Flamsteed to Sharp, 15 May 1711, in Forbes et al., eds., 1995–2002, 3:607 (letter 1296); Chapman 1982, 176 (on the "corrupted" catalog and observations at Halley's hands).

[35] Flamsteed to George, Prince of Denmark, November/December 1704 (draft), in Forbes et al., eds., 1995–2002, 3:122 (letter 970); Chapman 1982, 104; Flamsteed to [Newton], 24 October 1705 (draft), in Forbes et al., eds., 1995–2002, 3:230–31 (letter 1037); Flamsteed to Newton, 25 October 1705, in Forbes et al., eds., 1995–2002, 3:233–34 (letter 1038).

[36] Chapman 1982, 176 ("corrupted"); 177–78 ("I made a sacrifice of them to Heavenly Truth").

[37] Boulliau 1645, 12, 18.

Integrating "the history of observers and observations": this was a goal Boulliau pursued by compiling, collating, and evaluating ancient and modern observational data to investigate specific astronomical problems throughout his 1645 work, which Lalande praised over a century and a half later as "one of the best treatises of astronomy made."[38] Flamsteed clearly had this in mind in his preface, much of which was devoted to reconstructing the observational procedures and instruments deployed by his historical predecessors, with an eye toward assessing the empirical basis for their views.

After having published his own "celestial history" in 1741, Pierre Le Monnier reflected more than once on the history of astronomy, singling out Flamsteed and Boulliau as "the first to write the history of ancient and modern astronomy."[39] Le Monnier's juxtaposition served a similar purpose as Flamsteed's earlier pairing of Curtz and Boulliau. The systematic collecting and publishing of astronomical observations was one necessary component of "celestial history" as a mode of writing the history of astronomy. The other was the integration of "the history of *observers*"—how they worked, what instruments they used—and the resulting observations. As Le Monnier noted in the memoir he had read to members of the Academy a few years earlier (10 May 1738), he had projected a "celestial history" that would include all the observations made in France since 1666. The project's urgency was undeniable, as Le Monnier made clear from his opening sentence. "This work," he declared, "is absolutely necessary for the progress of physics and astronomy."[40]

Le Monnier illustrated what was at stake in a lengthy "preliminary discourse" to his 1741 *Histoire céleste*. He first provided "the history of different researches" on the problem of astronomical refraction, whose distorting effects on observed positions had long been "one of the greatest obstacles" to astronomy's progress, with no clear resolution in sight.[41] This history included a review of early Academy astronomers' efforts to increase the accuracy and precision of their empirical data, from improvements in instrumentation (seconds pendulum, micrometer for measuring apparent diameters, telescopic sights applied to quadrants/sextants) to observational protocols (corresponding altitudes for establishing meridian transits).[42]

Le Monnier then turned to his own work along the same lines: repeating observations made by his predecessors (sometimes even with the same instruments); comparing old observations to new ones; and using the results to assess attempts to correct for refraction, including his own.[43] In Le Monnier's view, the history of astronomy was neither a historical vindication of astronomy's excellence, nor a chronological sequence of peoples or individuals notable for their celestial knowl-

---

[38] Lalande 1803, 221.

[39] Le Monnier 1744, [v]; Le Monnier 1746, xxi. For Le Monnier's authorship of the "Introduction," see Montucla and Lalande 1799–1802, 4:213, and Diderot and D'Alembert 1751, 1:25.

[40] Le Monnier 1741, "Projet," [aiii]r.

[41] Le Monnier 1741, "Discours," ii.

[42] E.g., ibid., v–vi, xxxvi; 1–2.

[43] Ibid., ii–iii.

edge, discoveries, and inventions, but the iterative investigation of astronomy's past in light of present-day problems to propel its future progress.[44]

11.4 Reading Between the Bibliographical Lines

So where could a reader find "celestial history" in the pages of Lalande's *Bibliographie astronomique*? The generic boundaries and lineaments of the hefty volumes compiled by Curtz, Flamsteed, and Le Monnier had since dissolved in the welter of printed "observations," a scientific genre that had come into being alongside "celestial history."[45] But while publications of observations by individual astronomers grew in number and prestige over the course of the seventeenth century, the genre of celestial histories—more ambitious in scope, more vulnerable to conflict—foundered.

The title of Le Monnier's 1741 *Celestial History … a Collection of All the Astronomical Observations Made by the Order of the [French] King* promised far more than it delivered. The volume began well enough with the observations that the Paris Academy astronomer Jean Picard made starting in 1666, the year of the Academy's foundation, then at the Royal Observatory from 1673 to Picard's death in 1682, alongside those of his colleague, Philippe de La Hire, from 1678 through the end of 1685.[46] But Le Monnier admitted that those made by the early Academy astronomers Adrien Auzout and Ole Rømer were "entirely lost."[47] And while "a good part" of the observations made by the Academy's longtime leading astronomer, Jean-Dominique Cassini, had been published in the Academy's memoirs, many, many more were surely recorded in his private registers. These, Le Monnier remarked, "have not been communicated to me."[48] Though he promised that the next volume of the "celestial history of France" was forthcoming, it never appeared in print.[49]

Le Monnier's contemporaries and sometime colleagues floated rival proposals for publishing celestial histories: the young Joseph-Nicolas Delisle (in 1720), who thought his chances of advancement within the Paris Academy of Sciences and its observatory so slim that he accepted an offer from Peter the Great to join the academy in St. Petersburg; Alexandre Guy Pingré, whose "celestial annals" for the seventeenth century had to wait until 1901 to see publication; Jean-Dominique Cassini of the third generation of Cassinis to serve as astronomers at the Paris Academy of Sciences, who proposed a "Histoire céleste de l'observatoire royal de Paris" from its

[44] Ibid., xxxi.

[45] See Park 2011 and Pomata 2011.

[46] Le Monnier 1741, "Discours," i.

[47] Le Monnier 1741, "Projet," [aiv]r.

[48] Ibid.

[49] Le Monnier 1741, 233.

founding to the present, but managed to produce only about 15 years' worth of observations in print; and Lalande himself, whose *Histoire céleste français* of 1801 featured not a single observation made at the Paris royal—then national—observatory.[50]

Despite its ultimate failure as an independent genre, early modern "celestial history" articulated a new and compelling vision of an astronomical past sustained through generations of observational labor; an astronomical present animated by the questions it posed to its predecessors and precedents; and an astronomical future premised on the unlimited potential of such research.

References

Asiatic Society of Bengal. 1799. *Asiatic Researches, or, Transactions of the Society Instituted in Bengal, for Inquiring into the History and Antiquities, the Arts, Sciences, and Literature of Asia. … Printed Verbatim from the Calcutta Edition*, vol. 1–5. London: Printed for J. Sewell; Vernor and Hood; J. Cuthell; J. Walker; R. Lea; Lackington, Allen, and Co.; Otridge and Son; R. Faulder; and J. Scatcherd.

Bailly, Jean Sylvain. 1775. *Histoire de l'astronomie ancienne, depuis son origine jusqu'à l'établissement de l'école d'Alexandrie*. Paris: Chez les Freres Debure.

Bailly, Jean Sylvain. 1777. *Lettres sur l'origine des sciences, et sur celle des peuples de l'Asie, Addressées à M. de Voltaire par M. Bailly, & précédées de quelques lettres de M. de Voltaire à l'Auteur*. London/Paris: Chez M. Elmesly … Chez De Bure l'ainé.

Bailly, Jean Sylvain. 1779. *Lettres sur l'Atlantide de Platon et sur l'ancienne histoire de l'Asie. Pour servir de suite aux Lettres sur l'origine des Sciences, adressées à M. de Voltaire*. London/Paris: Chez M. Elmesly … Chez les Freres Debure.

Bailly, Jean Sylvain. 1779–1782. *Histoire de l'astronomie moderne depuis la fondation de l'école d'Alexandrie, jusqu'a l'époque de M.D.CC.XXX*. Paris: Chez les Freres de Bure.

Bailly, Jean Sylvain. 1787. *Traité de l'astronomie indienne et orientale, Ouvrage qui peut servir de suite à l'Histoire de l'Astronomie ancienne*. Paris: Chez Debure l'aîné.

Bernoulli, Johann. 1771–1776. *Recueil pour les astronomes*. Paris/Berlin: Chez l'Auteur… chez Desaint.

Boistel, Guy. 2022. *Pour la gloire de M. de La Lande: Une histoire matérielle, scientifique, institutionnelle et humaine de la* Connaissance des temps, *de 1679 à 1920*. Paris: Institut de mécanique céleste et de calcul des éphémérides.

Borri, Cristoforo. 1631. *Collecta astronomica, ex doctrina P. Christophori Borri; mediolanensis, ex Societate Iesu*. Lisbon: Apud Matthiam Rodrigues.

Boulliau, Ismael. 1645. *Astronomia philolaica*. Paris: Sumptibus Simeonis Piget.

Brahe, Tycho. 1598. *Astronomiæ instauratæ mechanica*. Wandesburg: Philippi de Ohr.

Cassini, Jacques. 1740. *Éléments d'astronomie*. Paris: De l'Imprimerie royale.

Chapman, Allan. 1982. *The Preface to John Flamsteed's* Historia coelestis britannica *or* British Catalogue of the Heavens *(1725)*. Translated by Alison Dione Johnson. London: Trustees of the National Maritime Museum.

Considine, John. 2008. *Dictionaries in Early Modern Europe: Lexicography and the Making of Heritage*. Cambridge, UK/New York: Cambridge University Press.

Copernicus, Nicolaus. 1617. *Astronomia instaurata*. Edited by Nicolaus Mulerius. Amsterdam: Excudebat Wilhelmus Iansonius.

[50] See Hsia 2017, 34–37.

Costard, George. 1746. *A Letter to Martin Folkes, Esq; President of the Royal Society, Concerning the Rise and Progress of Astronomy amongst the Antients*. London: Printed by Jacob Ilive, for T. Osborne … and J. Hildyard.

Costard, George. 1748. *A Further Account of the Rise and Progress of Astronomy amongst the Antients, in Three Letters to Martin Folkes, Esq; President of the Royal Society*. Oxford: Printed at the Theatre, for Richard Clements.

Cotte, Louis. 1803. Table alphabétique des matières. In Joseph-Jérôme Lalande, *Bibliographie astronomique*, [916]–966. Paris: de l'Imprimerie de la République.

Curtz, Albert [as Lucius Barrettus], ed. 1666. *Historia coelestis ex libris commentariis manuscriptis observationum vicennalium viri generosi Tichonis Brahe Dani*. Augsburg: Apud Simonem Utzschneiderum.

De Chales, Claude-François Milliet. 1674. *Cursus seu mundus mathematicus*. Lyon: Ex officina Anissoniana.

Diderot, Denis, and Jean Le Rond D'Alembert, eds. 1751–1772. *Encyclopédie, ou dictionnaire raisonné des sciences, des arts et des métiers, par une société de gens de lettres*. Paris: Chez Briasson … David l'ainé … Le Breton … Durand….

Estève, Pierre. 1755. *Histoire generale et particuliere de l'astronomie*. 3 vols. Paris: Chez Ch. Ant. Jombert.

Estienne, Robert. 1538. *Dictionarium latinogallicum*. Paris: Ex officina Roberti Stephani.

Estienne, Robert. 1543. *Dictionarium, seu Latinae linguae thesaurus*. Paris: Ex officina Roberti Stephani.

Fabricius, Johann Albert. 1796. *Bibliotheca graeca sive notitia scriptorum veterum graecorum*, vol. 5. Edited by Gottlieb Christoph Harless. 3rd ed. Hamburg: Apud Carolum Ernestum Bohn.

Forbes, Eric, et al., eds. 1995–2002. *The Correspondence of John Flamsteed, the First Astronomer Royal*. Bristol/Philadelphia: Institute of Physics.

Gesner, Conrad [as Euonymus Philiatrus]. 1555. *Tresor de Evonime Philiatre des Remedes secretz*. Lyon: Chez Balthazar Arnoullet.

Goguet, Antoine-Yves. 1758. *De l'origine des loix, des arts, et des sciences; et de leurs progrès chez les anciens peuples*. Paris: Chez Desaint & Saillant.

Grafton, Anthony T., and Noel M. Swerdlow. 1986. The Horoscope of the Foundation of Rome. *Classical Philology* 81:148–153.

Gruber, Doris. 2020. *Frühneuzeitlicher Wissenswandel: Kometenerscheinungen in der Druckpublizistik des Heiligen Römischen Reiches*. Bremen: Edition Lumière.

Hsia, Florence. 2017. Astronomy after the Deluge. In *Science in the Archives: Pasts, Presents, Futures*, ed. Lorraine Daston, 17–52. Chicago: University of Chicago Press.

Jardine, Nicholas. 1984. *The Birth of History and Philosophy of Science: Kepler's* A Defence of Tycho against Ursus *with Essays on Its Provenance and Significance*. Cambridge, UK: Cambridge University Press.

Jerratsch, Anna. 2020. *Der frühneuzeitliche Kometendiskurs im Spiegel deutschsprachiger Flugschriften*. Stuttgart: Franz Steiner Verlag.

Keill, John. 1721. *An Introduction to the True Astronomy: or, Astronomical Lectures, Read in the Astronomical School of the University of Oxford*. London: Printed for Bernard Lintot.

Kepler, Johannes. 1627. *Tabulae Rudolphinae, quibus astronomicae scientiae, temporum longinquitate collapsae restauratio continetur*. Ulm: typis … Jonae Saurii.

Kremer, Richard L. 1980. Bernard Walther's Astronomical Observations. *Journal for the History of Astronomy* 11:174–191.

Kremer, Richard L. 1981. The Use of Bernard Walther's Astronomical Observations: Theory and Observation in Early Modern Astronomy. *Journal for the History of Astronomy* 12:124–132.

Kremer, Richard L. 2004. Text to Trophy: Shifting Representations of Regiomontanus's Library. In *Lost Libraries: The Destruction of Great Book Collections since Antiquity*, ed. James Raven, 75–90. Houndmills/New York: Palgrave Macmillan.

Lalande, Joseph-Jérôme. 1764. Remarques sur le choix des meilleurs livres d'astronomie. In *Connoissance des mouvemens célestes, pour l'année commune 1766*, ed. Joseph-Jérôme Lalande, 228–237. Paris: de l'Imprimerie royale.

Lalande, Joseph-Jérôme. 1803. *Bibliographie astronomique*. Paris: de l'Imprimerie de la République.

Lalande, Joseph-Jérôme. 1803. *Connaissance des tems, à l'usage des astronomes et des navigateurs, pour l'an XIII [1805] de l'ère de la République française*. Paris: de l'Imprimerie de la République.

Lalande, Joseph-Jérôme. 1804. *Connaissance des tems, ou des mouvemens célestes, à l'usage des astronomes et des navigateurs, pour l'an XIV [1806] de l'ère de la République française*. Paris: de l'Imprimerie de la République.

Le Gentil de la Galaisière, Guillaume Joseph Hyacinthe Jean-Baptiste. 1779–1781. *Voyage dans les mers de l'Inde, fait par ordre du roi, à l'occasion du passage de Vénus, sur le disque de soleil, le 6 juin 1761, & le 3 du même mois 1769*. Paris: de l'Imprimerie royale.

Le Monnier, Pierre-Charles. 1741. *Histoire celeste, où Recueil de toutes les observations astronomiques faites par ordre du roy; avec un discours préliminaire sur le progrès de l'astronomie, où l'on compare les plus récentes observations à celles qui ont été faites immédiatement après la fondation de l'Observatoire royal*. Paris: Chez Briasson, Libraire.

Le Monnier, Pierre. 1744. Introduction où l'on donne une histoire abrégée de l'astronomie ancienne, & de son usage dans la navigation & dans la géographie. In *Traité complet sur l'aberration des étoiles fixes, avec une histoire générale de l'astronomie*, ed. Pierre Fontaine des Crutes, [v]–xxxii. Paris: Chez G. F. Quillau, Pere.

Le Monnier, Pierre. 1746. *Institutions astronomiques , ou leçons élémentaires d'astronomie ... précedées d'un essai sur l'histoire de l'astronomie moderne*. Paris: Chez Hippolyte-Louis Guerin, & Jacques Guerin.

Levera, Francesco. 1663. *Prodromus universae astronomiae restitutae*. Rome: Ex Typographia Angeli Bernabò.

Longomontanus, Christen Sørensen. 1622. *Astronomia danica*. Amsterdam: Ex officina Typographica Guiljelmi I. Caesii.

Mizauld, Antoine. 1549. *Cometographia ... insuper catalogum visorum cometarum & ignitorum aeris spectrorum, vsque ad annum 1540 cum portentis & euentis quæ secuta sunt*. Paris: Excudebat Christianus Wechelus.

Montucla, Jean Étienne. 1758. *Histoire des mathematiques, dans laquelle on rend compte de leurs progrès depuis leur origine jusqu'à nos jours; où l'on expose le tableau & le développement des principales découvertes, les contestations qu'elles ont fait naître, & les principaux traits de la vie des mathématiciens les plus célebres*. Paris: Chez Ch. Ant. Jombert.

Montucla, Jean Étienne and Joseph-Jérôme Lefrançais Lalande. 1799–1802. *Histoire des mathématiques, dans laquelle on rend compte de leurs progrès depuis leur origine jusqu'à nos jours; où l'on expose le tableau et le développement des principales découvertes dans toutes les parties des mathématiques, les contestations qui se sont élevées entre les mathématiciens, et les principaux traits de la vie des plus célèbres*. Paris: Chez Henri Agasse.

Mosley, Adam. 2013a. Past Portents Predict: Cometary Historiae and Catalogues in the Sixteenth and Seventeenth Centuries. In *Celestial Novelties on the Eve of the Scientific Revolution: 1540–1630*, ed. Dario Tessicini and Patrick Boner, 1–32. Florence: L. S. Olschki.

Mosley, Adam. 2013b. Reading the Heavens: Observation and Interpretation of Astronomical Phenomena in Learned Letters circa 1600. In *Communicating Observations in Early Modern Letters (1500–1675): Epistolography and Epistemology in the Age of the Scientific Revolution*, ed. Dirk van Miert, 115–134. London/Turin: Warburg Institute and Nino Aragno Editore.

Moxon, Joseph. 1665. *A Tutor to Astronomy & Geography, or, The Use of the Copernican Spheres; In Two Books*. London: Printed for Joseph Moxon.

Park, Katharine. 2011. Observation in the Margins, 500–1500. In *Histories of Scientific Observation*, ed. Lorraine Daston and Elizabeth Lunbeck, 15–44. Chicago: University of Chicago Press.

Pingré, Alexandre Guy. 1783. *Cométographie ou traité historique et théorique des comètes*. Paris: de l'Imprimerie royale.

Pomata, Gianna. 2011. Observation Rising: Birth of an Epistemic Genre, 1500–1650. In *Histories of Scientific Observation*, ed. Lorraine Daston and Elizabeth Lunbeck, 45–80. Chicago: University of Chicago Press.

Raina, Dhruv. 2003. Betwixt Jesuit and Enlightenment Historiography: Jean-Sylvain Bailly's History of Indian Astronomy. *Revue d'histoire des mathématiques* 9 (2): 253–306.

Remmert, Volker R. 2011. *Picturing the Scientific Revolution: Title Engravings in Early Modern Scientific Publications*. Translated by Ben Kern. Philadelphia: Saint Joseph's University Press.

Riccioli, Giovanni Battista. 1651. *Almagestum novum astronomiam veterem novamque complectens*. Bologna: Ex Typographia Hæredis Victorij Benatij.

Richards, Joan L. 2006. Historical Mathematics in the French Eighteenth Century. *Isis* 97 (4): 700–713.

Savérien, Alexandre. 1765. *Histoire des philosophes modernes ... Tom. V. Histoire des mathematiciens*. Paris: Chez François ... et chez les Libraires ordinaires.

Schöner, Johann, ed. 1544. *Scripta clarissimi mathematici M. Ioannis Regiomontani*. Nuremberg: apud Ioannem Montanum & Vlricum Neuber.

Sherburne, Edward. 1675. *The Sphere of Marcus Manilius Made an English Poem: With Annotations and an Astronomical Appendix*. London: Printed for Nathanael Brooke.

Smith, Edwin B. 1954. Jean-Sylvain Bailly: Astronomer, Mystic, Revolutionary, 1736–1793. *Transactions of the American Philosophical Society* 44 (4): 427–538.

Souciet, Étienne, ed. 1729–1732. *Observations mathématiques, astronomiques, geographiques, chronologiques et physiques, tirées des anciens livres chinois, ou faites nouvellement aux Indes et a la Chine, par les Peres de la Compagnie de Jesus*. Paris: Chez Rollin Libraire.

Špelda, Daniel. 2017. The History of Science as the Progress of the Human Spirit: The Historiography of Astronomy in the Eighteenth Century. *Studies in History and Philosophy of Science Part A* 63:48–57.

Stadius, Johannes. 1560. *Tabulae Bergenses aequabilis et adparentis motus orbium coelestium*. Cologne: Apud haeredes Arnoldi Birckmanni.

Swerdlow, Noel. 1993. Montucla's Legacy: The History of the Exact Sciences. *Journal of the History of Ideas* 54 (2): 299–328.

van Lansbergen, Philips. 1632. *Tabulae motuum coelestium perpetuae*. Middelberg: Apud Zachariam Romanum.

von Zach, Franz Xaver, ed. 1798–1799. *Allgemeine geographische Ephemeriden*. Vol. 1–4. Weimar: im Verlage des Industrie-Comptoirs.

von Zach, Franz Xaver, ed. 1800–1813. *Monatliche Correspondenz zur Beförderung der Erd- und Himmels-Kunde*. Gotha: im Verlage der Beckerischen Buchhandlung.

Weidler, Johann Friedrich. 1741. *Historia astronomiae sive de ortu et progressu astronomiae liber singularis*. Wittenberg: sumtibus Gottlieb Heinrici Schwartzii.

Weidler, Johann Friedrich. 1755. *Bibliographia astronomica*. Wittenberg: Sumtu Samuelis Godofredi Zimmermanni.

Willmoth, Frances. 1997. Models for the Practice of Astronomy: Flamsteed, Horrocks and Tycho. In *Flamsteed's Stars: New Perspectives on the Life and Work of the First Astronomer Royal, 1646–1719*, ed. Frances Willmoth, 49–75. Woodbridge/Rochester: Boydell Press.

Wilson, Curtis A. 1980. Perturbations and Solar Tables from Lacaille to Delambre: The Rapprochement of Observation and Theory, Part I. *Archive for History of Exact Sciences* 22 (1/2): 53–188.

Wolloch, Nathaniel. 2007. 'Facts, or Conjectures': Antoine-Yves Goguet's Historiography. *Journal of the History of Ideas* 68 (3): 429–449.

The manufacturer's authorised representative in the EU is Springer Nature Customer Service Centre GmbH, Europaplatz 3, 69115 Heidelberg, Germany. If you have any concerns regarding our products, please contact ProductSafety@springernature.com

Printed and bound by CPI Group (UK) Ltd, Croydon, CR0 4YY

17/09/2026

02218432-0005